The Future of US Warfare

This book provides an interdisciplinary analysis of the future of US warfare, including its military practices and the domestic and global challenges it faces. The need to undertake a comprehensive analysis about the future of warfare for the US is more pressing today than ever before. New technologies and adversaries, both old and new, have the potential to revolutionize how wars are fought, and it is imperative that policy makers, military planners, and scholars engage with the latest analyses regarding these new threats and weapon systems. The primary aim of this book is to provide a clear and comprehensive depiction of the types of conflict that the United States is likely to become involved with in the future, as well as the methods of warfare that it may employ within these struggles. While a number of scholarly books have previously considered some of the potential features of US warfare in the future, many of these writings are either outdated or have limited their focus to just one or two of the main types of warfare that may occur and omitted consideration of the others. This book intends to remedy this deficiency in the literature. The volume consists of thematic chapters which address the key issues relevant to the future of US warfare, including cyber warfare, asymmetric conflicts, drone warfare, and nuclear strategy. Through the provision of a series of analyses by leading international academics, the volume provides an important interdisciplinary examination of the different areas of warfare that the United States is expected to use or encounter in the future.

This book will be of great interest to students of US foreign policy, military studies, strategic studies and International Relations in general.

Scott N. Romaniuk is a PhD candidate in International Studies at the University of Trento, Italy.

Francis Grice is visiting assistant professor at McDaniel College, USA.

Military Strategy and Operational Art
Edited by Professor Howard M. Hensel, Air War College, USA

The Routledge Series on Military Strategy and Operational Art analyzes and assesses the synergistic interrelationship between joint and combined military operations, national military strategy, grand strategy, and national political objectives in peacetime, as well as during periods of armed conflict. In doing so, the series highlights how various patterns of civil–military relations, as well as styles of political and military leadership influence the outcome of armed conflicts. In addition, the series highlights both the advantages and challenges associated with the joint and combined use of military forces involved in humanitarian relief, nation building, and peacekeeping operations, as well as across the spectrum of conflict extending from limited conflicts fought for limited political objectives to total war fought for unlimited objectives. Finally, the series highlights the complexity and challenges associated with insurgency and counter-insurgency operations, as well as conventional operations and operations involving the possible use of weapons of mass destruction.

Also in this series:

The Failure to Prevent World War I
The Unexpected Armageddon
Hall Gardner
ISBN 978 1 4724 3056 4

Air Power in UN Operations
Wings for Peace
Edited by A. Walter Dorn
ISBN 978 1 4724 3546 0

The Swedish Presence in Afghanistan
Security and Defence Transformation
Edited by Arita Holmberg and Jan Hallenberg
ISBN 978 1 4724 7409 4

Culture and Defence in Brazil
An Inside Look at Brazil's Aerospace Strategies
Edited by Maria Filomena Fontes Ricco
ISBN 978 1 4724 7183 3

The Future of US Warfare

Edited by Scott N. Romaniuk
and Francis Grice

Routledge
Taylor & Francis Group
LONDON AND NEW YORK

First published 2017 by Routledge

2 Park Square, Milton Park, Abingdon, Oxfordshire OX14 4RN
52 Vanderbilt Avenue, New York, NY 10017

Routledge is an imprint of the Taylor & Francis Group, an informa business

First issued in paperback 2018

British Library Cataloguing-in-Publication Data
A catalogue record for this book is available from the British Library

Library of Congress Cataloging-in-Publication Data
Names: Romaniuk, Scott Nicholas, editor of compilation. |
Grice, Francis, editor of compilation.
Title: The future of US warfare / edited by Scott N. Romaniuk and Francis Grice.
Description: Milton Park, Abingdon, Oxon, New York, NY: Routledge, [2017] | Series: Military strategy and operational art | Includes bibliographical references and index.
Identifiers: LCCN 2016047668 | ISBN 9781472484048 (hardback) |
ISBN 9781315557892 (ebook)
Subjects: LCSH: National security–United States. | Military art and science–United States. | United States–Military policy–History–21st century. | Operational art (Military science) | War–Forecasting.
Classification: LCC UA23 .F8825 2017 | DDC 355/.033573–dc23
LC record available at https://lccn.loc.gov/2016047668

ISBN: 978-1-4724-8404-8 (hbk)
ISBN: 978-1-138-39369-1 (pbk)

Typeset in Bembo
by Deanta Global Publishing Services, Chennai, India

Contents

Contributors

Scott N. Romaniuk is a PhD candidate in International Studies at the University of Trento and is a recipient of the Fondazione Bruno Kessler (FBK) PhD grant focused on "International Politics and Conflict Resolution." He specializes in international relations, security studies, asymmetric warfare, terrorism and political violence.

Dr Francis Grice is an Assistant Professor in Political Science and International Studies at McDaniel College. He holds a PhD in defense studies from King's College London. His areas of research specialization include Asian security and foreign policy, international relations, security studies, leadership, and political violence, and civil wars and intrastate conflict.

William A. Taylor is an assistant professor of Security Studies at Angelo State University in San Angelo, Texas. Taylor has won grants from the Society for Military History and Harry S. Truman Library Institute and a George C. Marshall/Baruch fellowship to research *Every Citizen a Soldier: The Campaign for Universal Military Training after World War II* (Texas A&M University Press, 2014).

Frank Jacob received a PhD in Japanese studies from Erlangen University in 2012. He is an assistant professor (tenure track) for world and global history at the City University of New York (QCC). He has published numerous books on German, Japanese, and military history. He is also the editor of three academic journals and three academic series.

Glen M. E. Duerr is an assistant professor of International Studies at Cedarville University in southwest Ohio, US. He received his PhD in political science from Kent State University. His research focuses on nationalism and secessionist movements, counterterrorism, and counterinsurgency. He is the author of *Secessionism and the European Union* (Lexington Books, 2015).

Łukasz Kamieński is an associate professor at the Faculty of International and Political Studies of Jagiellonian University in Krakow, Poland. His work concentrates on military technology and military transformation, biotechnological human enhancement, the future of war, military profession, and strategy. His most recent book is: *Shooting Up: A History of Drugs in Warfare* (London and New York, 2016).

Roger Chapman is a professor of History at Palm Beach Atlantic University. He received his PhD in American culture studies and history from Bowling Green State University. He is editor of the two-volume *Culture Wars: An Encyclopedia of Issues, Viewpoints, and Voices* as well as *Social Scientists Explain the Tea Party Movement*. He served in the US Army as a member of the 1st Ranger Battalion and was a participant in the ill-fated Operation *Eagle Claw*.

Martin S. Catino is the associate dean of Doctoral and Strategic Studies at Henley Putnam University and a specialist in US strategic security and cultural and human terrain analysis. He is a US Fulbright Scholar (2007–09), US Dept. of State, visiting scholar to Saudi Arabia (2009), visiting scholar to Beijing University (2004), Sasakawa Fellowship scholar in Japanese Studies at San Diego State University (2005), and visiting scholar to the University of Asia and the Pacific, Manila, Philippines (2005).

Tobias J. Burgers is a doctoral candidate at the Otto Suhr Institute, Free University Berlin, from which he holds a diploma in political science. His research interests include the impact of cyber and robotic technology on security dynamics, East Asian security relations, maritime security, and the future of conflict.

Senia Febrica received her PhD from the University of Glasgow. She is a researcher at the American Studies Center, Universitas Indonesia, and a Gerda Henkel Stiftung Post-Doctoral Scholar. She is a lecturer in the Faculty of Social and Political Sciences, Universitas Indonesia, and is a visiting fellow in the Department of Asian Studies, University of Edinburgh.

Tony Tai-Ting Liu is an assistant research fellow at the Contemporary Center for China Studies, National Chung Hsing University. He is a PhD candidate at the Graduate Institute of International Politics, National Chung Hsing University. His research interests include international relations theory, Chinese foreign policy, and East Asian political economy.

B. D. Mowell is a tenured senior professor at Broward College and has taught at the University of Georgia and Florida Atlantic University. He has over 50 publications to date. He has authored or edited several books, including two titles published by the National Council for Geographic Education, and he has written numerous chapters in other books.

Abdelrahman Rashdan is an academic staff member of the Department of Political Science at the Future University in Egypt, where he specializes in national security theory. He is an internationally published columnist covering Arab world politics and US foreign policy.

Suzie Sri S. Sudarman is a lecturer in the International Relations Studies Department, Faculty of Social and Political Science, University of Indonesia,

Jakarta, Indonesia. She has been the director of the American Studies Center (University of Indonesia) since August 2003 and was the director of the Center for International Studies (CIReS) from May 2007 until 2010.

Antônio Sampaio is research associate for security and development at the International Institute for Strategic Studies (IISS) in London. He specializes in the intersection of urban governance and armed violence, with a particular interest in the responses to insecurity in cities of the developing world. His analysis has been featured in several media outlets such as *Foreign Policy*, *The Financial Times*, *World Economic Forum Agenda*, Phoenix TV (China), *El Universal* (Mexico), *Época* and *O Globo* (Brazil).

Jorge E. Delgado received his PhD in War Studies from the Department of War Studies at King's College London, where he worked on the history of Colombia's counterinsurgency operations. His research on Colombia has been published in *Small Wars and Insurgencies* and *The Journal of Strategic Studies*. He worked in a renowned Colombian think tank called *Fundación Ideas para la Paz* (FIP) in different consultancy projects for the Ministry of Defense and the Colombian Air Force.

John M. Callahan is an instructor of political science at New England College, focusing on US foreign and public policy. He received his PhD in International Studies from Old Dominion University in 2015 with research focused on national security decision making and framing of policy decisions. He is a contributor to the UK Defense Forum's Defense Viewpoints.

Ronald Lorenzo is an adjunct lecturer at Prairie View A&M University in Texas. He is the author of *The Puritan Culture of America's Military: US Army War Crimes in Iraq and Afghanistan*. His research interests are in the areas of social theory, the sociology of science and technology, and the sociology of religion.

Shih-Yueh Yang received his PhD from National Chengchi University and is an associate professor at the Department of International Affairs and Business, Nanhua University. His research interests include international relations theory, strategic studies, regional security, and Chinese foreign and defense policy. He has contributed to *The China Review*, *The Korean Journal of Defense Analysis*, *Issues and Studies*, and *Defense and Security Analysis*.

Emeka T. Njoku holds a master's in political science from the University of Ibadan, Nigeria, where he is currently a PhD candidate in the Department of Political Science. He is a fellow at the Institute of French Research in Africa and was a fellow of the Social Science Research Council in New York.

Introduction

Lessons of the past for future war

Francis Grice

In August 1914, over a million French soldiers marched determinedly into Germany as part of the now infamous Plan XVII. Clad in bright red trousers and dark blue coats, the French proved easy targets for the German machine gunners, who repulsed them bloodily. Following the war, the French built the massive series of fortifications along the Franco–German border that came to be known as the Maginot Line. When military operations against Germany began in 1940, the mainstay of the French forces lingered behind these walls, waiting patiently for a massive enemy assault. That offensive never came, for the Germans undertook a daring armored thrust through Belgium instead and encircled the hapless defenders from behind. In both cases, the French demonstrated in a most painful fashion the importance of predicting the future of war. In the former, the French attempted to fight using the same kinds of offensive strategy that had defined the Franco–Prussian War of 1870–1871, in which rapidly advancing columns of Prussian soldiers had triumphed over the more defensively oriented French. In 1940, the French again attempted to use the strategies of the past, this time relying upon the defensive strategies that had come to dominate the First World War. In both cases, the French suffered heavily for their unwillingness to plan based upon what war might be like in the future rather than upon how it had been fought in the past. Their German counterparts, in contrast, made substantial efforts to forecast the nature of future warfare and made use of their predictions to shape their military strategies and tactics.

These and many other examples help to underscore the importance of considering and making informed predictions regarding the future of warfare. Of course, such an endeavor brings with it significant peril: no one can ever dependably *know* the future, and all forecasts must, inevitably, include some errors. Nevertheless, it is possible to identify what current technological, sociological, and political trends seem likely to exist in the coming decades and to make informed predictions regarding the ways that these trends are likely to shape the characteristics of warfare in the future.

Attempts to predict the future of warfare have also featured regularly throughout history but have often failed due to oversimplification. The Hellenistic successors to Alexander the Great placed great stock in trying to foretell the future character of their military campaigns and battles, yet they synthesized their forecasts into a single driving theme: warfare would become more technologically sophisticated, with increasingly longer pikes, gargantuan ships, and colossal war machines. Over the next two centuries, massed legions of Romans, armed primarily with short spears and swords, proved them catastrophically wrong. Similarly, the British after 1914 endeavored to predict how subsequent conflicts might be waged, but the mainstream military establishment became convinced that "colonial warfare" would be the only future and so stymied other military advances. When war with Germany broke out in 1939, the British lacked both the volumes of tanks and the armored expertise needed to initially counter the coming German *blitzkrieg*. Both examples demonstrate the risks and problems created by presuming that warfare will advance in one area only and overlooking other possibilities.

The need to undertake a comprehensive analysis about the future of warfare for the United States (US) is more pressing today than ever before. Multiple technologies and adversaries, both old and new, have the potential to revolutionize how wars are fought, and it is imperative that policy makers, military planners, and scholars engage with the latest analyses regarding these new threats and weapon systems. With this in mind, this edited volume sets out to provide a multifaceted response to the question: *What will US warfare look like in the future?* While previous studies on this topic have tended to focus on just one or two potential features of US warfare in the future, this interdisciplinary book brings together authors from all over the world to comment on multiple aspects of how US warfare may appear in the future. Each author discusses a different dimension of warfare, fitting in with three primary themes: strategy and joint warfare, nontraditional warfare, and new and reinvigorated technology. These are discussed briefly below. Some of these chapters present mainstream scholarly, policy, and military perspectives; others are more radical and controversial. One aim for this volume is that these contributions will stimulate new conversations and debates about the future of US warfare, so that we can learn from the lessons of the French and avoid creating our own Plan XVII and Maginot Line.

Strategy and joint warfare

Modern wars are increasingly interconnected and involve extensive planning and coordination at levels that dwarf those of earlier eras. Even the briefest of military involvements abroad now require massive levels of strategic planning, financial and logistics preparation, collaboration between different government agencies and branches of the armed services, coordination with allies, and negotiations with other states. The US intervention into Iraq in 2003 serves as an example of the huge amounts of planning and funding that can be needed for undertaking military initiatives and the kinds of problems that can occur when insufficient investment is made in these two areas (Grable, 2015).

As a result of the burgeoning costs of preparing for and carrying out military operations, states are looking for ways to streamline their expenses, achieve greater cost efficiency, and get "more bang for their buck." One way that is gaining popularity is the drawing together of individual service branches into a single framework, known as "joint warfare." Measures to improve "jointery" in the US include efforts to integrate the different branches of the armed forces under a single unified command, the design and production of weapons and equipment that can be used to fulfill the needs of multiple different services, the alignment of strategic efforts under a single joint doctrine, enhanced collaboration between both uniformed and nonuniformed agencies, and increased cross-agency training programs. These and other measures are designed to aid the transition of the military forces of the country from being formed of multiple separate entities, all acting independently to one another, into a singular flexible "Joint Force with the agility to rapidly transition from low-end to high-end missions when directed to do so by our commander in chief" (Porter & Mykleby, 2012, 59).

To address these important trends and their likely continued development in the decades ahead, the book begins by examining the theme of strategy and joint warfare. In the opening chapter, Scott N. Romaniuk and Emeka T. Njoku assess the orientation and reorientation of US foreign policy and identity in a changing security environment as well as in the context of US warfare at various points over previous decades. It is precisely within the area of US foreign policy and the country's identity relationship with its wars at home and abroad that we find a critical step in the framing of American public opinion in future wars. In the following chapter, William Taylor appraises the ties between US national security and grand strategy, including analysis of how new threats are influencing American approaches to these two areas. In Chapter 3, Scott N. Romaniuk and Tobias Burgers examine how the US military has evolved from single-service to joint warfare and "full spectrum dominance." Romaniuk and Burgers also consider the emergent threats that face the US and discuss the need for both enhanced domestic interagency working and international multilateral activities to meet these challenges.

Nontraditional warfare

The age of conventional warfare seems very much in decline, if indeed it ever existed at all. The largest risk facing the US is no longer the possibility of massed columns of Soviet armor crossing over the Fulga Gap into Western Europe; the threats it now faces stem instead from less overt sources. Over the past 15 years, the US has encountered a host of asymmetric challenges across the globe, beginning with the resistance to its occupations of Iraq and Afghanistan and continuing through to such phenomena as the rise of the insurgent group Islamic State in Syria and Iraq, piracy around the coast of Somalia, and covert Russian aid for pro-Russian rebels in the Ukraine. These kinds of threat are likely to grow yet more pervasive and prolific in the future (Dixit, 2010).

This trend is occurring, in part, because the unrivalled sophistication, budget, and force projection capabilities of the US armed forces means that the country is unlikely to be challenged by adversaries employing conventional warfare any time soon. Even China and Russia, who are probably the closest contenders for possessing military forces that rival the US, are suspected to have developed a doctrine of hybrid rather than conventional warfare for their operations (Weitz, 2016). Moreover, with the world becoming increasingly urbanized, there appears to be a growing transition from the practice of warfare in the open countryside toward fighting in the cities (Ward, 2015). In fact, urban warfare has remained with us in war and has been a recurrent theme over the past several years in cities across the Middle East, including Aleppo, Basra, Fallujah, and Damascus, and with the spread of tactics used by notorious extremist groups like the Islamic State, urban warfare is becoming bloodier.

This all makes the study of nontraditional warfare in its present and potential future incarnations a crucial theme in this book. To this end, in Chapter 4, Frank Jacob appraises the current and future challenges posed for the US by asymmetric adversaries, particularly Islamic State, as well as the tools and capabilities that the US could deploy to tackle these threats. The question of whether or not the Western world employs a different kind of warfare from the non-Western world is highly contentious and much debated. In Chapter 5, John Callahan advances one side of this debate by arguing that the non-Western world uses a more flexible, decentralized, and technologically adept approach to warfare than the West and that the West must adapt to meet this threat. For Chapter 6, Glen M. E. Duerr leverages the case studies of Yugoslavia, Afghanistan, Iraq, and Libya to argue that creating military peace in combination with economic incentives is the most effective way to counter virulent insurgent movements. In Chapter 7, Roger Chapman analyzes the rise in size and usage of Special Operations Forces since September 11, 2001. This includes discussion about how Special Operations Forces are about to embark upon a period of renewed training and education in preparation for tackling Islamic extremists, weapons of mass destruction, and proxy wars. Next, in Chapter 8, Martin S. Catino addresses the rise of organized crime and drug trafficking in the US, as well as the failings of the country in addressing these threats, through the eyes of ranchers in Southern Arizona. Suzie Sri Suparin S. Sudarman and Senia Febrica continue this section in Chapter 9 by discussing the use of psychological warfare to win the hearts and minds of both civilians and enemy combatants in the future, with emphasis placed upon the lessons that can be drawn from case studies in Indonesia and Iraq. In Chapter 10, Jorge E. Delgado draws upon the case of American involvement in Columbia since 1999 to rebut the notion that "small footprint" advising should replace "big footprint" counterinsurgency of the kind used in Iraq and Afghanistan as the primary method of US intervention abroad. Antônio Sampaio rounds off this section with discussion in Chapter 11 about the challenge of operating in complex city environments, which is becoming increasingly prevalent, including discussion about recent peacekeeping and anticrime operations in South America.

New and reinvigorated technology

Technology has always affected the ways that wars are prepared for and fought, and this trend has heightened during the end of the twentieth and start of the twenty-first centuries. Over the past several decades, new scientific breakthroughs and developments to existing discoveries have dramatically altered the way that wars are prepared for and fought and promise to continue to do so long into the future (US Army Signal Corps OCS Association, 2012). Moreover, new technologies have regularly brought with them substantial moral and ethical quandaries and shifts, a process that is also likely to continue in the years ahead (Algaze, 2015). As the global scientific and military leader, the US is particularly influenced by the advent of new technological developments and improvements, including those that have been fostered by its own scientists and engineers and those that have been produced by foreign powers, to which it must respond (Flournoy & Lyons, III, 2016).

As a result, the impact of new and reinvigorated technology upon the future of US warfare cannot be underestimated. One new form of warfare that has already arisen and promises to continue growing as a medium for the US is unmanned drone warfare, which is already becoming an increasingly prevalent staple in the US arsenal (Hasian Jr., 2016: 1–18). This new type of weapons platform has been used heavily in the Middle East with a range of successes, including the killing of numerous militant leaders and destruction of hostile strategic assets. Drones have come to be viewed as one of the most controversial of weapons ever created as they allow their pilots to kill and destroy enemy targets without placing themselves in any danger (Enemark, 2014). Moreover, their small size and absence of a pilot who can be captured allows them to be used with substantially lower risk of being detected by the adversaries they pursue and, even if they are destroyed, it is harder to identify their source. Another area of technology that is crucial for the US is its nuclear weapons, which continue to be viewed in many quarters as representing an essential part of the country's security toolkit against threats from nuclear rivals and potential adversaries such as China, Russia, Iran, and North Korea (Perry & Scowcroft, 2009). A third is the field of stealth warfare, which entered into mainstream use with the revealing of the F-117 Nighthawk stealth fighter in 1988 and which has been re-energized by the rolling out of the F-35 Lightning IIs in July 2016 (Cunningham, 1991; Seligman, 2016). All of these technologies stand as notable milestones in the path toward future conflicts being fought by the US using what Edward Luttwak so famously dubbed "Post-heroic warfare" (Luttwak, 1995). Drones and stealth technologies eliminate or reduce the threat to their pilots of being shot down during combat, while nuclear weapons continue to offer states the ability to destroy enemy armies and vast swathes of their infrastructure with the use of unmanned missiles that can be launched from thousands of miles away.

Beginning the book's discussion of these technological developments and their impacts, Łukasz Kamieński outlines in Chapter 12 how drone and robotic warfare is positioned to become ever more sophisticated in the future. He then analyzes some of the complex military, political, and moral dilemmas that the US faces in making use of these weapons platforms. B. D. Mowell continues in Chapter 13 by examining how indications that the US might curtail its nuclear weapons and doctrine after the collapse of the Soviet Union in 1992 have failed to materialize. Instead, he argues that the US is being challenged to reinvest in its nuclear arsenals and doctrine due to nuclear challenges from both old and new rivals, especially Russia, China, North Korea, and Iran. Turning toward China, Abdelrahman Rashdan argues in Chapter 14 that it is in the strategic interest of the US to curtail Chinese access to its defense technologies. The author hones in on Sino–Israeli relations and the degree to which these have facilitated China's access to American military secrets. The theme concludes with an examination of stealth technology from two different angles. In Chapter 15, Ronald Lorenzo and Tony Tai-Ting Liu take a different tack by examining the concepts of barbarism, force, chicanery, novelty, and pecuniary damage in relation to stealth in order to illuminate the social, ethical, and moral aspects of stealth technology. In Chapter 16, Shih-Yueh Yang discusses the military effectiveness of stealth technology in the US, including the array of challenges that the technology faces and the alternatives that should be considered to mitigate these weaknesses.

References

Algaze, B. (2015). "Weapons of World War III: How new technology will affect the next great conflict," *ExtremeTech* [online]. Available at: <http://www.extremetech.com/extreme/201542-weapons-of-world-war-iii-how-new-technology-will-affect-the-next-great-conflict>.

Cunningham, J. (1991). "Cracks in the Black Dike: Secrecy, the media, and the F-117a," *Airpower Journal* [online]. Available at: <http://www.airpower.maxwell.af.mil/airchronicles/apj/apj91/fal91/cunn.htm>.

Dixit, K. C. (2010). "The challenges of asymmetric warfare," *IDSA Comment*. The Institute for Defence Studies and Analyses, New Delhi, India [online]. Available at: <http://www.idsa.in/idsacomments/TheChallengesofAsymmetricWarfare_kcdixit_090310>.

Enemark, C. (2014). *Armed Drones and the Ethics of War: Military Virtue in a Post-Heroic Age*. Oxford, UK: Routledge.

Grable, S. (2015). "Why strategy matters," *Air Force Comptroller* (48)1, 5–8.

Flournoy, M. A. & Lyons III, Robert R. (2016). "Sustaining and enhancing the US military's technology edge," *Strategic Studies Quarterly* 10(2): 3–14.

Hasian Jr., M. (2016). *Drone Warfare and Lawfare in a Post-Heroic Age*. Tuscaloosa, AL: University of Alabama Press.

Luttwak, E. N. (1995). "Toward post-heroic warfare." *Foreign Affairs* (May/June). Retrieved August 24, 2016 from <https://www.foreignaffairs.com/articles/chechnya/1995-05-01/toward-post-heroic-warfare> .

Perry, W. J. & Scowcroft, B. (2009). "U.S. Nuclear Weapons Policy." *Independent Task Force Report* 62. New York: Council of Foreign Relations.

Porter, W. & Mykleby, M. (2012). "Rethinking America's joint force: Strength and credibility in a constrained fiscal environment," *American Foreign Policy Interests* 34, 57–68.

Seligman, L. (August 5, 2016). "F-35 marks new era for stealth," *Aviation Week* [online]. Available at: <http://aviationweek.com/defense/f-35-marks-new-era-stealth>.
US Army Signal Corps OCS Association (2012). "Technology and war" [online]. Available at; <http://www.armysignalocs.com/veteranssalultes/tech_on_society_2.html>.
Ward, A. (March 24, 2015). "Battleground metropolis: The future of urban warfare," *The National Interest* [online]. Available at: <http://nationalinterest.org/feature/battleground-metropolis-the-future-urban-warfare-12467>.
Weitz, R. (2016). "The US faces rival powers waging hybrid warfare," *Yale Global Online* [online]. Available at: <http://yaleglobal.yale.edu/content/us-faces>.

Part 1

Strategy and joint warfare

1 Speaking identity

Security practice and US foreign policy

Scott N. Romaniuk and Emeka T. Njoku

Introduction

The United States (US) has reoriented its foreign policy since the American Revolution from a noninterventionist approach to a policy of a global hegemon[1] built on the idea of American exceptionalism. During the Cold War, Lyndon B. Johnson's (1965) discourse stressing the "value of America's word" and "the value of an American commitment" was intertwined with American foreign policy. After 1990, President George H. W. Bush (1991) represented the US as taking a "firm stance against inhumane aggression" by acting to "check that aggression" and "defend civilized values around the world." After 9/11, the discourse of President George W. Bush offered an acute illustration of speaking to American core values and the defense of freedom when he stated that Iraq was a "grave and gathering danger" to "free nations." President Barack Obama (2009) spoke of the US having "entered a new era" in which the US would "need new tools to protect the American people," and that the US would once again "have the strength and standing to truly lead the world." He did so by describing American efforts as the "cause of securing our country," "determination of the American people," and "because of who we are" (Obama, 2009). These representations of identity form part of a unique relationship with US foreign policy.

These narratives depicting America's exceptionalism have been consistently employed by US policy makers to "engineer consent" from the American public in the face of unpopular wars. This was case in the Vietnam War, where the US advanced that Vietnam posed an existential threat to the security of the US interest as a result of its adoption of communism, hence garnering support from the American people during the early period of the war. In the period leading to the 1990 and 2003 wars in Iraq, the US military misled the American people into believing that Saddam Hussein posed both economic and major security threats to the US interests in the Middle East and North Africa (MENA) region in order to justify both wars. The Iraq War facilitated the construction of binary opposites between the West and parts of the MENA region such as "us" against "them" and "good" versus "evil." Such identity constructions aided in sustaining the Global War on Terror (GWoT) as the US and her allies framed public discussions that viewed and securitized states such as Iraq, Libya, and Iran as supporters of terrorism and thus as threats to global political and economic stability. In contemporary times, the responsibility to protect (R2P) principle, though a controversial concept, has become a tool that has helped in polishing US identity as world police—charged with the responsibility of intervening in humanitarian crises in undemocratic states where the fundamental rights of its citizens have been infringed on. R2P, which echoes the need for humanitarian intervention in conflict states, will continue to aid in framing American public opinion in future wars.

Identity in words

An abundance of literature on identity within international relations (IR) provides a range of definitional accounts: personal, corporate, relational, social, role, national, collective, and self-identity (Koslowski & Kratochwil, 1994; Mattern, 2001; Gries, 2005). The concept of *national identity* is a good starting point. Discourse has been widely used to support policy ends, to condition public opinion, and to recast subjects and objects to fundamentally alter understandings of identity. It served a special function in maintaining and strengthening the coherence of a unique Japanese (monocultural and homogenous) identity in the face of increasing numbers of migrants (Burgess, 2004). Dominant discourses within the European Union (EU) have competed with each other to formulate different visions regarding the question of state accession (i.e., Turkey) and the potential for wider enlargement (other states within the EU's current "Enlargement Agenda") (MacMillan, 2013).

Epstein (2008) examines how a powerful anti-whaling discourse (not changing material interests) brought about core changes in worldwide attitudes toward whaling that recast the image and needs of whales. These and other cases offer detailed discussions about the relationship between identity, discourse (dominant, oppositional, and media), and foreign policy. Foreign policy, in each of these cases, relies on the representations of identity, which are then produced and reproduced through those same policies. In each context, there is no prediscursive or stable identity. Policies can be seen as discursive practices that bring together ideas, meaning, and material factors that bring about the institutionalization of an understanding of a single identity or multiple identities.

The US has been considerably reliant on these same practices of representation for drawing on (national) identity. Between 1945 and 1991, American identity was structured around articulations of Communist danger and of "civilized" and "barbaric," with Washington's Containment Policy featured as an extension of US identity (Campbell, 1992). Events following 9/11 point directly to the creation of narratives of "enemies" through the Othering of Muslims—part of the concept of national security and the idea that the killing of "them" is done to avoid more killing of "us" (Brooks, 2013). American society has shown reluctance toward the acceptance of discourse attempting to manipulate public support for military activities, hostilities, and violent resistance and has not been readily accepting of everything it reads. A display of the US public's reception of discourse from its government supporting violent war fits neatly with the emergence of new security threats to the US in the past 15 years.

The discursive and constructed character of the US is both visible and accessible through discourse analysis and its relational configurations. First, despite its reluctance to buy into anti-American discourse, the US public appears to be more receptive of discourse campaigns that use scare tactics and fear as the core of those campaigns. Second, pro-war sentiment in the US no longer arises independent of elite-level efforts to create or stoke it, which shows a significant deviation from the past. Third, dominant discourses over the last decade deviate sharply from others in US history. Fourth, by comparing recent US wars with its past wars, what begins to show is that the use of optimistic discourse has been just as successful in rallying the US public and Congress by appealing to the their fear and hesitation rather than their *principles* and *core values*.

Articulating identity

National identity is both a constructed and public national self-image supported through an actor's acceptance into a specific community like a political group and is interconnected with the understanding of security (Lipschutz, 1995). Other factors like an actor's history, symbolism, political forces, language, and cultural norms found inherently among members of a state fulfill this role (Schlesinger, 1991; Smith, 1971, 1986; Gellner, 1983; Anderson, 1991; Niebuhr, 1967; Hobsbawm, 1990; Deutsch, 1953). These factors serve as a "system of orientation for self-reference and action" (Ross, 1997: 42). The modernist stance on national identity assumes that, regardless of how nations come into being, identity needs to be *constructed* and repeatedly *reconstructed*. Therefore, accepting the existence or construction of a single American national identity is problematic, if not entirely impossible (Calabrese & Burke, 1992). American presidents and administrations since the formation of the American state (nation) have expressed American identity and its uniqueness in various ways (i.e., civil and religious beliefs and core values and principles of liberty, democracy, freedom, equality, and self-government) (Beasley, 2001). We cannot necessarily expect to explain just one American identity.

Certain values and core beliefs remain prominent over time, even if they have not been entirely supported by the behavior of the US government. The notion of individual freedom conflict with US history but has also remained a central factor supporting American idealism and identity (Calabrese & Burke, 1992). National identity is also dependent on the position and description of other nations in the national narrative of the US. Much of the animus of the US as a contemporary military, economic, and cultural power is formulated by comparing it to other states and their positions in the world. Ignatieff (1998: 18) explains that "[n]ational identity is not fixed or stable: it is a continuing exercise in the fabrication of illusion and the elaboration of convenient fables about who 'we are.'" The role of the US as compared to that of the Soviet Union, China, and nations comprising the Middle East has helped to construct a core part of US identity repeatedly over time. It is also reinforced not merely by its oppositional position to other nations but also by the roles that the US assumes in relation to the roles of other nations. The spread of Communism during the Cold War and the threat of militant Islam do not just represent the policies of the US. The US government and political elites are able "to manipulate national discourses and symbols in order to engender and mobilize support among the mass public for specific political goals" (Hutcheson, Domke, Billeaudeaux & Garland, 2004: 29). However, to a large extent, this process has been taken into account in order to analyze the impact that it has on the public for the purpose of using discourse as a foreign policy instrument. But *how* and *to what extent* it features or operates at a level of forming US "nationhood" should also be considered.

Foreign policy and identity

Looking at the literature on foreign policy and identity reveals that one of the greatest challenges within the field is *interpretation*. Scholars have gone about addressing foreign policy and identity by establishing what sort of beliefs and emotions lie behind the decision-making and policies of states rather than establishing the relevant social and political contexts in which those decisions and their subsequent actions occur (Tsygankov, 2012). In the context of foreign policy and identity, action has predominantly been considered in terms of "why" and not so much in terms of "how" or "in what specific context(s)" (Tsygankov, 2012). A significant feature of the literature is that the ability to explain "objective" factors has been taken for granted in attempting to understand foreign policy decisions and what they mean for identity formation. Mattern (2001), Barnett and Duval (2005), and Guzzini (2005) argue that, as part of foreign policies, the capabilities of any number of factors like material power can only fulfill this type of role if they are socially rooted. Barnett (2002: 61) writes that even though "cultural foundations make possible and desirable certain actions, they cannot fulfill their roles so long as they are missing the necessary *institutional* and *ideational* (cultural) setting."

Context specificity and actor sensitivity has not been extensively written into the literature. Scholars place a great deal of constraints on the application of grand theories that are used to help understand action and policy. Meaningful context is important and should be foundational in studying multiple identities of actors like the US, especially when such a study would span multiple historical periods and moments. Hoffmann (1977) shows the importance of applying International Relations (IR) theory and knowledge in the right contexts and of not operating with political, ideological, or epistemological bias. Competing interpretations of identity and its relationship with foreign policy is made evident through ongoing debate. Hixson (2008), on one hand, looked extensively at linguistic discourse on US identity and showed that certain narratives or "stories" of America emerge over time and impact US foreign policy. This suggests that the narrative of identity feeds the policies that are carried out. On the other hand, foreign policy, "with all the symbolic trappings of sovereignty and statehood," has a considerable effect on identity (Anderson, 1991). Foreign policy is precisely what feeds the construction of national identity as it relates to customs, institutions, territory, myth, rituals, and values/principles.

US foreign policy and identity

Campbell (1992) reworked conventional understandings of foreign policy along the lines of what foreign policy *is* and what it *does*. He shows that foreign policy is not merely an action or response by a particular subject in the world. Instead, he points to foreign policy as the reproduction of subjectivity and holds that these are state practices of representation that, in turn, constitute a "voice" (Campbell, 1992). Foreign policy is "retheorized as one of the boundary-producing practices through which the United States, for example, is performatively reproduced as a subject of global political life" (Campbell cited in Laffey, 2000: 431). A salient feature of Campbell's (1992) work is the concept of "performativity." He operates in the interest of avoiding two forms of reductionism: the rendering of agency as an unexplainable quality of a sovereign subject, and seeing the subject as something completely determined by its social context (Laffey, 2000). Campbell's treatment of foreign policy and identity is unique, and in his work he demonstrates what it means to say that foreign policy is about the construction of the Other. "The construction of identity," writes Campbell (1992: 259), "is not the deconstruction of politics; rather it establishes as political the very terms through which identity is articulated."

Campbell's work presents a thick description of foreign policy. In doing so, a continuous network of discourse and political practice through foreign policy comes together from a series of performances over a long historical period. Contrary to this, the US as Self can be seen as a narrative but not as a narrative structure or as vital prediscursive foundation essential to the existence of all Selves (Campbell, 1992: 11). Like Connelly (1991), Campbell (1992) argues that practices of representation need to be implemented and carried out without the practice of Othering; this, claims Neumann (1996: 158), "is exactly what the United States has failed, and is failing, to do. Campbell (1992: 195) argues that the US is *constantly* looking to implement a practice of Othering and states:

> [i]f we take the cold war to be a struggle related to the production and reproduction of identity, the popularly heralded belief that we are witnessing the end of the cold war embodies a misunderstanding: while the objects of established post-1945 strategies of otherness may no longer be plausible candidates for enmity, their transformation has not by itself altered the entailments of identity which they satisfied.

An issue to bring attention to in this work is not one of "imaginedeness" but one of setting the US Self apart from the Other. After the Soviet Union collapsed, the US Self did not change; instead, there was a search for a new object/actor to Other (Neumann, 1996: 158–9). There is a need to consider the continuity of the US Self in Campbell's work, which presents the US Self as a smooth case—a single story with one voice. Meaning performs through its place in language, which in turn finds its strength in specific contexts. If this is the case, then there is reason to argue that the preservation of stable identity over a long period of time is dubious. Instead of looking for the US Self, we might look within the discontinuities that form part of that Self (i.e., the other Selves). Ashley (1987, 1989) and Walker (1993) highlight the idea that the US Self cannot be fully understood without seeing what the impact of competing political interests and aims has on both Self and Selves. Hansen (1996) has examined the relationship between Self and Other with a specific focus on the conflict that results from different discursive practices (Neumann, 1996). Campbell (1992), on the other hand, took the US and portrayed the US Self as a single continuing discursive practice.

Instantiations of "America" through conflict and war

The subsequent "events" have a striking political saliency. They represent critical periods in which the US administration expressed why the nation faced a crisis to its national security and how certain policy choices were framed. The policy decisions were both preemptive and responsive to internal/external acts and effects. Noticeable effects on US identity present in these historical periods are also disrupted and reshaped by previously established arrangements of US identity. The use of historical periods or events like this is a viable approach to the construction and understanding the construction of identity. Events offer rich pronouncements of identity for use in social science research. They offer platforms for discursive competitions and challenges to the formulation of identity. The historical periods and courses of action taken compelled elites to examine US policy options and responses to threats in a way that might not have happened if these specific events did not occur.

Vietnam war

The Vietnam War showed divisions in US public and government sentiments toward national security and foreign policy. Intense changes in values, commitments, and objectives by both the public and government toward war can be observed. Vietnam can be regarded as the defining moment that brought about the collapse of unrivalled support in the US public toward the bipolar security arrangement during the Cold War. Vietnam helps to conceptualize how a link is created between identity and policy. Through articulation, a process of differentiation and linking took place. US discourse constructed multiple identities, which described the objective of North Vietnam as "total conquest" fueled by an "appetite for aggression" and the "deepening shadow of Communist China" that has "destroyed freedom in Tibet," "attacked India," and was "condemned by the United Nations for aggression in Korea" (Lyndon B. Johnson, 1965).

Among the various narratives constructed to sustain the Vietnam War was the fear of communist expansion. This was used to entrench fear of insecurity and thus frame public opinion and support during the early stages of the war. Presidents John F. Kennedy and Lyndon B. Johnson claimed that the adoption of communism by one state could influence others; hence, they argued that the fall of South Vietnam to communism could potentially influence the rapid spread of communism in Southeast Asia and other parts of the continent (Gardner & Young, 2007). They thus reinforced the Cold War ideology of military containment of communism as a justification for the war in Vietnam (Karnow, 1997). Moreover, in order to successfully control public opinion during the early period of the conflict, the state established institutions such as the Committee for Peace with Freedom in Vietnam and the Vietnam Information Group, which aided the proliferation of their propaganda.

The American military also influenced the media, as information on the war was sieved or distorted in order to control the media's view and how it reported about the war to the American people. This explains the support of the media during the beginning of the conflict, which ultimately influenced public opinion of the war (Herring, 1986). Furthermore, Hallin (1986) contended that although the media played significant role in aiding the state in the war, the position of the media changed when the key administration officials began to question the need for the continuation of the war. Specifically, he argues that the Tet offensive of 1968 was a turning point that influenced the outcome of the Vietnam War, as both the media and the American people became critical about government's position on the conflict. The construction of identities was both enabling and constraining. Limitations like institutional pressures, military capabilities, and the media were engaged through creative aspects in foreign policy and discourse. US identity construction was hinged on notions of the "value of America's word," "the value of an American commitment," and a "national pledge" to "enrich the hopes and existence" that "we have helped to defend" (Lyndon B. Johnson, 1965).

The Gulf war

Between 1990 and 2000, the international structure and US security discourse changed substantially. In 1990, the global structure was drifting in the wake of the aftermath of Soviet collapse. American discourse still operated in Cold War style and did not yet regard potential territorial threats to the US. The Gulf War is one of the first televised events where people around the world watched US and coalition forces' military operations unfold via global TV satellite networks (Kellner, 1992, 1995). It was complemented by one of the most successful public relations campaign in US political history. Bush Sr.'s construction of the identities of the Saddam regime focused on the "Iraqi dictator" and his use of a "powerful Iraqi army" that was used to invade a "trusting neighbor" to eventually "threaten Saudi Arabia." Bush Sr. represented the US as taking a "firm stance against inhumane aggression" by acting to "check that aggression" and "defend civilized values around the world," "to protect freedom, protect our future, and protect the innocent." That construction positioned "Iraq against the world," which "shared our [America's] concern for principle" (George H. W. Bush, 2009: 129).

Furthermore, Lakoff (1991) argues that these metaphorical statements put forward by President George H. W. Bush as justification for the Gulf War influenced the thought processes of the American people and their reactions to the war. The Iraq invasion was constructed by the US government as posing both security and economic threats to US interest in the region. Consequently, President Bush launched Operation Desert Shield on August 7, 1990, as a defensive strategy to prevent Iraq from invading Saudi Arabia. However, this operation was abandoned on August 8, 1990, following the declaration of Kuwait as the 19th province of Iraq by Saddam Hussein. The US deployed its military might in the Saudi, Kuwait, and Iraq border to prevent Iraqi military from further encroachment (Miles, 2006). Furthermore, the US also influenced other states into forming a coalition against Saddam Hussein, and on January 17, 1991, the US and the coalition forces launched a military offensive tagged "Operation Desert Storm" on the Iraqi forces, forcing them to abandon their objectives of annexing Kuwait (Global Security, 2016).

It is important to state that the US government, through its military, distorted and even fabricated information that was made available to the American people in order to raise support for the war through it propaganda strategies. For instance, Peterson (2002) reported:

> When George H. W. Bush ordered American forces to the Persian Gulf to reverse Iraq's August 1990 invasion of Kuwait part of the administration case was that an Iraqi juggernaut was also threatening to roll into Saudi Arabia. Citing top-secret satellite images, Pentagon officials estimated in mid-September that up to 250,000 Iraqi troops and 1,500 tanks stood on the border, threatening the key US oil supplier. But when the St. Petersburg Times in Florida acquired two commercial Soviet satellite images of the same area, taken at the same time, no Iraqi troops were visible near the Saudi border just empty desert.

Furthermore, organizations such as the "Citizens for a Free Kuwait" were established and funded by the Kuwaiti Government to tilt Congress and the US public in favor of an intervention. This organization embarked on a wide campaign in colleges and through media to influence public opinion. It fabricated images of gross human-rights violations perpetrated by Saddam Hussein, specifically the killing of children who were pulled out from an incubator and left to die on the floor before the senate committee. This was a turning point that influenced the congressional support for the Gulf War through a 52–47 vote (MacArthur, 1992).

The Iraq war

By 2003, the US was the sole and unrivalled unipolar state, but the events of 9/11 brought international terrorism "deterritorialized subject[s]" as threats to the US mainland into American security discourse (Hansen, 2006). US discourse during the Iraq War offers an acute illustration of speaking to American values and principles, the defense of freedom, and the protection of free and democratic societies while also playing on US fears and apprehension about war and violence. "Iraq" became a construction on the basis of a dictatorial "regime" that committed "atrocities" and violated international law and an "oppressed Iraq people" who were denied "freedom," "liberty," and democracy (George W. Bush). The Saddam regime was seen as an "enemy" characterized as a "grave and gathering danger" to "free nations" and part of an "axis of evil" that used "weapons of mass destruction" to carry out "mass murder" and the ability to "plot another September the eleventh" (George W. Bush cited in Hansen, 2006: 25). This construction can be analyzed on different levels: textual construction of identities and their linkage with foreign policy, and responses to official policy and dominant discourse.

Similar to the strategy used by the state to garner support from the people in the Gulf War, these constructions used to depict Saddam Hussein's Iraq became the justification for the US military intervention in Iraq in March 2003. In the process of building support for and congressional approval of the war, the US government reconstructed Saddam's Iraq not only as dictatorial regime that infringed on the fundamental rights of it citizens but also as a key supporter of terrorism and an impediment to its GWoT. In fact, the government subtly alleged that the Iraqi government under Saddam Hussein participated in the September 11, 2001 terrorist attacks on US homeland and further that Iraq possesses weapons of mass destruction, which it could share with terrorist groups, hence posing a major security threat to the US and her allies. Thus, the US government exploited the extant fear of terrorism that emerged globally following the 9/11 attacks. President Bush emphasized the role that the American people, regarded as bastions of the principles of free democratic society, must play to ensure that those who opposed freedom and the rights and civil liberties of others were stopped.

Global War on Terror (GWoT)

The Iraq War triggered a wave of "orientalist" mentality (Said, 1978) and Othering of Muslims and Islam to construct binaries of "us" against "them," and "good" versus "evil." took place almost immediately after 9/11. "We had entered *a new era*," where "enemies who did not abide by any law" attacked the US. President Obama (2009) states that not sacrificing security for values is the "unique genius of America." Rather than ensuring the US public of a lower risk to terrorist threats under the new security paradigm, the linguistic discourse that is not used and the things that are not shown by the US government would appear to also be part of concrete policy action in this period. Technological applications to fighting new threats reveal themselves when examining US foreign policy and the identity construction of the US attached to it in this period.

The construction of countries such as Iraq, Libya, and Iran as supporters of terrorism by the US and her allies in the context of the GWoT aided large public support of its counterterrorism policies. The policies of the US in countering terrorism or going after those it believes sponsor terrorism emphasize security, support extrajudicial executions and repression, and, ironically, infringe on the fundamental rights of individuals in these countries. This depicts an image of countering terrorism with terrorism, as these counterterrorism measures equally breed an atmosphere of fear and deaths and destructions. As stated earlier, these policies receive support from the American people not only because the state feeds on the fear of the people to derive legitimacy for its policy, but also because the information they receive on the activities of the government in its GWoT program is also largely filtered or misleading.

Interestingly, the Neo Gramsci theory of hegemony as advanced by Robert Cox finds more relevance here. Cox (1981) states that

> hegemony within the historical structure is constituted in three spheres of activity: the social relations of production encompassing the totality of social relations in material, institutional and discursive forms that engender particular social forces; forms of state, consisting of historically contingent state civil society complexes; and world order, which not only represent phases of stability and conflict but also permit scope of thinking about how alternative forms of world order might emerge.

He argued that production should be seen in terms of the proliferation of ideas, morals, and institutions, which usually comes before material production (Cox, 1989). These social relations of productions of ideas, morals, and institutions are facilitated by social forces, which direct patterns of production, subject to the dynamism of social relations. Secondly, forms of state emerge as a result of the activities of the social forces by proselytizing citizens within their domain into accepting the produced ideas, morals, and institutions, thus establishing a consensus of ideas locally by the integration of varied interest of groups into a common national ideological position (Hoare & Nowell-Smith, 1971).

This establishes hegemony. In other words, orthodox forces and institutions engender the dominance through the systematic method of common knowledge articulation and aggregation. Hegemony is thus understood as an opinion—molding activity in the society (Cox, 1981). Chomsky further elucidates these hegemonic processes with particular emphasis on the US strategy of adopting measures aimed at restraining "freedom of thought," even though freedom of speech is guaranteed. That is, the hegemons within a state must ensure that where the people can express themselves freely, they must be guided to say the right thing. According to him, this approach has been termed "manufacture of consent by Walter Lippman or engineering of consent by Edward Bernays" Chomsky (2002:19). Furthermore, Bieler and Morton (2004: 93) argue that "once hegemony has been consolidated domestically, it may expand beyond a particular social order to move outward on a world scale and insert itself through the world order."

Future direction of US identity and war

The practice of identity construction as a justification for wars by US policy makers may continue to define the US foreign policy formulation and implementation. In other words, the US government may continue to project itself to the American people as the world's police, having the onerous responsibility of ensuring democratic values, freedom, fundamental rights, and civil liberties through its military might. Moreover, among other states in Asia, the MENA region will continue to be the key focus of the US's projection of its power and values. Future US presidents are likely to continue the policy of securitizing the MENA region before the US public as justification for their foreign policy actions. A key concept that has aided in defining the US interventions in recent wars in the MENA region is the R2P. The R2P, which has been seen as an emerging international norm, not only defined President Barack Obama's foreign policy choices (Rodriguez 2012) but will also define US identity construction as justification for future wars. According to the International Commission on Intervention and State Sovereignty (ICISS), the term R2P has two key components. The first is a reconstruction of the concept of sovereignty from "sovereignty as control to sovereignty as responsibility." Succinctly, this entails that the concept of sovereignty has moved beyond the idea that a state can perform any actions it so desires within its latitude of authority, meaning instead that the state or sovereigns are responsible for ensuring that the fundamental rights of their citizens are protected from acts of violence within their territory. Another key component of the concept of R2P is that, in the event that a state has shown an inability to protect its citizens and uphold their fundamental rights, the responsibility moves beyond that state to the international community (ICISS 2001:9).

Weiss (2011: 287) asserts that

> no idea has moved faster in the international normative arena than the "Responsibility to Protect," which was formulated in 2001 by the International Commission on Intervention and State Sovereignty (ICISS). Friends and foes have pointed to the commission's conceptual contribution to reframing sovereignty as contingent rather than absolute, and to establishing a framework for forestalling or stopping mass atrocities via a three-pronged responsibility—to prevent, to react, and to rebuild.

However, the US, China, and other states were reluctant to accept the principles of R2P after it was adopted by the UN General Assembly in 2005 on the grounds that it contravenes the principle of noninterference and utter the national security policy, and many other criticisms have trailed the R2P doctrine as a result of the belief that world powers might use the guise of humanitarian intervention to further their capitalist ideals (Bellamy, 2005). However, Rodriguez (2012: 13) argues that "normative pressures have had the effect of compelling states to frame and justify their behavior according to the norms it advances." Reiterating this further, a group of professionals who examine the idea of humanitarian intervention concluded it is "illegal but legitimate." They contend that while the concept contravenes international law principles. it is "sanctioned by its compelling moral purpose" (Clark, 2005). This reinforces the general understanding that states in the international community have a moral right to intervene in humanitarian crises (Bellamy, 2005). This was further underscored by Weiss (2011), who asserts that "when humanitarian and strategic interests coincide, a window of opportunity opens for coalitions of the willing to act on the humanitarian impulse in the Security Council, or elsewhere."

The principles of R2P were reinforced by President Barack Obama as justification of its involvement in the 2011 Libyan Civil War. The War became a source of intense debate among stakeholders within the US. The debate centered on the constitutionality of the US intervention in Libya by the setting up of a no-fly zone and the ascribed role of the protection of foreign citizens in other territories (Rodriguez, 2012). Senator Paul Rand questioned the legality as well as the necessity of the US involvement in the war, having stated that:

> President Obama has failed to heed his own advice. He has ignored our constitution and engaged us in a military conflict without congressional debate and approval. What imminent threat did Gadhafi or Libya pose to the United States? Obviously, the decision to take military action of this magnitude is something that should not be taken lightly, and should first require determining whether it is in the United States' vital national interest.
>
> (The New American, 2011)

Although President Obama's actions were not a clear departure from the narratives projected by his predecessors as justification for involvement in previous wars, Obama told the American people that the US has a responsibility to use its military might to protect its interests and values at all times, and not merely in situations where they are threatened. He stated that it is "un-American" to look away or do nothing in the face of the suffering of people in foreign lands under dictatorial regimes where the fundamental human rights of foreign citizens are taken away or where the state has failed to protect people's lives. He further advanced that the "common humanity" that both the American and Libyan people share makes it incumbent on the US to intervene and protect these people on humanitarian grounds. President Barack Obama's foreign policy decision in Libya was regarded as the first major "unambiguous military enforcement of the Responsibility to Protect norm" (Patrick, 2011).

Therefore, we argue that the construction of the idea of R2P to protect may continue to resonate in US foreign policy actions in spite of the objections raised by some states. The principles of R2P will further aid the construction of narratives that reinforce American exceptionalism before the American public and, consequently, justification for future wars or American intervention in civil wars in other regions. The words of Thomas Weiss (2011: 135, 149) underline this claim:

> In spite of the current preoccupations, in the United States and in the United Nations, with the wars on terrorism and the occupation in Iraq, humanitarian intervention remains an important policy option. Future debates and action are framed by the International Commission on Intervention and State Sovereignty, whose report entitled The Responsibility To Protect and an accompanying research volume were published in December 2001. Future humanitarian crises will arise in conjunction with the need for military force to protect human beings ... Whether US power will underpin or undermine humanitarian intervention is uncertain. But one thing is clear. It will be decisive. If the responsibility to protect is to flourish, the United States must be on board. The current moment is dark, but that is not to say that humanitarian intervention will not dawn again.

Conclusion

The use of Self and Other in identity studies has also been prominently used in order to grasp how particular ideas about how entities like states are constructed and how their identities come about. The sociological approach is useful for understanding a relationship that takes place, but there is still an underlying assumption that a reciprocal relationship exists (Stryker, 1980). Groups, organizations, networks, and institutions are formed when the Self influences society, while that Self is influenced through factors such as language and culture that enable it to become the Other. The Other also takes part in social interaction and can consider itself an "object." Existence of the Other is not concrete. Rather than staying constant, the Other reconfigures and becomes redefined and in doing so destabilizes and reformulates the state of the Self. Several concerns stem from the Self/Other nexus. In the first place, there exists another assumption that the Self is in constant need of an Other. Second, there exists the idea that the Other needs to be treated as a threat to the Self. Third, Othering implies not only that a distinction is being made but is also cast in negative light, whereby the Other implies a level of conflict or confrontation (Stryker, 1980). Creating binaries in this way can lead to researchers accepting reality as something completely devoid of a middle ground or grey zone. By using the Self and Other in such (restrictive) ways, a further assumption is being made about the nature of identity or the relationships that exist within a given identity, and in the case of this chapter, we apply this reasoning to the US and its wars of the past, present, and future. As Hopf (2002) explains, the possibility of characterizing a relationship between Self "as opposed to another idea or history or place" should not be discounted (Hopf, 2002: 3).

Note

1 We contend that, while many argue that American hegemony is here to stay, the US is in fact no longer a global hegemon and departed from this position roughly half a decade ago. We find that, in line with the reasoning of Hadar (2009), US political and economic influence has significantly waned in the past several years and along with it, US military power, which was stretched to a new limit during the contentious years of the so-called "Global War on Terror" (what has recently been switched to "Overseas Contingency Operations" [OCO] under the Obama administration) and has also ebbed, at least relatively speaking. Washington's failure to meet its albeit impossible GWoT-related objectives, and failure to contain the rise of competitor states, among other factors and issues, is part and parcel of America's departure from its hegemonic seat. For those claiming still that the US occupies its more archaic position as a global hegemon, "such assumptions about US omnipotence," argue Hadar (2009), "are woefully out of touch with reality."

References

Anderson, B. (1991). *Imagined Communities: Reflections on the Origin and Spread of Nationalism.* New York, NY: Verso.

Ashley, R. (1987). "The geopolitics of geopolitical space: Toward a critical social theory of international politics," *Alternatives*, 12(4): 403–34.

Ashley, R. (1989). "Living on Border Lines: Man, Poststructuralism, and War," in Der Derian, J. & Shapiro, M. J. (eds.), *International/Intertextual Relations: Postmodern Readings of World Politics*. Lexingtom, MA: Lexington, 259–321.

Barnett, M. N. (2002). "The Israeli identity and the peace process: Re/Creating the un/thinkable," in Telhami, S. & Barnett, M. N. (eds.), *Identity and Foreign Policy in the Middle East*. New York, NY: Cornell University Press, 58–87.

Barnett, M. & Duval, R. (2005). "Power in International Politics," *International Organization*, 59(1): 39–75.

Beasley, V. (2001). "The rhetoric of ideological consensus in the United States: American principles and American pose in presidential inaugurals," *Communication Monographs*, 68: 169–83.

Bellamy, A. J. (2005). "Responsibility to Protect or Trojan horse? The crisis in Darfur and humanitarian intervention after Iraq," *Ethics & International Affairs*, 19(3): 31–54.

Bieler, A. & Morton A. D. (2004). "A Critical Theory Route to Hegemony, World Order and Historical Change Neo-Gramscian Perspectives in International Relations," *Capital & Class*, 28(1): 85–113.

Brooks, S. (2013). "The 'Othering' of Muslim Americans," *The Islamic Monthly* [online]. Available at: <http://theislamicmonthly.com/the-othering-of-muslim-americans/>.

Burgess, C. (2004). "Maintaining identities: Discourses of homogeneity in a rapidly globalizing Japan," *Electronic Journal of Contemporary Japanese Studies*, 4(1).

Bush, G. H. W. (2009). *Speaking of Freedom: The Collected Speeches*. New York, NY: Scribner.

Calabrese, A. & Burke, B. (1992). "American identities: Nationalism, the media and the public sphere," *Journal of Communication Inquiry*, 16(2): 52–73.

Campbell, D. (1992). *Writing Security: United States Foreign Policy and the Politics of Identity*. Manchester, UK: Manchester University Press.

Chomsky, N. (2002). *Pirates and Emperors, Old and New: International Terrorism in the Real World*. London, UK: Pluto Press.

Chomsky, N. (2015). *Pirates and Emperors, Old and New: International Terrorism in the Real World*. London, UK: Pluto Press.

Clark, I. (2005). *Legitimacy in International Society*. Oxford, UK: Oxford University Press.

Connelly, W. (1991). *Identity\Difference: Democratic Negotiations of Political Paradox*. Ithaca, NY: Cornell University Press.

Cox, R. (1981). "Social forces, states, and world orders: Beyond international relations theory," *Millennium: Journal of International Studies*, 10(2): 126–55.

Cox, R. (1989). *Production, the State, and Change in World Order*. Toronto, Canada: Lexington Books.

Deutsch, K. W. (1953). *Nationalism and Social Communication: An Inquiry into the Foundations of Nationality*. Cambridge, UK: Cambridge University Press.

Epstein, C. (2008). *The Power of Words in International Relations: Birth of an Anti-Whaling Discourse*. Cambridge, MA: MIT Press.

Gardner, L. C. & Young, M. B. (eds.) (2007). *Iraq and the Lessons of Vietnam*. New York, NY: The New Press.

Gellner, E. (1983). *Nations and Nationalism*. Oxford, UK: Blackwell.

Global Security (2016). "Operation Desert Storm" [online]. Available at: <http://www.globalsecurity.org/military/ops/desert_storm.htm>.

Gries, P. H. (2005). "Social psychology and the identity-conflict debate: Is a 'China-threat' inevitable?" *European Journal of International Relations*, 11(2): 235–65.

Guzzini, S. (2000). "A Reconstruction of Constructivism in International Relations," *European Journal of International Relations*, 6(2): 147–82.

Guzzini, S. (2005). "The Concept of Power: A Constructivist Analysis," *Millennium: Journal of International Studies*, 33(3): 495–521.

Hadar, L. T. (Feb. 3, 2009). "The US is no longer a global hegemon," CATO Institute, Washington, DC [online]. Available at: <http://www.cato.org/publications/commentary/us-is-no-longer-global-hegemon>.

Hallin, D. C. (1986). *The Uncensored War: The Media and Vietnam*. Los Angeles, CA: California University of California Press.

Hansen, L. (1996). "R. J. B. Walker deconstructing a discipline," in Neumann, I. B. & Wæver, O. (eds.), *Twelve International Relations Theorists Reassessed: Masters in the Making?* London, UK: Routledge.

Hansen, L. (2006). *Security as Practice: Discourse Analysis and the Bosnian War*. London, UK & New York, NY: Routledge.

Herring, G. C. (1986). *America's Longest War, The United States and Vietnam, 1950–1975* (2nd ed.). New York, NY: Newbery Award Records.

Hixson, W. L. (2008). *The Myth of American Diplomacy*. New Haven, CT: Yale University Press.

Hoare, Q. & Nowell-Smith, G. (eds.) (1971). *Selections from the Prison Notebooks of Antonio Gramsci*. London,UK: Lawrence & Wishart.

Hobsbawm, E. (1990). *Nations and Nationalism Since 1780: Program, Myth, Reality*. London, UK: Clarendon.

Hoffmann, B. (1997). "Terrorists and WMD: Some preliminary hypotheses," *Nonproliferation Review*, 4(3): 45–53.

Hoffmann, S. (1977). "An American Social Science: International Relations," *Daedalus*, 106(3): 41–60.

Hopf, T. (2002). *Social Construction of International Politics: Identities & Foreign Policies, Moscow, 1955 & 1999*. Ithaca, NY and London, UK: Cornell University Press.

Hutcheson, J., Domke, D., Billeaudeaux, A., & Garland, P. (2004). "US national identity, political elites, and a patriotic press following September 11," *Political Communication*, 21: 27–50.

Ignatieff, M. (1998). *The Warrior's Honor: Ethnic War and Modern Conscience*. New York, NY: Holt Paperbacks.

International Commission on Intervention and State Sovereignty (ICISS) (2001). "The Responsibility to Protect," Ottawa, Canada [online]. Available at: <http://responsibilitytoprotect.org/ICISS%20Report.pdf>.

Johnson, L. B. (Apr. 7, 1965). "Address at Johns Hopkins University: 'Peace without Conquest,'" The American Presidency Project [online]. Available at: <http://www.presidency.ucsb.edu/ws/?pid=26877>.

Karnow, S. (1997). *Vietnam: A History* (2nd edition). New York, NY. Penguin.

Kellner, D. (1992). *The Persian Gulf TV War*. Boulder, CO: Westview Press.

Kellner, D. (1995). *Media Culture*. London, UK & New York, NY: Routledge.

Koslowski, R. & Kratochwil, F. (1994). "Understanding change in international politics: The Soviet empire's demise and the international system," *International Organization*, 48(2): 215–47.

Laffey, M. (2000). "Locating identity: Performativity, foreign policy, and state action," *Review of International Studies*, 26(3): 429–44.

Lakoff, G. (1991). "Metaphor and war: The metaphor system used to justify war in the Gulf," *Peace Research*, 23(2/3): 25–32.

Lipschutz, R. (1995). "Negotiating the boundaries of difference and security at the millennium's end," in Lipschutz, R. (ed.), *On Security*. New York, NY: Columbia University Press.

MacArthur, John, R. (1992). *Second Front: Censorship and Propaganda in the Gulf War*. Berkeley, CA: University of CA Press.

Macmillan, C. (2013). *Discourse, Identity and the Question of Turkish Accession to the EU: Through the Looking Glass*. Farnham, UK: Ashgate.

Mattern, J. B. (2001). "The power politics of identity," *European Journal of International Relations*, 7(3): 349–97.

Miles, D. (2006). "15 Years after Desert Storm, US commitment to region continues" [online]. Available at: <http://archive.defense.gov/news/newsarticle.aspx?id=14792>.

Neumann, I. (1996). "Self and Other in international relations," *European Journal of International Relations*, 2(2): 139–74.

Niebuhr, R. (1967). "The social myths of the Cold War," in Farrell, J. & Smith, A. (eds.), *Image and Reality in World Politics*. New York, NY: Columbia University Press.

Obama, B. (May 21, 2009). "Remarks by the President on national security," The White House, Washington, DC [online]. Available at: <https://www.whitehouse.gov/the-press-office/remarks-president-national-security-5-21-09>.

Patrick, S. (2011). "Libya and the future of humanitarian intervention: How Qaddafi's fall vindicated Obama and RtoP," Council on Foreign Relations (CFR), Washington, DC [online]. Available at: <https://www.foreignaffairs.com/articles/libya/2011-08-26/libya-and-future-humanitarian-intervention>.

Peterson, S. (Sept. 6, 2002). "In war, some facts less factual," *Christian Science Monitor* [online]. Available at: <http://www.csmonitor.com/2002/0906/p01s02-wosc.html>.

Rodriguez, P. M. (2012). "Constructing responsibility: Sovereignty and intervention in the wake of Libya," *E-International Relations* [online]. Available at: <http://www.e-ir.info/2012/08/01/constructing-responsibility-sovereignty-and-intervention-in-the-wake-of-libya/>.

Ross, M. (1997). "Culture and identity in comparative political analysis," in Lichach, M. & Zuckerman, A. (eds.), *Comparative Politics: Rationality, Culture and Structure*. Cambridge, UK: Cambridge University Press.

Said, E. (1978). *Orientalism*. New York, NY: Random House.

Schlesinger, P. (1991). "Media, the political order and national identity," *Media, Culture and Society*, 13: 297–308.

Smith, A. (1986). *The Ethnic Origins of Nations*. Oxford, UK: Blackwell.

Smith, A. (1971). *Theories of Nationalism*. London, UK: Duckworth.

Stryker, S. (1980). *Symbolic Interactionism: A Social Structural Version*. Menlo Park, CA: Benjamin Cummings.

The New American (Mar. 29, 2011). "Rand Paul Rebuts Obama on Libya." [Online]. Available at: <http://www.thenewamerican.com/usnews/politics/item/5679-rand-paul-rebuts-obama-on-libya>.

Tsygankov, A. P. (2012). "Assessing cultural and regime-based explanations of Russia's foreign policy: 'Authoritarian at heart and expansionist by habit?'" *Europe-Asia Studies*, 64(4): 695–713.

Walker, R. J. B. (1993). *Inside/Outside: International Relations as Political Theory*. Cambridge, UK: Cambridge University Press.

Weiss, T. G. (2004). "The sunset of humanitarian intervention? The Responsibility to Protect in a unipolar era," *Security Dialogue*, 35(2): 135–53.

Weiss, T. G. (2011). "RtoP alive and well after Libya," *Ethics & International Affairs*, 25(3): 1–6.

2 US national security strategy and threats

William A. Taylor

Introduction

This chapter analyzes the important relationship between United States (US) grand strategy and national security. First, the chapter outlines the purposes of national strategy, instruments of power, and grand strategy. Second, the chapter details the historical context of the National Security Act of 1947 and the Goldwater-Nichols Department of Defense Reorganization Act of 1986. These two watershed laws provide the foundation for the contemporary US national security apparatus and illuminate both the promises and challenges of American civil–military relations and national security reform today. Third, the chapter analyzes current US national security strategy. It surveys key policy documents such as the quadrennial defense review, national defense strategy, national security strategy, national military strategy, and quadrennial diplomacy and development review to elucidate US grand strategy. Fifth, the chapter evaluates the role that alliances play in national security strategy. Sixth, the chapter considers the interagency process and potential for reform. The chapter concludes with an assessment of likely future threats to US national security.

Definitions

When discussing strategy, it is always useful to define terms at the outset. This chapter explores national strategy, which is the rational correlation of ends (objectives) to means (resources) and ways (activities). At its core, national strategy involves setting priorities among constrained options (Lucas, 2015, p. 5). Catherine Dale observed, "Critical to defense strategy, and to national security strategy in general, is the linkage between the key trajectories in the global security environment and U.S. national interests" (Dale, 2014, p. 7). Among the many constraints on national strategy, domestic politics and fiscal limitations are two key ones. Sometimes they combine, as in the Budget Control Act. On August 2, 2011, President Obama signed the act into law, resulting in an austere fiscal environment with specific limits on defense spending for a decade and mandatory cuts of 10 percent in the absence of consensus.

Instruments of power are the means that fuel national strategy. Tami Davis Biddle explained, "If they are sufficient in quality and quantity, an actor's resources can serve as 'instruments of power'—leverage mechanisms—that can help it to sustain itself, and perhaps even thrive, in a competitive and often dangerous world" (Biddle, 2015, p. 9). A common construct to frame instruments of power is the DIME model, referring to diplomatic, informational, military, and economic resources. Instruments of power are most effective for national strategy when used together. President Obama stated, "Our influence is greatest when we combine all our strategic advantages." He further warned, "The use of force is not, however, the only tool at our disposal, and it is not the principal means of U.S. engagement abroad, nor always the most effective for the challenges we face" (Obama, 2015, p. 4).

Grand strategy is national strategy that takes into account all instruments of power. Definitions of grand strategy abound. Paul Kennedy characterized it as a process, involving "the constant and intelligent reassessment of…means and ends" (Kennedy, 1991, p. 6). According to Colin Gray, grand strategy is the "purposeful employment of all instruments of power available to a security community" (Gray, 2007, p. 283). R. D. Hooker Jr. emphasized its value as a deterrent: "At its best, grand strategy is not always or fundamentally about fighting or the military application of force, but rather creating in the adversary's mind an appreciation of its potential, along with the other instruments of power" (Hooker, 2015, p. 330). Biddle went further: "At the top of the definitional hierarchy, 'grand strategy' identifies and articulates a given political actor's security objectives at a particular point in time and describes how they will be achieved using a *combination* of instruments of power—including military, diplomatic, and economic instruments" in a manner that is feasible, acceptable, and suitable (Biddle, 2015, pp. 5, 13–14). Overall, grand strategy seeks to achieve vital national interests by incorporating all instruments of power, including diplomatic, informational, military, and economic means. In this sense, US national security strategy is the tangible manifestation of American grand strategy.

Background

Historical context illuminates the foundations of the contemporary US national security strategy organization and process. World War II revealed limitations of the structure at that time. In the postwar era, civilian and military policy makers sought to apply the lessons of that war to improve both the composition and method undergirding US national security. Some significant changes had already occurred. In February 1942, President Franklin D. Roosevelt issued an executive order to establish the joint chiefs of staff, although the group's preference to reach "unanimous agreement" (Locher, 2001, pp. 96–97) and the lack of any formal guidelines from Roosevelt regarding the cluster's specific "roles or powers" (Stevenson, 2008b, p. 13) handicapped it throughout the war and beyond.

After the war, America's newfound role as a world superpower resulted in a plethora of additional responsibilities, evidenced by significant US involvement in the Bretton Woods and San Francisco conferences, creating the

International Monetary Fund, World Bank, and General Agreement on Tariffs and Trade; and the United Nations, respectively. Domestic discussions about unification of the US military underscored differences between the various services, with the army strongly supporting unification and the navy and Marine Corps fiercely opposing it (Stevenson, 2008a, p. 130). After much contentious debate, Congress passed and President Truman signed on 26 July 1947 the National Security Act of 1947, known formally as Public Law 80-253. The act resulted in a National Military Establishment comprised of three separate military departments and headed by a secretary of defense, although that position remained weak for some time. The act also established the National Security Council, the Central Intelligence Agency, an independent air force, and a National Security Resources Board to coordinate mobilization. The National Security Act continued the joint chiefs of staff, albeit still without a chairman. In 1949, Congress revised the act, converting the National Military Establishment to the Department of Defense, strengthening the secretary of defense, and creating the position of chairman of the joint chiefs of staff, although it was a nonvoting position at that time. Revisions in 1953 and 1958 sought to bolster the authority of the secretary of defense and also attempted to strengthen the unified commanders *vis-à-vis* the service secretaries and chiefs (Locher, 2001, p. 99).

Overall, the National Security Act sought to expand national security strategy from only involving military issues to incorporating of all elements of national power into a coherent plan. This was evidenced by the boards that the act created, although the results often fell far short of desired goals. Stevenson indicated, "National security policy involves the integration of domestic as well as foreign and military policies. This was one of the original functions of the National Security Council listed in the 1947 law." He conceded, "But prior to 1993, with the creation of the National Economic Council, and 2002, with the creation of the Department of Homeland Security, little was done to include domestic and economic matters as part of national security policymaking" (Stevenson, 2008a, p. 132).

The National Security Act of 1947 provided the underpinning of US national security for the next four decades. During the 1980s, civilian and military leaders reexamined the national security organization and process and sought to reform them. The result was the Goldwater-Nichols Department of Defense Reorganization Act of 1986. Its purpose was simple; it aspired to promote joint military planning. James R. Locher III perceived, "The overarching objective of Goldwater-Nichols as it was ultimately formulated was to balance joint and service interests" (Locher, 2001, p. 105). Passage of the act was an intense and lengthy struggle. As Locher recorded, "That bitter battle lasted for four years and 241 days—a period longer than U.S. involvement in World War II—and it pitted two former allies, Congress and the services, against each other" (Locher, 2001, p. 96).

The two major sponsors of the act, Barry M. Goldwater (R-AZ) and William F. Nichols (D-AL), were veterans. Goldwater was a major general in the US Air Force Reserve, and Nichols was a World War II combat veteran. After several years of exhaustive work on military reform, on 4 February 1986, the Senate Armed Services Committee began markup of the Goldwater-Nichols Act amidst quarrelsome deliberation. Senators and representatives offered a total of 80 written amendments and another 60 oral amendments to the bill. The vast majority of the service secretaries and service chiefs argued that increasing the role of unified commanders would undermine US national security[1] (Locher, 2002, pp. 39, 44). After three weeks of final negotiations, the bill moved forward and annually required a national security strategy in both classified and unclassified versions.

One of the main goals of the Goldwater-Nichols Act was increased civilian control, although actual improvements failed to reach stated ambitions. One of the key outcomes was improved strategy formulation regarding national security, as evidenced by the new national security strategy requisite and better military advice to the president and secretary of defense. Ultimately, the bill strengthened both the chairman of the joint chiefs of staff and the unified combatant commanders. The chairman became the principal military adviser to the president, while the act placed combatant commanders directly in the operational chain of command under the National Command Authority comprised of the president and secretary of defense (Gorman and Krongard, 2005, pp. 51, 56).

The current US national security strategy configuration and practice resulted from these reforms. As discussed, Congress required a periodic national security strategy in the Goldwater-Nichols Department of Defense Reorganization Act of 1986. Specifically, Section 3043 mandated, "The President shall transmit to Congress each year a comprehensive report on the national security strategy of the United States" (Lucas, 2015, p. 1). Title 50, US Code, §3043(b) obliged the national security strategy to address five major components: "worldwide interests, goals, and objectives"; "foreign policy, worldwide commitments, and national defense capabilities"; "proposed short-term and long-term uses of the political, economic, military, and other elements of the national power"; "adequacy of the capabilities"; and "such other information as may be necessary to help inform Congress on matters relating to the national security strategy of the United States" (Lucas, 2015, p. 1).

Current US national security strategy

Presidential administrations have sporadically and unevenly produced national security strategies since it became a requirement beginning in 1987. Prior to 2000, presidents published a national security strategy every year except 1989 and 1992. Since 2000, however, there have only been four: 2002, 2006, 2010, and 2015. Although originally conceived as both a classified and unclassified document, the national security strategy is routinely published as an unclassified document detailing US national security strategy in its broadest contours. Observers debate whether or not the national security strategy should be classified. Some specialists argue that a classified version allows more detailed planning, whereas other analysts contend that a classified national security strategy precludes valuable public debate (Lucas, 2015, p. 6).

The Fiscal Year 1997 National Defense Authorization Act compelled a quadrennial defense review, and the Fiscal Year 2000 National Defense Authorization Act made permanent the requirement. The quadrennial defense review is a Department of Defense process that evaluates the likely international security environment for the next two decades and accordingly crafts US national defense strategy to meet it. As the name implies, the process results in a quadrennial defense review report once every four years. The year after the review, the secretary of defense must provide Congress, specifically the House and Senate Armed Services Committees, with the report prior to the submission of the president's budget request. Dale documented, "The mandate requires that the review be conducted every four years, during the first year of every presidential administration. The review is required to take a 20-year outlook, and to be fiscally unconstrained by the concurrent President's budget request" (Dale, 2014, p. 2). The quadrennial defense review evaluates the strategy and priorities of the Department of Defense. It also shapes the national defense strategy and supports the national security strategy. The Pentagon has produced every one required: 1997, 2001, 2006, 2010, and 2014. The quadrennial defense review also entails the formation of a National Defense Panel that reviews the report from an independent perspective and the members of which are chosen by the secretary of defense, chairmen, and ranking members of the House and Senate Armed Services Committees.

On 4 March 2014, Secretary of Defense Chuck Hagel presented the 2014 Quadrennial Defense Review to Congress. Hagel explained that the current quadrennial defense review rested upon "three pillars," including "Protect the homeland… Build security globally… Project power and win decisively" (Hagel, 2014, p. v). He also noted the impact of resources on defense strategy: "Overall, the Department can manage these risks under the President's FY2015 Budget plan, but the risks would grow significantly if sequester-level cuts return in FY2016, if proposed reforms are not accepted, or if uncertainty over budget levels continues" (Hagel, 2014, p. v). Hagel outlined such "dynamic and unpredictable" regional threats as North Korea and Iran and emphasized the importance of addressing violent extremism, sectarian conflict, weak and failing states, and cyber security. He reinforced that doing so must occur within a domestic context of significant fiscal constraints (Hagel, 2014, p. iii).

Many actions contribute to the national defense strategy, which, in turn, supports the national security strategy. In January 2012, the Defense Strategic Guidance established the basis for the Department of Defense's military strategy for the next decade. In March 2013, the Strategic Choices and Management Review analyzed the impact of budget cuts on the Department of Defense. The national defense strategy is the Pentagon's military strategy; its purpose is to buttress the national security strategy. Since the quadrennial defense review is a legislative requirement, sometimes the national defense strategy is part of the quadrennial defense review (e.g., 1997, 2001, and 2014) and at other times the national defense strategy is a separate document (e.g., 2005 and 2008). For the most recent iteration, Secretary of Defense Hagel included the national defense strategy as part of the 2014 Quadrennial Defense Review. He emphasized that the Pentagon's focus was "[r]ebalancing for a broad spectrum of conflict," a departure from the prior stress on stability operations and counterinsurgency in Iraq and Afghanistan (Hagel, 2014, p. vii). He also underscored downsizing the force, reiterated the military's rebalance to the Asia-Pacific region, and highlighted "key capability areas," including cyber security, missile defense, nuclear deterrence, space operations, and precision strike (Hagel, 2014, vii–xi).

On 6 February 2015, President Obama published a new national security strategy, focusing on "strong and sustainable leadership" primarily "within a rules-based international order" (Obama, 2015, p. 1). Significant additions to the previous 2010 US National Security Strategy included increased import on climate change, global health security, cyber security, and energy security. President Obama articulated his administration's assessment of the mutable nature of the contemporary international security environment: "Five recent transitions, in particular, have significantly changed the security landscape, including since our last strategy in 2010" (Obama, 2015, p. 4). Obama explained that "power among states is more dynamic" as demonstrated by the Group of Twenty (G-20); "power is shifting below and beyond the nation-state" as exhibited by substate and nonstate actors; there is "increasing interdependence" as revealed by climate change, cyber security, health security, and transnational issues; "a struggle for power" exists in the Middle East and North Africa as displayed by the Arab Spring in 2011; and "the global energy market has changed dramatically" as confirmed by the ascendancy of the US as the world's leading oil and natural gas producer (Obama, 2015, pp. 4–5).

The 2015 National Security Strategy was more explicit geographically by articulating strategy within each region and shifting the US's primary focus to Asia and the Pacific by listing it first among all world regions and emphasizing it throughout the strategy. President Obama declared, "We are rebalancing toward Asia and the Pacific" (Obama, 2015, p. 1). Congressional Research Service Analyst Nathan J. Lucas confirmed, "The 2015 report retains much of the underlying thought of the 2010 version." He added, "However, its emphasis appears to shift away from the U.S. role in the world being largely a catalyst for action by international institutions to one that reflects more involved leadership both inside those institutions and between nations" (Lucas, 2015, p. i). The 2015 National Security Strategy focuses on such state actors as Russia and China and also emphasizes increased engagement with Africa, partly to counter the prevalence of weak and failing states on that continent. It also highlighted nontraditional security issues such as climate change (e.g., reducing greenhouse gas emissions and carbon pollution) and health security (e.g., the February 2014 Ebola outbreak in West Africa).

The Fiscal Year 2004 National Defense Authorization Act required a national military strategy every other year that articulates the military means to achieve the political ends outlined in the national security strategy. Therefore, the national military strategy resides below the national security strategy and the national defense strategy and seeks to articulate the military strategy to achieve both. In June, Chairman of the Joint Chiefs of Staff Martin E. Dempsey issued the 2015 National Military Strategy. Dempsey admitted in his foreword, "Today's global security environment is the most unpredictable I have seen in 40 years of service" (Dempsey, 2015, p. i). The national military strategy illustrates such complexity by focusing on both state and nonstate actors. It highlights that Russia "has repeatedly demonstrated that it does not respect the sovereignty of its neighbors and it is willing to use force to achieve its goals" (Dempsey, 2015, p. 2). Throughout the document, Dempsey conveyed a clear shift toward state actors after more than a decade of emphasis on nonstate actors, primarily within the context of the Iraq and Afghanistan Wars. The national military strategy created a dichotomy of threats consisting of such "revisionist states" as Russia, Iran, North Korea, and China and "violent extremist organizations" (VEOs) (McInnis, 2015, p. 1). Dempsey identified three overall goals, termed *National Military Objectives*: "to deter, deny, and defeat state adversaries; to disrupt, degrade, and defeat VEOs; and to strengthen our global network of allies and partners" (Dempsey, 2015, p. 6).

Alliances

The Congressional Research Service reported that "the [National Military Strategy] states—for the first time in a major strategy document produced in the past twenty years—that there is a distinct possibility that the U.S. may find itself at war with another great power, although it notes that the probability of that actually happening is 'low but growing'" (McInnis, 2015, p. 1). With a return in strategic emphasis on state actors, alliances have also ascended in importance in US grand strategy. A common thread weaving through all these documents is an amplified appreciation of the value of alliances. "Cooperation with allies adds their military forces to ours and secures forward basing and other rights we need to secure U.S. interests around the world," Hooker concluded. "We do not enter into them altruistically, but because they serve U.S. interests" (Hooker, 2015, p. 321). Of course, allies' interests often diverge from American ones. Biddle discerned, "Additionally, a political actor's ends will very rarely, if ever, align completely with its allies' preferred ends.... Allies may share one or two overarching objectives, but they are likely to differ over the nature of more specific ends and the methods required to achieve them" (Biddle, 2015, p. 18). Dempsey reinforced the significance of alliances in the national military strategy: "America's global network of allies and partners is a unique strength that provides the foundation for international security and stability" (Dempsey, 2015, p. 9). Europe remains a region vital to American interests, especially given the continued relevance of the North Atlantic Treaty Organization (NATO) to countering Russia's recent aggression. The worth of alliances is also reinforced by the rebalance to the Asia-Pacific region, including agreements with Australia, Japan, Thailand, the Republic of Korea, and the Philippines.

Another key development has been an attempt to link more closely civilian and military strategic planning. In 2010, Secretary of State Hillary R. Clinton first applied the quadrennial defense review process to the US State Department and US Agency for International Development (USAID), converting the Pentagon's review of military strategy and priorities to the State Department's emphasis on diplomacy and USAID's core competency of development. Secretary of State John F. Kerry continued and expanded this process. On 22 April 2014, Kerry began to formulate the 2015 Quadrennial Diplomacy and Development Review (QDDR). Kerry remarked, "Building on the spirit of innovation of the previous quadrennial review, the QDDR examines the management of our resources and people, the flexibility for experimentation, the modernization of our activities, and the actions that will guide a modern State Department and USAID" (Kerry, 2015, p. 4). Based on guidance in the 2015 National Security Strategy, Kerry articulated four "global policy priorities for the State Department and USAID" (Kerry, 2015, p. 10), including countering violent extremism, fostering democratic societies, expanding economic growth, and moderating climate change. Kerry contended, "Pursuing these priorities and reforms will ensure that American diplomatic and development efforts are better able to advance U.S. interests and priorities as articulated in the 2015 National Security Strategy" (Kerry, 2015, p. 9).

On 17 November 2011, before a crowded audience, President Obama addressed the Australian Parliament and made the first of many calls for increased emphasis in US national security strategy toward the Asia-Pacific region. "As we consider the future of our armed forces, we've begun a review that will identify our most important strategic interests and guide our defense priorities and spending over the coming decade," he remarked. "So here is what this region must know. As we end today's wars, I have directed my national security team to make our presence and mission in the Asia Pacific a top priority" (Obama, 2011, p. 6).

His goals were simple. President Obama sought to strengthen existing economic engagement throughout the region and increase American military presence there, including deployments to Australia, Singapore, the Philippines, and the Indian Ocean, among other areas. As a team of Congressional Research Service analysts wrote, "The fundamental goal underpinning the shift is to devote more effort to influencing the development of the Asia-Pacific's norms and rules, particularly as China emerges as an ever-more influential regional power" (Manyin *et al.*, 2012, p. i). At its core, the Obama administration's rebalance toward Asia seeks to prioritize the region as critical to US national security, with attendant implications and consequences. Mark E. Manyin and his team reported, "Underlying the 'pivot' is a conviction that the center of gravity for U.S. foreign policy, national security, and economic interests is being realigned and shifting towards Asia, and that U.S. strategy and priorities need to be adjusted accordingly" (Manyin *et al.*, 2012, p. i). The implicit rationale also included a reassessment of US national security strategy in the midst of concluding a decade of costly wars in the Middle East.

Such a shift had been building for some time. Dale recorded, "The 2012 [Defense Strategic Guidance] signaled geographic prioritization more emphatically than most guidance documents—with a strong new emphasis on Asia and the Pacific region, some continued attention to the Middle East, and a reasonably explicit intent to decrease the scope and scale of U.S. commitment and engagement everywhere else" (Dale, 2014, pp. 8–9). This "geographic prioritization" also resulted in US policy makers protecting naval forces from the larger budget cuts that both army and Marine Corps ground forces experienced, since many analysts and military planners consider the Asia-Pacific region a primarily naval theater of operations. The vast maritime area includes two major oceans, the Indian and Pacific, as well as the critical shipping choke point located at the Straits of Malacca, through which more than 25 percent of the world's goods and supplies pass.

Reforms

Just as previous reforms set the context for the contemporary US national security organization and procedure, some observers have argued that improvements must be made to the current interagency process in order for future US national security strategy to be implemented effectively. Gorman and Krongard have contended "that a fundamental mismatch exists between the international threat environment and the current national security structure and that the lack of national-level joint interagency organizations undermines the ability of the United States to develop appropriate policies and implement comprehensive strategies" (Gorman and Krongard, 2005, p. 52). Other experts have added their own unique suggestions. The Atlantic Council Combatant Command Task Force evaluated current national security organization and process and advocated significant changes, including assigning an ambassador-level civilian deputy to each geographic combatant command, shifting the emphasis from geographic to unified regional commands, and integrating prepositioned supplies. They argued that the existence of the geographic combatant command as the sole government entity implementing foreign and defense policy at the regional level emphasizes military power at the expense of an effective interagency process. Chairman James L. Jones warned, "The US geographic combatant commands are priceless in strategic value, but their structure, function, and organization are increasingly relics of a bygone era. A purely military approach to the myriad of national security challenges that the nation faces will no longer be enough" (Jones, 2014, p. vi).

Threats

The US faces a plethora of potential future threats, including Iran's nuclear ambitions and state-sponsored terrorism, North Korea's proliferation of weapons of mass destruction and missile technology, Islamic State of Iraq and Syria (ISIS)'s violent extremism resulting in regional instability, and Russia's aggression in Europe. There are also vast positive possibilities, including improving relations with China; increasing engagement with India, the world's largest democracy; and strengthening cyber security. Regarding cyber security, Obama mentioned in his national security strategy, "Prosperity and security increasingly depend on an open, interoperable, secure, and reliable internet" (Obama, 2015, p. 12).

At one minute past 9 p.m. on 10 September 2014, President Obama clarified US strategy toward ISIS.[2] He referred to an alternate name for the group, Islamic State of Iraq and the Levant (ISIL) and vowed, "Our objective is clear: We will degrade, and ultimately destroy, ISIL through a comprehensive and sustained counterterrorism strategy" (Obama, 2014, p. 3). Dempsey, in his 2015 National Military Strategy, characterized the group as "working to undermine transregional security, especially in the Middle East and North Africa." He added, "Such groups are dedicated to radicalizing populations, spreading violence, and leveraging terror to impose their visions of societal organization" (Dempsey, 2015, p. 3). In order to counter ISIS, the US has employed a multipronged approach. The Congressional Research Service affirmed, "Stated U.S. strategy to achieve this objective consists of a number of 'lines of effort.'" The report for Congress catalogued many of them: "direct military action, support for Iraqi and Syrian partner ground forces, intelligence gathering and sharing, and efforts to restrict flows of foreign fighters and disrupt the Islamic State's finances" (Blanchard and Humud, 2015, p. 22). To highlight the threat of ISIS, Director of National Intelligence James R. Clapper testified before the Senate Armed Services Committee that the group "has become the preeminent terrorist threat because of its self-described caliphate in Syria and Iraq, its branches and emerging branches in other countries, and its increasing ability to direct and inspire attacks against a wide range of targets around the world" (Clapper, 2016, p. 4).

Russia, the former Cold War adversary of the US, has increasingly demonstrated aggression as of late, occupying Crimea and supporting separatists in Ukraine, among other provocative actions. Director of the Russia and Eurasia Program at the Center for Strategic and International Studies, Andrew Kuchins, remarked that heightened Russian hostility has led to "incidents at sea" and "dangerous military activities" that have made determining Russian leaders' intentions paramount, especially "What is the [Russian] threshold for military use of nuclear weapons?" (Grady, 2015, pp. 1–3). Russia also prominently appears in the 2015 National Security Strategy. Obama cautioned, "In the realm of inter-state conflict, Russia's violation of Ukraine's sovereignty and territorial integrity—as well as its belligerent stance toward other neighboring countries—endangers international norms that have largely been taken for granted since the end of the Cold War" (Obama, 2015, p. 10). Of course, European dependence on Russian energy is another cause for concern, as it involves both US allies and adversaries.

China plays the most conspicuous role in potential future threats to the US. Many strategic analysts characterize China as a rising power and include as a major issue the resolution of controversies in the South and East China Seas. Other potential problems between the US and China include trade disputes, maritime security, human rights, military modernization, and cyber security. China's perceived ascent in the Asia-Pacific region also triggered the previously discussed shift toward the region in the 2015 US National Security Strategy. Hooker explained, "The U.S. 'Rebalance to Asia' and opposition to Chinese territorial moves in the East and South China Seas can be seen as an attempt to counter the rise of China in a manner consistent with long-standing U.S. grand strategy" (Hooker, 2015, p. 320). In fact, one of the underlying rationales for the Obama administration's rebalance toward Asia is "China's growing military capabilities and its increasing assertiveness of claims to disputed maritime territory, with implications for freedom of navigation and the United States' ability to project power in the region" (Manyin *et al.*, 2012, p. 2). Foremost among these military concerns are China's efforts to build a deep-water navy and a robust anti-access/area denial (A2/AD) capability. Writing for the *US Naval Institute Proceedings*, US Navy Lieutenant John F. Tanalega observed, "The changing face of China's surface fleet is proof of its growing ambitions on the open ocean." He explained, "It is developing not only carrier-based naval aviation, but area air-defense ships such as the Luyang II- and III-class destroyers to defend it—roughly equivalent to U.S. *Arleigh Burke*-class destroyers" (Tanalega, 2016, p. 34). Of course, conflict is never preordained. In the 2015 National Security Strategy, Obama hoped: "We seek cooperation [with China] on shared regional and global challenges such as climate change, public health, economic growth, and the denuclearization of the Korean Peninsula. While there will be competition, we reject the inevitability of confrontation" (Obama, 2015, p. 24). Even so, he did evidence growing alarm over China. Dempsey reiterated many of the same concerns in the 2015 National Military Strategy. "China's actions are adding tension to the Asia-Pacific region," he warned. "For example, its claims to nearly the entire South China Sea are inconsistent with international law" (Dempsey, 2015, p. 2).

Based on past trajectories of threats to US national security, it is likely that future threats will continue to evolve. US national security strategy will have to anticipate such changes and adapt accordingly. Probable future threats include more sophisticated cyber attacks, revealing advanced offensive cyber capabilities among nations and nonstate actors and an increased willingness to employ them. In addition, international terrorism will continue to spawn both coordinated and lone-wolf attacks aimed at the US and its key allies. Resurgent nationalism will escalate tensions in key regions, especially Europe and Asia. This trend will be especially volatile in the maritime environment.

Conclusion

Effective grand strategy is pivotal to ensuring US national security, both today and in the future. The National Security Act of 1947 and the Goldwater-Nichols Act of 1986 provide the historical foundations for contemporary

US national security strategy. Such key policy documents as the quadrennial defense review, national defense strategy, national security strategy, national military strategy, and quadrennial diplomacy and development review all seek to correlate American national ends to available means and ways. Alliances play a major role, especially in an era of heightened globalization. The US rebalance toward Asia provides glimpses into the prioritization of American interests, while reform of the interagency process continues to be an important arena for strategic improvement. Future threats are numerous and reinforce the necessity of cogent and prescient US national security strategy to counter them into the next decade of the twenty-first century.

Notes

1 The lone exception was US Air Force Chief of Staff General Richard Gabriel.
2 Analysts and policy makers alternatively refer to ISIS as Islamic State of Iraq and Syria (ISIS), Islamic State of Iraq and the Levant (ISIL), Islamic State (IS), or *Daesh* in Arabic.

References

Biddle, T. D. (2015). *Strategy and Grand Strategy: What Students and Practitioners Need to Know*. Carlisle Barracks, PA: US Army War College Press.

Blanchard, C. M. and C. E. Humud. (2015). *The Islamic State and U.S. Policy*. Washington, DC: Congressional Research Service.

Clapper, J. R. (2016). "Statement for the Record: Worldwide Threat Assessment of the US Intelligence Community, Senate Armed Services Committee." February 9. Available at http://www.dni.gov/files/documents/SASC_Unclassified_2016_ATA_SFR_FINAL.pdf (accessed 10 February 2016).

Dale, C. (2014). *The 2014 Quadrennial Defense Review (QDR) and Defense Strategy: Issues for Congress*. Washington, DC: Congressional Research Service.

Dempsey, M. E. (2015). *The National Military Strategy of the United States of America*. Washington, DC: US Government Printing Office.

Gorman, M. J. and A. Krongard. (2005). "A Goldwater-Nichols Act for the U.S. Government: Institutionalizing the Interagency Process." *Joint Force Quarterly*, *39*(1), 51–58.

Grady, J. (2015). "Panel: Moves from Putin Administration Ending 'Strategic Holiday' for U.S., NATO." *US Naval Institute News*, September 8. Available at http://news.usni.org/2015/09/08/panel-moves-from-putin-administration-ending-strategic-holiday-for-u-s-nato (accessed 7 February 2016).

Gray, C. S. (2007). *War, Peace and International Relations—An Introduction to Strategic History*. New York: Routledge.

Hagel, C. T. (2014). *Quadrennial Defense Review*. Washington, DC: U.S. Government Printing Office.

Hooker Jr., R. D. (2015). "Understanding U.S. Grand Strategy." *Orbis*, *59*(3), 317–330.

Jones, J. L. (2014). *All Elements of National Power: Moving Toward a New Interagency Balance for US Global Engagement*. Washington, DC: Atlantic Council of the United States.

Kennedy, P. (Ed.) (1991). *Grand Strategies in War and Peace*. New Haven, CT: Yale University Press.

Kerry, J. F. (2015). *Enduring Leadership in a Dynamic World: Quadrennial Diplomacy and Development Review*. Washington, DC: US Government Printing Office.

Locher III, J. R. (2002). "Goldwater-Nichols: Fighting the Decisive Battle." *Joint Force Quarterly*, *31*(1), 38–47.

Locher III, J. R. (2001). "Has It Worked? The Goldwater-Nichols Reorganization Act." *Naval War College Review*, *54*(4), 95–115.

Lucas, N. J. (2015). *The 2015 National Security Strategy: Authorities, Changes, Issues for Congress*. Washington, DC: Congressional Research Service.

Manyin, M. E. et al. (2012). *Pivot to the Pacific? The Obama Administration's 'Rebalancing' Toward Asia*. Washington, DC: Congressional Research Service.

McInnis, K. J. (2015). *The 2015 National Military Strategy: Background and Questions for Congress*. Washington, DC: Congressional Research Service.

Obama, B. H. (2015). *National Security Strategy*. Washington, DC: US Government Printing Office.

Obama, B. H. (2014). "Statement by the President on ISIL." The White House, Office of the Press Secretary. September 10. Available at https://www.whitehouse.gov/the-press-office/2014/09/10/statement-president-isil-1 (accessed 8 February 2016).

Obama, B. H. (2011). "Remarks by President Obama to the Australian Parliament." The White House, Office of the Press Secretary. November 17. Available at https://www.whitehouse.gov/the-press-office/2011/11/17/remarks-president-obama-australian-parliament (accessed 29 January 2016).

Stevenson, C. A. (2008a). "Underlying Assumptions of the National Security Act of 1947." *Joint Force Quarterly*, *48*(1), 129–133.

Stevenson, C. A. (2008b). "The Story Behind the National Security Act of 1947." *Military Review*, *88*(3), 13–20.

Tanalega, J. F. (2016). "A Sling for Goliath." *US Naval Institute Proceedings*, *142*(2), 34–39.

3 "Full spectrum dominance"

US national security doctrine in the new global security environment

Scott N. Romaniuk and Tobias J. Burgers

Introduction

"Full-spectrum dominance" is an overarching feature of the United States (US) Department of Defense's (DoD) "Joint Vision 2020," referring to the ability of US forces to confront, engage, and defeat adversaries in the twenty-first-century security environment along or in tandem with military partners and allies. The strategic security environment of the twenty-first century presents a unique set of challenges that differs considerably from the past. Nonstate actors, the emergence of new states, nontraditional security issues (NTS)[1] (which has no clarity of definition) (Bedeski, 1992; Walsh, 2011), advanced technology, rising and potentially declining powers, all contribute to make "the world a more dangerous place, and have caused vibrant debate in national security establishments around the world (New York University, 2014). The emergence of such threats and their ability to impact the nature of local, national, and global security environments present serious obstacles for national militaries and military power. A few years ago, President Barack Obama stated in a Department of Defense (DoD, 2012: n. p.) strategic guidance publication that "[t]his country is at a strategic turning point after a decade of war, and we are shaping a Joint Force for the future that will be smaller and leaner, but will be agile, flexible, ready, and technologically advanced." Despite the president's optimism concerning the US's ability to adapt to a wide spectrum of budding security challenges, US military pre-eminence—vital for the protection of US strategic interests—is not guaranteed in this new environment.

Recent security events and dynamics, such as the rise of the Islamic State, also known as ISIS, ISIL, or Daesh, plus an host of other nonstate terrorist organizations across the Middle East and Africa; Russia re-emergence as security actor of significance in East-Europe; Assad's continuing survival; and China's ever-increasing military power seem to indicate that the pre-eminence of the US military is not dawning but could be limited. Maintaining and strengthening joint force integration; downsizing the force, including the directed reduction of civilian workforces; and reduced operating budgets have become salient features of current debates, all predicated on the nature of warfare and how the US military can adapt to the twenty-first century. The escalation of hybrid threats in recent years has illustrated how the security of the twenty-first environment could look. This picture is one of highly complex security threats, often interwoven, with a wide array of various security actors, state and nonstate, who have been genuinely testing the US military's leadership and institutional ability to field the right force with the right strategies and tactics to confront and mitigate them.[2] At the same time, and as illustrated above, the US armed forces have in recent years struggled to maintain their budgets, capabilities, and size, which has caused significant discrepancies between what is expected from the US military and what it can provide.

It has sought to adapt to this new security environment, as well as the new economic and resource environment, by adapting a new future framework for strategy and operations. Current US publications (see Joint Operating Environment [JOE], 2010; JP 3-0, 2011; JP 3-03.3, 2012; JP 1, 2013) engage with broad domains related to the new security environment. JOE (2010) contextualizes geopolitical issues such as the broad domain of globalization and connects them to US perceptions of global security. In the post-9/11 era, the US is strengthening its commitment to joint forces operations and command while simultaneously grappling with its wide spectrum of strategic commitments as outlined in JOE. Partnerships have proven crucial for maintaining and strengthening security and are able to add value to particular actors where weaknesses may exist. Nuclear states, for example, bring a net of security to those that do not possess such capabilities (or if they do, then those states that do not possess nuclear weapons). Combining US DoD capabilities and extending cooperative efforts to the defense-oriented establishments of other countries also aids in reinforcing special security interests (DoD, 2007: 20). This willingness to partner is reflected in the most recent US National Security Strategy (NSS, 2015), which likewise emphasizes the need for collaboration, partnership, and joining forces in order to meet the new security environment.

Despite the view that the future of security maintenance and strengthening by means of joint force operations is all around us, the US has only been gradually moving to this end. Many conflicting views on "jointness" still resonate within US public and private circles. The US does not just face conventional military forces, which largely defined its role in the past (such as bipolarity during the Cold War). Weak and failing states, unconventional threats (nonstate actors and armed groups), radical ideological and acts, weapons of mass effect (WMEs), urbanization, and battles of discourse come together further still to present a very comprehensive and far-reaching threat matrix to the US. What is the future of US joint warfare given the constraints placed upon the US military in the strategic security environment of the post-9/11 period? Adherents of force modernization might be quick to point out that preservation of US military pre-eminence may fall in the face of increased defense spending, expanding contractor personnel, extensive military headquarters, and greater focus on the singular roles of service branches. The future of US joint warfare, however, is more likely to be predicated on two distinct and mutually supportive factors. First, future operations are expected to be full-spectrum with respect to military services (whole-of-government); and second, the US will likely be increasingly required to operate as part of an integrated team with allied and partner countries (whole-of-partners).

This chapter approaches military "jointness" by examining factors that are able to shape this concept now and in the long-term future. The first aim is to examine the NSS of the US and its ability to portray the global security environment and its many fluctuations over time. We refer to this as a "rolling process." After exploring existing literature on the conceptualization of joint warfare, we turn to the evolution of the US from single-service to joint warfare. What follows is an analysis of the contemporary state of US joint warfare to illustrate how nascent challenges to the US and its interests in the "new" security environment require US military services to function as a unified team that brings its major military services and branches closer together, including multinational partners.

As the nature of the international security environment continues to undergo strident change, there is a growing need to change the manner in which states and nonstate authorities approach conflict and the practice of war. Some of the factors that have contributed to the dramatic need for change in this regard include emerging and more challenging national security threats, increased pressures to operate in line with a relatively restrictive national economy for the US, and the enhanced interconnectivity of the contemporary geopolitical world. These factors alone have placed the US in a position that requires the nation to deeply reorient its national defense strategies and doctrines, with the manner in which the US operates militarily being no exception. Given the confluence of factors that are increasingly shaping the character of US national defense and its military relationships with its friends and allies, it may prove difficult to dispel any doubt about the need to move further toward the concept of joint military operations. In this, the concept of joint military operations, as both interservice and multinational, could be regarded as logical consequences of the broader globalization that has developed during the last two decades. This, albeit primarily economic in nature, too emphasized joint undertakings and operating with partners, beyond national and regional borders, and is probably best manifested through the ever-increasing number of multinational cooperation. Therefore, it is essential to examine the conditions that are necessary for enhanced (improved) joint operations concerning the US and its military forces. We stress the interaction of multiple contexts for the purposes of this chapter because understanding the interaction of different factors and contexts, the variables of the environment influencing and affecting the subject in question, is critical for richer intersubjective treatment and analysis. That is, it should take place not only in its character, capabilities, and viability but also in the context.

National security threats: Extant and predicated threats[3]

Near-term threats

The two most recent NSSs were in May 2010 and February 2015. Albeit with a five-year hiatus, both documents have a strong similar focus, and the 2015 version can be regarded as an updated version of the 2010 document, adapted to changes that took place in the security environment during this hiatus and largely discussing/following the same threats. Indeed, the 2015 document explicitly references the national interests outlined in the 2010 edition (NSS, 2015: 2). The focus, as Goldgeier and Suri (2016) note in both documents, has been on reacting to prior (perceived) mistakes made by the prior Bush administration. As such, both NSSs—interestingly enough, as they should emphasize future visions—seem to be haunted by ghosts from the past.

Within both documents, President Barack Obama articulated a number of near- and long-term *threats*. Weapons of mass destruction (WMDs) are described as the "gravest danger to the American people and global security" (DoD, 2012: 5; Mullen, 2011: 15; NSS, 2010: 8). This was inclusive of nuclear weapons as a salient threat, especially when pursued by terrorists/extremists and "their proliferation by additional states (NSS 2010: 4). In the 2015 NSS, this threat was once more listed as the top threat (NSS, 2015: 1). The ongoing issue of terrorism (i.e., al-Qaeda and its affiliates) and danger presented through violent (militant) extremism was added to the persistent threat matrix as "top national security priorities" (DoD, 2012: 1–2; NSS, 2010: 4; Obama, 2011: 3). Here too we notice the continuity between the 2010 and 2015 documents: The 2015 edition uses nearly the same language in defining the threat of terrorism, with the major difference being that now ISIL is explicitly mentioned (NSS, 2015: 1). Obama also noted that cyber security was of great importance for the safety of the US (DoD, 2012: 5; NSS, 2010: 18; NSS, 2015: 1–2). Additionally, the US is currently challenged by the hardships of war and conflict, ongoing economic crisis, and a continued reliance on fossil fuels—although the 2015 document has a more positive tone toward economic stability and progress (NSS, 2010: 1, 8; NSS, 2015: 1, 3). The 2015 NSS furthermore points to the risks of global diseases and their threat to global stability, such as, foremost, Ebola (NSS, 2015: 2).

These factors can be taken as both direct and indirect (or latent) threats). They are stress points for the US including allies, not merely in and of themselves, but also in conjunction with other states seeking to secure thriving economic and energy positions. The NSS focuses significantly on abstract concepts that represent a wide range of security challenges on the horizon. It is difficult to discriminate between what constitutes immediate threats and what will be enduring challenges (i.e., threats that either lie just over the horizon or that will become prevalent in just a few years' time). With some of the illusory concepts and tactics mentioned, the NSS underscores more substantive areas of concern, namely those of North Korea and Iran, with their need to be responsible international players, foremost with regard to their nuclear ambitions; Russia's increasingly aggressive behavior, which limits international cooperation with it; and constructive partnerships with as well as cooperation by China and India (both of which are considered rising powers) (DoD, 2012: 2–5; Johnson, 2009; NSS, 2010: 3–4; NSS, 2015: 1–2, 4, 11–12, 24–5).

Long-term threats

Both the 2010 and 2015 NSSs illuminate numerous key points not traditionally associated with high politics in a security context. This is evident as they call attention to "new global actors," "fragile democracies," "non-democratic regimes," "corruption," "transnational crime," and anti-government and governance effects," all of which represent long-term challenges and potential threats (NSS, 2010, 2015). The emergence of new actors in the international system suggests that new forms of state engagement should be expected. All actors (state and nonstate actors alike) seeking to build influence and military prowess to support their influence and defend their interests bring uncertainty to the table. This is particularly the case with overseas interests in areas that overlap with those of the US. Fragile democracies are vulnerable points in US security interests because they represent the same values upheld by the US (as a liberal democracy). Democracy, human rights, and development are innate elements in US democratic identity, and unless they are defended abroad, the US stands to see its democratic ideals undermined at home (DoD, 2012: 2; NSS, 2010: 37; NSS, 2015: 19 and 27). Nondemocratic regimes are therefore part and parcel of this perceived threat because nondemocratic (i.e., authoritarian) states might by injurious to delicate democratic movements that require US support via several (including military) means. Corruption contributes to an unhealthy international system in multiple ways (DoD, 2012; NSS, 2010: 21; NSS, 2015: 1 and 21). By neglecting the issue of corruption, the US stands to see weak, unstable, and failing social, political, and economic systems emerge (in such states as Afghanistan, Pakistan, Iraq, and elsewhere in democratic structures that are either embryonic, weak, or virtually nonexistent) (NSS, 2010: 29; NSS, 2015: 7).

National security threats: Existing and predicted strategies

Near- and long-term responses

US strategic approaches in dealing with immediate threats to the US and US security and interests resemble a tripartite relationship; it includes (1) the domestic element; (2) state-level arrangements; and (3) nonstate or interorganizational cooperation. Inclusive of this relationship is the Joint Force's capacity extension and reformatting to elevate its preparation for the "multi-nodal" climate that Mullen (2011) refers to in "The National Military Strategy of the United States of America." There is no clear delineation between US approaches to near- and long-term threats. The national security documents hone in on different domains of threats and necessity on two distinct levels: these include *substantive* and *conceptual* elements. The NSS and DSG (Defense Strategic Guidance) are elite-level documents produced for public spheres within the US and overseas. They are intended to outline the general approach the US adopts in addressing globally dynamic threats, but both documents are rationally disconnected from the global scope of strategic affairs that they address. Even as broad strategic illustrations of the US, the documents somewhat fail to consider the development of security systems elsewhere in the world. Europe (particularly the European Union [EU]) and Asia, for example, are two notable areas where considerable changes have been taking place—especially in recent years. The most recent NSS does discuss the upheaval in Eastern Europe (e.g., Ukraine, Baltic states security), and possible conflicts in East Asia (e.g., the South China Sea), but it fails to put this in a framework of changing, alternative security systems. However, in order to discuss the formulation of international security arrangements, systems must be examined independent and collectively.

Collectivity also concerns individual states. It is necessary to consider how the US can and will employ its available resources and whether it will do so wisely. The first condition for "success," therefore, covers all three domains. "A whole governmental approach" is comprised of "integrating all of the tools of American power and work with our allies to do the same" (NSS, 2010: 14). The statement touches on renewed investment in diplomacy and development capabilities as well as institutions so as to complement and "reinforce our global partners," and this is once more reinforced in the 2015 edition (Canada and Mexico are predominant mechanisms in the defense of the North American "homeland" (Mullen, 2011: 11; NSS, 2010: 14; NSS, 2015: 1–2, 4). This renewed focus on diplomacy and development is the most visible element of the efforts of the Obama administration to change, to fix perceived mistakes from the prior government and its national security strategy, and to redevelop US national security strategy (Goldgeier & Suri, 2016). In the field of intelligence, this has become most visible in terms of the renewed effort to work together with other states' intelligence communities to ensure accuracy in threat assessment and security procedures, with the intention of detecting conventional and asymmetric threats. The NSSs (2010, 2015) identify eight fields that will receive attention: Defense, Diplomacy, Economic, Development, Homeland Security, Intelligence, Strategic Communications, The American People, and the Private Sector (14–16).

These points cover a full spectrum of threats both near and distant, and are, albeit in a different language, once more strongly reaffirmed in the 2015 document. International cooperation and engagement underscores the need to work closely with other countries in the present (i.e., intelligence efforts). Shared values and interests that serve US and international security drive this approach. Rather than focusing on building relationships with states that have traditionally been viewed as primary actors and states that have been more dominant in international affairs, the US will also extend its partnerships with states such as Brazil, South Africa (NSS, 2010: 44; NSS, 2015: 15, 27), China, and India (the world's two most populous nations), which require special recognition as major players in the international domain. This is a significant change from years past as modernization is China's number one strategic priority (politically, militarily, and economically) (Johnson, 2009: 13). States such as these will be constructively engaged to facilitate cooperation in detailed ways. A dual purpose is fulfilled here: to include these states increasingly in the international system and to maintain America's involvement beyond its borders. By expanding its cooperation with other states, the US stands to open pathways in new regions.

Mullen (2011: 4) presents four specific points regarding US national interests and national military objectives that have become running themes throughout the previous four US NSSs since 2002:

1. The security of the US, its citizens, and US allies and partners
2. A strong, innovative, and growing US economy in an open international economic system that promotes opportunity and prosperity
3. Respect for universal values at home and around the world
4. An international order advanced by US leadership that promotes peace, security, and opportunity through stronger cooperation to meet global challenges

Understandably, Iran receives more attention than others. Both as an immediate and long-term threat because of its regional (Middle East) geopolitical aspirations and—albeit subdued for the moment—its potential nuclear weapons capabilities, as well its continuing support for the Assad regime and Hezbollah (as well as religious mechanisms), the Iranian government demonstrated behavior that can foster volatility in existing situations but also trigger unpredictable events (Mullen, 2011: 11; Terrill, 2011). The US hopes, through the nuclear deal and the opening up of Iran's society, to change the political discourse (foremost among young Iranians) and hopefully encourage Iranian leadership to assume a different political course. The promotion of democracy and human rights is expected to foster a favorable environment for US interests. Interestingly, US values are attached to the idea of "advancing universal values," thereby ensuring a distinct relationship between the promotion of international development and US foreign policy. Alliance building involves strengthening the US relationship with European states (i.e., the EU and regions immediately adjacent to EU borders); developing a new strategic concept for the North Atlantic Treaty Organization (NATO); advancing stability in the Balkans, the Caucasus, and Cyprus; and engaging further with Turkey (Mullen, 2011: 12; NSS, 2010: 42; NSS, 2015: 27).

Preparing for future conflict

War: Past, present, and future

Addressing the issue of preparing for future conflict can be done through a number of assumptions. The military plays a dominant role. It can be assumed that conventional forces provide the basis for US preparation and readiness regarding predicted and unpredicted wars and the general application of American power in the global commons (Gray, 2011). Accordingly, we connect with the primary missions of the US Armed Forces given that they are equipped with the requisite tools for confronting conflict—past, present, and future. The US will remain the preeminent economic and military power in the foreseeable future (Mullen, 2011). This does not absolve the US, however, of necessary adjustments to account for "stresses of sustained combat operations" and "potential adversaries" (Mullen, 2011: 2).

US Armed Forces (the Joint Force) is expected to "recalibrate its capabilities and make selective additional investments" to meet the demands of national security. It also emphasizes leadership as expedient in how the US exercises its power (Mullen, 2011). The Joint Force faces many obstacles. Overcoming some of its traditional military doctrinal mentality is one such hurdle. Additionally, there is the need to take account of new and emerging dimensions of warfare, to become increasingly cognizant of customary modalities of war and conflict, and to adhere to the international security environment in a "multi-nodal world" (Mullen, 2011). The US has set out to make several fundamental changes to the Joint Forces by 2020. Multifaceted program development orbits around eight central points.

1 Maintain a broad portfolio of military capabilities
2 Differentiate between investments that should be made now and those that can wait
3 Maintain a ready and capable force even amid force reduction
4 Cut costs by relying on partnerships and making cost-effective and strategically viable decisions
5 Take stock of limited resources and the manner in which the US Joint Forces uses its current resources
6 Reassess (by the DoD) the mix of Active Component (AC) and Reserve Component (RC) elements and fit them with strategic aims effectively and efficiently
7 Advance in networked-warfare
8 Maintain the necessary industrial base and investment in science and technology

(Obama, 2011: 6–8)

Conventional practices of warfare no longer dominate the standing role of the US Armed Forces today. Notions of irregular warfare now receive more attention in the NSS (2010, 2015) than does conventional warfighting (although the documents do not treat the issue of conflict in such austere terms). There is now an emphasis on "global counterterrorism efforts" that are formulated by way of "direct action" and "security force assistance" (DoD, 2012: 3). The National Strategy for Counterterrorism (2011) is in line with the NSS (2010, 2015), which calls for adherence to "US core values," "building security partnerships," "applying CT tools and capabilities appropriately," and "building a culture of resilience" (Obama, 2011: 4). Just as the security environment of the past has called for US reliance on partnerships, so do current and future environments. However, the difference rests in the nature of the threat—from conventional to asymmetric. Previously, the US maintained deterrence, but mention is now made of the concept of "credible deterrence" (DoD, 2012: 4). The integrative element is here emphasized with the statement: "[o]ur planning envisages forces that are able to fully deny a capable state's aggressive objectives in one region by conducting a combined arms campaign across all domains—land, air, maritime, space, and cyberspace" (DoD, 2012: 4; Mullen, 2011: 18–9).

The denial of access and freedom of US military forces and force projection is limiting the effectiveness of US military force in general and thus represents a partial failure (or area for improvement) in the US's primary objective of national and international security. The US, for that reason, seeks to maintain its ability to operate in areas that have proven challenging and where access has been limited due to a wide array of conditions. Given the projection of China and Iran and future threats using asymmetric measures to counter US power, the US military will pursue investment as a primary strategic approach "to ensure its ability to operate effectively in anti-access and area (A2/AD) environments" (DoD, 2012: 4). In this vein, the US will have to compensate for the changing military balance of East Asia and other regions (Johnson, 2009: 2). As such, the Joint Operations Concept is deeply engrained within the perceived success of this strategy. Specifically, Washington projects to sustain its undersea capabilities (i.e., nuclear and attack submarines), retain its surface maritime capabilities, start the development of a next-generation stealth bomber, invest in advanced missile defenses (for protection of its friends and own forces stationed in such regions as [Central and Eastern] Europe and the Korean peninsula), and expand upon so-called "critical space-based capabilities" (DoD, 2012: 5).

The emphasis of measures to be taken in the national documents is a strong indication of the difference between wars of the past and those projected into the future (including those of the present day, which ultimately define the future). From this, one can extrapolate that states will continue to use conventional force capabilities but will combine them with elements of asymmetric force to meet desired ends. Nuclear, biological, and chemical weapons (NCBWs) remain a constant from wars of the past to the present and forthcoming. They have become part of a wider portfolio of military means, which now and will continue to include the element of cyber warfare. Missile defense and armed forces for space-based operations are becoming increasingly evident in the operational and strategic domain of US and international security. Moreover, counterinsurgency (COIN) operations figure flagrantly in US military and military-to-military doctrine (Obama, 2011: 6). With increased mention and prediction of violent militant extremism and asymmetric threats, US forces will prepare for COIN and what has been referred to increasingly as "stability" operations at any time, if required (DoD, 2012: 6; Obama, 2011: 6). This importance of COIN and countering asymmetric terror threats has been validated itself once more in recent years. Following the 2010 NSS, the Obama government scaled back its COIN efforts and decreased its troops in Afghanistan and Iraq, only to see new asymmetric terror groups rise once more. As a result, in the last year and a half, the US COIN effort has gradually increased again with growing numbers of US advisors in Iraq, a nondeparture of US forces from Afghanistan, and increased presence through the US Special Operations Command on the African continent. These events and the US security reaction to them have proved the opposite of prior strategies as outlined five years earlier. Then, the national security documents stated that the US approach will lessen its intention of conducting long-term and prolonged stability operations (DoD, 2012: 6; Obama, 2011: 6).

Irregular threats: Transnational Criminal Organizations (TCOs)

Near- and Long-term threat and projected strategies

TCOs have stood against US efforts to foster international security for decades, and the US will likely face this threat for decades to come (NSS, 2015: 15). TCOs retain unprecedented wealth and international ties and markets to continue operations. Much of their work is covert and is able to penetrate deep into American society (i.e., narcotics trafficking in US communities that enter the US from Mexico) (NSS, 2010: 43, 49; Williams and Fellab-Brown, 2012: ix, x). Good governance and US security cannot share a relationship with TCOs. These organizations are underscored in the new threat assessment, which notes that TCOs are expanding, becoming richer, growing in influence, and posing a large threat today and in the future. No single TCO is mentioned. The strategy is therefore one that projects to expand well beyond the traditional confines and ensure that the US is able to operate *carte blanche*.

A "multidimensional approach" is noted as the requisite condition for overcoming this challenge. Mitigating TCO financial assets and thus their strength, disrupting illicit trafficking, tempering corruption, strengthening the rule of law, "bolstering judicial systems," and ensuring transparency comprise a collective strategy, combined with other states' strategies to overcome the threat of TCOs today and tomorrow (NSS, 2010: 49; NSS, 2015: 15, 28). Optimizing efforts in combating TCOs requires the coordination of departments and affiliations, hence the "multidimensional" label. With this, the US looks to "safeguarding the global commons"—working with US friends and allies to bring oversight to "shared sea, air, and space domains" (Mullen, 2011: 3; NSS, 2010: 49; NSS 2015: 15). Much of the operations of TCOs rely on the lack of concentration by states on the governance of regions that fall beyond the exclusivity of national boundaries. The NSS (2010: 49) regards these spaces as critical and upon which "all nation's security and prosperity depend."

As part of the strategic approach to dealing with TCOs, the US will focus increasingly on access points to the essential markets that fund TCO operations. Access points are vital links in TCO resource movements in multiple domains and include border crossing and international ports of call (NSS 2010: 50). Multilateral cooperation with other national governments and bodies of authority is aimed at enhancing the awareness of what comes "in" and what goes "out" of territories. Constant monitoring is key to knowing and understanding what sort of traffic takes place in all places at all times. Norms and standards (i.e., anticorruption efforts) are appendages of the anti-TCO approach (Mullen, 2011: 4; NSS, 2010: 50; NSS, 2015: 10, 21). Through establishing a preferred and agreeable set of norms and standards in line with those of the US, a healthier routine of action and conduct in political, economic, and social domains is expected to take place. This in turn, as inferred from the documents, will hinder the operations of TCOs, which would find conducting "business" under such conditions increasingly difficult.

Commerce is an intrinsic part of the health of TCOs. Much of TCO commerce takes place in maritime contexts (Mullen, 2011). Consequently, the US aims at detecting emerging maritime threats and patrolling maritime theaters and regions in order to meet the criteria established in "knowing" what moves where at any given time and place. US attention to TCOs shares ground with its attention to preventing and ensuring its response to other state- and nonstate threats.

Military transitions: Hybrid and "compound" warfare

Although warfare has been practiced and has changed over the course of human history, many of the changes that have taken place and subsequent demands placed upon militaries worldwide have occurred in recent years, such as the rise of unmanned systems, cyber technology, and an increasing digitalization of conflict. General Raymond Odierno, former US Army's Chief of Staff (CoS), commented that, the US military needed to take stock of and positively utilize "one of the most costly lessons it has learned over the last decade: how to deal with the challenge of hybrid warfare" (quoted in Hoffmann, 2012). Hoffmann (2012), and Murray and Mansoor (2012) have also noted that the nature of warfare is changing so as to include both regular and irregular elements operating together on the battlefield. What these authors describe as hybrid warfare is also referred to as "compound warfare" and has far-reaching effects on the structure, organization, and application of military forces (Hoffman 2009). These authors have underscored the impact that hybrid or compound warfare has had on the US by forcing the US government (USG) to reassess the nature of threats it faces in the contemporary world, its resources, and its strategic, operational, and tactical capabilities.

A great deal of changes took place in the aftermath of Soviet dissolution in 1991. The US was no longer faced by a singular or monolithic threat in the form of the Soviet Union and its close allies (i.e., the former Warsaw Pact countries). Resultantly, a surfeit of changes took place in the international strategic environment that ultimately led to a much broader framework in which the US would thence have to operate. The National Military Strategy of the United States of America (NMS) (1995: I; Avance, Ceplecha, Clay, Featherston, Grantham, Kelleher, Kelly, Pendleton, Rupp & Yelder, 1996: 12) devoted a great deal of attention to the view that, "challenges to our [US] global interests did not disappear with the end of the Cold War. Today we face a world in which threats are widespread and uncertain, and where conflict is probable, but too often unpredictable." The US will have to be prepared to contend with threats on multiple fronts and will be required to perform a robust body of tasks and missions.

The Rand Corporation (2011) has engaged in military transitions, that put the US in a position of vulnerability and opportunity. The "US military," notes Rand (2011: n. p.), "was largely designed for major combat operations against state actors but over the past several years it has made significant changes to its training, organizing, and equipping paradigms to adapt to nonstate, irregular threats in Afghanistan and Iraq." Iraq and Afghanistan are marked as watersheds in the history of US military operations for demonstrating radically new approaches that the US's military services will need to adopt in order to acclimatize to a changing threat matrix. Israel is also undergoing this process. The Second Lebanon War, according to Rand (2011), served as a "wake-up call" for the Israeli Defense Forces (IDF) by demonstrating that new forms of weaponry employed by different nonstate actors in multiple environments (including across borders) will demand significant changes of the US as well as changes in the way it approaches conflict. Such changes have given profound permanence to new and potentially unfamiliar domains of security and warfare, whereas "the IDF focused [...] almost exclusively on irregular combat, [it] became generally incapable of the joint combined arms fire and maneuver capabilities associated with major combat operations" (Rand, 2011: n. p.). The US has undergone the same growing pains in Iraq and Afghanistan. As noted by Rand (2011: n. p.), "[t]he ability of an adversary to transition from irregular to hybrid is relatively simple, requiring only the sponsorship of a state actor able to provide sophisticated weapons and training."

Murray and Mansoor (2012) have selected a healthy body of cases to represent the imminent challenges posed by hybrid warfare and the pressures that it will increasingly place on the US to assume or evolve toward a joint position in order to deal with them. "There is no overwhelming consensus," writes Hoffman (2009), "about US strategy in the emerging security environment. History suggests that betting on point predictions or single scenarios is imprudent." Indeed, the term *evolution* has proven highly problematic within this field of studies. Snider (2003) argues that the term *evolution* is misleading and is implicit of a concrete direction and commitment by the US to address a changing security environment. The management approach developed through the Goldwater Nicholas Act, reasons Snider (2003), has done little more than produce a cadre of managerial professionals based on an institutionalized rhetoric of what it takes to adapt a military to new and emerging threats in the modern age. Snider (2003: 19) continues by explaining that the lack of "jointness" that can be measured in US military practice has become "critically damaging to the intended defense transformation." Military transformation cannot simply be measured in terms of emerging and sophisticated technologies, hardware and software; rather, it has much more to do with the "mindset of military and civilians DOD [Department of Defense] professionals, the vision and commitment to carry within their profession, and the service those professional render to the American people."

Empirically, the study of joint warfare and the "jointness" of the US military profession assumed a different image. Kumar (2010), who examines and compares existing literature on joint warfare, claims that the changing dynamic of American "jointness" can be measured in accordance with the changes that have taken place within the Chinese military complex. In this vein, any adaptations to the practice of warfare made by China are a direct result of changes implemented by the US. Therefore, if China is changing its practice, then the explication is made that the US is doing the same. This sentiment, consequently, negatively confronts the conjecture put forward by Snider (2003) that changes in US military practice are not as prominent as originally suggested nor are they exclusively evolutionary in nature. By looking at post-1986 developments, Kumar (2010: 7) employs key events (i.e., the Gulf War, the Somalian crisis, the Kosovo conflict, the recent war in Afghanistan, and the Iraq War) to show that, "the degree of unity and interoperability in warfare has been enhanced."

The level of praise proffered by Kumar (2010) regarding US military interoperability has attracted criticism. In spite of the apparent success of US forces working closely together in the aforementioned conflicts, only one notable theater (the Gulf War theater in 1991) in which all three major military elements (land, sea, and air) were brought under the direct command of a single military leader (that of General Norman Schwarzkopf) can be noted. Skelton (1993–4) approbated the manner in which operations were carried out ("jointly") in 1991. Notwithstanding the high level of acclaim that the US "victory" received post-1991 (with the fall of Soviet Communism), some analyses level sharp criticism against the conduct of military commandership and therefore operations overall during the Gulf War. Kumar (2010: 8) explains that Washington contributed to operations by identifying targets intended for attack, which contradicts the necessity and placement of the senior joint forces commander.

"Hybrid" and "joint" warfare

The US's military forces were originally designed to engage in major combat operations in which traditional state actors would be involved. As a result of the changing security environment of the twenty-first century as described above, the US and its various military services have implemented a number of reforms with the aim of bringing about positive change in the areas of training, organization, equipment, technological applications, and the introduction and use of sophisticated weaponry. Preparations in these areas signal some of the most critical measures to respond to the conditions imposed upon the US and its forces with the emergence of hybrid warfare and a wide range of opponents.

As noted, hybrid warfare brings together aspects of conventional warfare, irregular warfare, and cyber warfare (see Manea, 2016; Monaghan, 2016). What should be underscored is the potential for a combination of nuclear, biological, and chemical weapons; improvised explosive devices (IED); and information attacks. Hoffman (2007: 28) describes hybrid wars simply as "a blend of lethality of state conflict with the fanatical and protracted fervor of irregular warfare." Irregular warfare, according to the US Joint Chiefs of Staff (JCS) (2006–8: GL-16), is a:

> [v]iolent struggle among state and non-state actors for legitimacy and influence over the relevant population. Irregular warfare favors indirect and asymmetric approaches, though it may employ the full range of military and other capacities, in order to erode an adversary's power, influence, and will.

A number of other definitions contribute to the evolving debate over the nature of hybrid threats and wars. Some offer support to others while some are undermining.

Basic concepts of regular and irregular warfare conjoin to form a general understanding of what hybrid warfare is. In essence, hybrid warfare is the potential for any type of attack at any point, anywhere, with the expectation that it will likely be protracted and demand a "joint" response whereby the full range of a nation's military and civilian capabilities is applied. Hoffman's (2009: n. p.) definition buttresses this one by defining a hybrid threat as "[a]ny adversary that simultaneously and adaptively employs a fused mix of conventional weapons, irregular tactics, terrorism and criminal behavior in the battle space to obtain their political objectives."

Historical dimensions of "jointness"

Effectively responding to what Hoffman (2009) describes as an "adversary that simultaneously and adaptively employs a fused mix of conventional weapons, irregular tactics, terrorism and criminal behavior in the battle space to obtain [its] political objectives," the US is moving to merge military services into a coherent body (such as joint force commands) so that one service can reinforce another and enhance overall strength through the principle of complementarity. "The Air Force," as explained by Rubel (2000–01: 46), "provide[s] the Army more air defense than ground forces can provide for themselves [and] [f]or the enemy to defend against one service it must become vulnerable to others; hence the dilemma."

The Goldwater-Nichols (DoD Reorganization) Act of 1986 (signed by former President Ronald Reagan) was an important step in building the requisite foundation for joint warfighting by US military services. The legislation, according to Widnall and Fogleman (1995: 15), "codified jointness and recent historical trends have reinforced this concept. Today, few dispute the efficacy of joint warfighting, which Desert Storm clearly validated." With the end of the Second World War, the US entered into a phase of decline in its military resources. It was not until shortly after the Goldwater-Nicholas Act came into force that the US extensively elaborated its principles of war and fundamentals of joint warfare to span a far more robust range of military operations.

Today, the US finds itself in the midst of its second largest reduction of military assets since 1945, with only the nuclear arsenals and the special operations forces command (JSCOM) unaffected by the reduction in forces and budgets. Each service of the US military possesses specific proficiencies required to respond to a range of dangerous threats to US national interests. No single service is able to fulfill the responsibility of the US to meet every challenge; "[o]nly those assets," states Avance et al. (1996: 18), "which are capable of complementing the capabilities currently in existence and acting as significant force multipliers should be acquired in the future." American military power is predicated on the nature of modern warfare, role of doctrine, values in joint warfare, fundamentals of joint warfare, and exercise of command.

To master multifaceted conditions, the US currently looks at "how to best employ [its] national military power to achieve strategic ends" (JCS, 1995: vi). It is guided by a set of values that include integrity, competence, physical and moral courage, teamwork, trust and confidence, delegation, and cooperation. These principles and the fundamentals of joint warfare are applied broadly enough so as to avoid literal or dogmatic construction. The fundamentals of joint warfare are comprised of unity of effort, concentration of military power, seizing and maintaining initiative, agility, operations extended to fullest breadth and depth, maintaining freedom of action, sustaining operations, clarity of expression, knowledge of self, and knowledge of the enemy (JCS, 1995: viii).

The "new" security environment

True to the tune of the United States Joint Forces Command (USJFCOM) 2010 JOE publication, the US and its military services are preparing for a variety of potential future threats. Joint concept development and experimentation are part and parcel of US readiness today and preparation for tomorrow's strategic environment. "Future trends, shocks, contexts, and implications for future joint force commanders and other leaders" are some of the central pillars of USJFCOM's efforts to understand and merge the concepts of US military "jointness" to the new security environment in which joint operations will increasingly take place (JFC, 2010). "Speculative in nature," JOE (2010) provides the US with an "intellectual foundation" on which to formulate ideas that will guide US future force deployment.

Much like the idea of "jointness," the nature of the new security environment in which the US operates in truly comprehensive and is guided by questions about trends and disruptions likely to affect the Joint Force as well as ways in which they will affect future contexts of joint operations (JFC, 2010: 5). Answers to these problems shape the contexts in which US forces will operate and ways in which joint operations are sure to unfold over the next 25 years. Forecasting the security environment takes place in the context of what the US describes as "enormous disruptions and surprises" found in a number of key areas (JFC, 2010: 12). Special attention is paid to areas that have produced no shortage of debate on what the future holds and how its uncertainties should impact the US. Specific nontraditional security issues are highlighted as the major trends that influence and will continue to impact the security environment in the foreseeable future. It is through an analysis of the nontraditional domains that the US is able to build assumptions and predictions about the security environment of the twenty-first century and link them to the necessity of constructing a twenty-first century joint force able to address them and other national and international stability and security concerns.

Toward a US joint force in the "new" security environment

Recent chairman of the Joint Chiefs of Staff (CJCS) General Martin E. Dempsey has been named the point man who is leading the US toward completely disaggregating the US military and shaping the Joint Force 2020. The purpose is to meet the "growing complexities of defense threats," and to achieve this, General Dempsey calls for "integration of joint command and control frameworks and specialized assets and creation of joint operating environment that is inherently flexible at its core" (quoted in Steele, 2012: 27). Globally integrated operations represent the keystone of the overall concept of Joint Force, and it prepares the US and its military services as parts. It then brings them together as a single body to address the new security environment confronting the US. Joint Force finds its strength in the capacity to merge unique services already in existence with the aim of constructing an overlapping structure in which each service is able to lend support to the other. No single service is expected to fulfill a strategic objective alone. A network of military services lend support to one another to fulfill the objectives as laid out in US military doctrine and strategy documents.

Integration is the core concept of the Joint Force. Here, the idea of integration hones in on the merging of services that exist and operate within the US and its numerous departments. The concept introduces narratives about "new ways of fighting and partnering" in the twenty-first century (JCS, 2012: 4). The idea of new methods of fighting shares a special relationship with the idea of effectiveness. Several elements lie at the very heart of the Joint Force doctrine. First, "globally integrated operations require a commitment to the use of mission command," as noted by the JCS (2012: 4). Rather than elevating mission command decisions, the US looks to empowering mission-level leadership with the freedom do exercise judgment with regard to how an operations unit will orchestrate their assigned tasks (JCS, 2012: 4). The human component is an essential piece of the overall success of this vision because of the critical importance of trust, force of will, intuitive judgment, and creativity as a means of satisfying strategic, operational, and tactical objectives, as well as demonstrating agility, flexibility, efficiency, and effectiveness as essential traits (JCS, 2012: 4). Mission command, however, can be criticized for its inappropriateness in certain situations. Some situations inextricably subordinate actors at the operational and tactical levels of warfare, especially when issues of the highest national interest are at stake (i.e., nuclear deterrence and authorization, air and littoral control, and strategic synchronization of military and civilian resources) (JCS, 2012: 4).

US military preeminence in the new security environment cannot be separated from retaining the initiative, as the NSS (2015) noted strongly. Regulating events on both micro and macro levels means the difference between success and failure in warfare. The US, as such, must be ready to take, retain, and exploit the initiative both temporally and spatially to maintain its position relative to state and nonstate threats (including hybrid threats) (Grothe, 2009: 2; NSS, 2015). Failing to achieve, hold, or take full advantage of and capitalize on different initiatives weakens the US military at the service level (single parts) and on the network level (the entire body). This second point shares a distinct relationship with the first. Grothe (2009: 8) states that "leaders can seize the initiative through opportunities generated by creative thought." An enemy bereft of initiative operates "off balance and [is] not [able to] adapt, or react to the tactics of" its adversary (Grothe, 2009: 5). Effects of holding the initiative can be assessed on multiple levels from the battlefield to capstone command, and they also imbue the US and military as institutions with operational ability that is difficult to measure. "Institutionally," argues Grothe (2009: 5–6), "this will enable the Army [US] to respond quicker to the nation's threats and maximize the potential benefits from new and emerging organizational designs resulting from the recent and ongoing Army transformation."

Conflict that is able to erupt at many points throughout the world accounts for the need to respond to events in much more agile ways. Operations today, much as they have in the past, are likely to extend initially from specific posture points (i.e., bases, stations, prepositioned stocks). Currently, the US operates between 700 and 800 bases and military installations around the globe and maintains a presence of US military personnel in well over 150 countries (Dufour, 2013). It can be expected that the number of bases globally, but foremost in the East Asian region, will increase even further, where the US has been seeking to spread its military assets through building various new bases in order to decrease vulnerability to a possible Chinese attack. Bases are used for various purposes, including training, preparation, and storage of military equipment, and are used by national armies.[4] Not only can conflict erupt at any given point in the world, it can also take place with little or no warning and develop at tremendous speed. Therefore, the need for force agility should be considered alongside the nature in which conflict occurs and the positioning of forces within the US and overseas. JCS (2012: 5) claims that "globally integrated Joint Forces can use capabilities such as cyber and global strike to rapidly bring combat power to bear," yet consideration will still need to be given to the way that force formations are conceived and what their operational reach will look like.

Because consideration has been given to the possibility of fewer massed formations and more nimble command and control, the US—in its future operations—will also have to reconfigure its strategic priorities, shifting from the desire to engage in long-term and large-scale conflict, as was the case with the Iraq War (Mar. 20, 2003–Dec. 2011) and US-led operations in Afghanistan (2001–ongoing) to more select and strategically shallow-investment operations such as the US's role in international military operations in Libya ("Operation Odyssey Dawn" [March 19–31, 2011]), with the aim of enforcing United Nation Security Council (UNSC) Resolution 1973 (United States African Command [AFRICOM], Mar. 2011; US AFRICOM Public Affairs, 2011; DoD, 2011). However, and at the same time, recent events in the Middle East have illustrated that despite a desire to shift away from long-term and large-scale conflict, this remains for the moment a future goal: The US still remains engaged in Afghanistan and Iraq and became engaged in a new conflict in Syria, with no short-term prospects for a quick victory. Furthermore, after a few years of declining numbers, in the last year and a half the number of soldiers active in the Central Command area of operations (the larger Middle East, including Afghanistan) has seen a rapid rise once more, with increasing numbers of soldiers in Iraq foremost, but as well as Special Operations Forces (SOF) operators in Syria and Libya among others, and a now stagnant number of forces in Afghanistan. As such, we could argue that despite the best intentions, the shallow investment operations of the US have somewhat backfired, with the US now invested in anything but shallow operations in the Middle East for a long time to come.

Given an understanding that maintaining a military presence in multiple theaters requires increased integration of current services branches, the US need to look to deeper partnering with other states. International partnering grants further the access of US forces to bases in other states' territory from which to launch operations such as short- and long-range strikes, surveillance and reconnaissance, and conduct training as well as other low-level security missions (Dufour, 2013). Partnering further serves US national interests by illustrating its commitment to working closely with its friends and allies and maintaining a military presence in order to support the existing forces of its allies. Another notable wave of partnering is the ability to demonstrate to potential enemies a nation's capacity for acting on a proactive military philosophy. Partnering is a logical response to the problematic that contemporary security challenges requires much more than the power of a single nation in addition to more than "the military instrument of national power" (JCS, 2012: 6). And indeed, over the course of the last few years, the US military has followed its mantra and doctrine and has increased its number of bases, with foremost Africa Command being extremely active in increasing its number of bases on the African continent. Where prior US military personal presence was limited, by now the African area of operations has become a key element in the US defense strategy, and deployment of forces has increased steadily in the last few years.

Global agility of US forces leads to the concept of globally integrated operations. Essentially two sides of the same coin, the concept of globally integrated operations brings together the abilities of different levels and spheres of command and operation. "Globally postured joint forces," states Combined Arms Support Command (CASCOM) (2013: 12), "quickly combine capabilities with mission partners across domains, echelons, geographic boundaries, and organizational affiliations." The core component of globally integrated operations is to surpass restrictions placed on functional forces by geographical features and force division based principally upon geography. According to the JCS (2012: 6):

> [g]eography remains the logical basis for conducting theater cooperative security, while some missions, such as strategic deterrence, remain functionally distinct. Rather, the intent is to explore hybrid command arrangements that provide greater flexibility in how Joint Forces accomplish their mission.

Geographic proximity is a core characteristic of the concept of "cross-domain synergy" (JCS, 2012). Closing the spatial gaps between joint forces is central to building an advantage over hybrid threats in the new security environment because all forces will be positioned to engage large and small threats. In doing so, existing capabilities of any given force because of the availability of employing adjacent military assets is augmented. Configuring US military forces in this manner produces offensive- and defensive-focused standing capabilities. "The full range of operations," according to DoD (2000: 6), "includes a posture of strategic deterrence. It includes theater engagement and presence activities." Thus, the US will be increasingly poised to meet threats that require decisive proactive operations as well as reactive or response-based operations (JCS, 2012: 6–7).

Informing US military advantage in the new security environment requires increasing integrated efforts with domains that leave little trace or evidence of US military activity (i.e., cyber-space operations, special forces operations, cross-regional strike capabilities, intelligence gathering, surveillance and reconnaissance [ISR], and other forms of discreet military activity) (JCS, 2012). Although some of these components have always been vital to successful military action in the past, the expectation is that action, based on accurate, timely, and accessible intelligence (for example), will enhance the power of a military force and its allies. Beyond accentuating military power, the potential also exists for extending the reach of US military forces with the mechanisms in place. These specific pieces of national military power will provide the US with much-needed flexibility in containing threats as they arise by allowing the US the option of noncommitment or overextending specific policy as it relates to a particular threat. A primary example can be seen in the US "drone" program, in which targets in high-risk and high-threat areas are selectively eliminated. Unintended consequences of alternative operations involving large-scale conventional forces can also be allayed. At the same time, such programs increase transparency and operational oversight due to the number of operating personnel and command structures involved—although there is always the potential for further unintended and unforeseeable consequences. At this given point, it seems that the drone program offers certain capabiltiies and advantages; however, much of the program, and its effectiveness in counterterrorist efforts, remains poorly understood. Kilcullen, a famed Australian COIN specialist formerly working for General David Patrues and considered one the world's leading authorities on insurgencies, argues that the use of drone and its program is actually counterproductive to US national security efforts. As such, it seems at this point difficult to determine to what extent the drone program is an effective tool in the lexicon of national security strategy.

Joint force, full-spectrum dominance, and UAVs

The concept of full-spectrum dominance is about the ability of the US to project its military capabilities toward any type of enemy that presents a threat to US national security or interests abroad. It is not only about engaging with enemies; rather, it is about "overcome[ing] an adversary in any operational environment" (Rees, 2016: 74). The concept of force projection is examined in the context of the Asia-Pacific region by Rees (2016: 74), who argues that "America would need to be able to project force and to damage an enemy with precision strikes (Cordesman & Hess, 2013). It would be neither realistic nor desirable to engage in a manpower-intensive land war in Asia because of the probability of facing numerically superior forces." Speaking of the use of precision strikes, the emphasis is placed on the ability to use long-range airpower in a variety of forms. Rees (2016) refers to the use of airpower in conjunction with naval power to control both the skies and the seas, referencing the necessity of carrying out strikes along the shores through amphibious operations. Naval power and amphibious components are removed when considering the issue of dominance in the case of inland operations such as Afghanistan and Iraq. Lessons from both of these campaigns highlight the critical need for the US and its military forces to be ready for full-spectrum combat operations including "from high-intensity war-fighting to counterinsurgency operations" (Rees, 2016: 73; Korb, 2009).

The US is caught in a growing dilemma. It is neither experienced in the strategies and tactics of counterinsurgency warfare through long-term campaigns and engagements, nor is it fully prepared for engaging in traditional warfare using armored vehicles in high-intensity conflicts. This dilemma is partly the result of the disagreement that has taken place between the military branches of the US and the government bodies of authority in Washington,

DC. The Obama administration has shown a marked reluctance to be drawn into long-term or major conflict in other countries that would see the US use its armed forces in high-intensity conflict, as can be witnessed by the language and goals of the 2010 and 2015 NSS. Instead, President Barack Obama has demonstrated the abilities of the US to pursue the idea of full-spectrum dominance by utilizing remotely piloted aerial systems (RPAS) or armed drones in areas that extend beyond the reach of US conventional, manned, military systems and to steer away from high-intensity conflict, perfectly reflecting the position of the US government at the highest levels. The use of RPAS is a defining feature for the capabilities of the US and its military forces because of how the technology itself allows for the US to engage in low-intensity conflict and project its military power in locations that would have remained inaccessible to its conventional forces. Extending the reach of its forces in this way also allows the US to avoid placing its service men and women in harm's way (Rees, 2016). Motivational logic of this kind transcends military and civilian practice. In both cases, the technology makes it possible to be present in areas that are "inhospitable or inaccessible to humans" (Vallor, 2015: 30).

Integrating RPAS into military operations was advocated as a valuable although risky venture in numerous papers and reports by Lieutenant-Colonel Andrew F. Krepinevich, Jr. (Ret.) (2002: 79), who reasoned that UAVs and robotics are important for allowing the US military to "conduct beyond-line-of-site reconnaissance and surveillance." In a comprehensive report, Krepinevich, Jr. (2004) explained that the "sanctuary factor" will not aid the US and its military operations and capabilities. The US, if it is to remain an effective fighting and security force in the contemporary security environment overseas, should take advantage of the information age and the availability of new technologies:

> New information systems, ranging from global positioning system (GPS) satellites, to unmanned aerial vehicles (UAVs), to tactical internets, promise similar leaps in effectiveness. Combined with the growing ability to conduct precision fires at ever-greater ranges, new information systems offer the prospect of an Army that can violate the principle of mass to disperse its forces (thereby reducing their vulnerability to detection and destruction) without suffering the traditional penalty of loss of effectiveness. Land warfare conducted in this manner presents two major competitive advantages. First, it moves the Army beyond the heavy, mechanical air-land era of warfare that began with the blitzkrieg, enabling the Expeditionary Army. Second, highly distributed and networked forces may be the antidote to the growing A2/AD threat (which will be elaborated upon presently). Thus, the Army's ability to develop distributed, networked forces will likely be critical to its long-term dominance.

Dominance, force transformation, and joint interoperability are not only a matter of priority within the US. Joint Vision 2020 exemplifies the need for the

US to incorporate new technologies, tactics, and techniques into its military practice; however, Joint Vision 2020 plays another role that is often overlooked when trying to understand how such measures are adopted by other military institutions in other states. Through the manner in which the US integrates its operations and brings its services and warfare centers together, other governments can either take notice or are actually encouraged by the US military to do the same. "Joint Vision 2020," writes Lieutenant-Colonel W. Bruce Rember (2000: 13), "emphasizes the importance of interoperability across not just service lines, but also among allies and governmental agencies." "Total force transformation requires families of systems to be built in overarching strategic architectures" that can take place simultaneously in the US and in other states facing similar (i.e., budgetary) constraints on their militaries while operating in the same uncertain and dangerous security environment, placing increasingly intensive and diversified demands on the state and its security forces.

Conclusion

The first half of this chapter has engaged with revisions and adaptations in the US NSS' and DSG since 9/11, and the second half has engaged with the concept of US joint forces and joint warfare in the post-9/11 period predicated on the idea of "full-spectrum dominance." By identifying and outlining its national security priorities and noting immediate and prospective (future) threats to US security, the NSSs and DSG are intentionally diverse in content. In noting these factors, the US calls attention to the need for a "comprehensive range of national actions and a broad conception of what constitutes our [US] national security" (NSS 2010: 51). Indeed, the US has looked at both utilitarian (substantive) and conceptual aspects of US security and threat assessment. It has focused intensively on such states as Iran, North Korea, China, Russia, and India and has looked to non-state-oriented entities, including violent militant extremists and TCOs. In less substantive terms, the US turns its attention to notions of "cooperation" and "engagement," "recognition" or new actors (rising powers), the pursuit of "constructive engagement," "promoting responsibility" in the international system, democracy building and strengthening human rights abroad, the practice of "principled engagement," "strengthening norms," and building "cooperation" (NSS 2010; NSS 2015). The strategies truly reflect the nature of current and projected threats, and the US has put forward strategic approaches that are a duality of hard and soft power (Gray 2011).

In maintaining its commitment to broad conceptions of security, the US is able to continue with the macrosecuritization program that arguably began under the Bush administration back in 2001 with the War on Terror (WoT). The national documents cover such a wide range of regions and factors that there remains little room to argue that nothing constitutes a security issue or has the potential for becoming a threat—especially in the face of US interests. Further implicit within the discourse of the NSSs and DSG is US foreign policy objectives and the ideological connotation that US foreign policy remains a priority above the interests of other nations in the world system. The US has properly extended its view of an ideal security arrangement in the international system while at the same time linking US values and interests to that system. Overwhelmingly, US strategic success is rooted in force projection on a global scale and in a manner that seeks to establish a new set of beliefs about the role of the US in the new security environment, which then acts as an enabler of new forms of US behavior in the context of the use of force. The US has extended itself a healthy range of operationality even if this might be at the expense of the Westphalian tradition.

Sustaining US military pre-eminence is part of a shared relationship with the idea that future operations will move toward full spectrum in terms of military services (whole-of-government) and integrated with friends and allies (whole-of-partners). In looking at the emerging security environment of the twenty-first century and the multiplicity of emerging threats within it, the US and its military forces are faced with a deceptively simple problem in an effort to yield a US force that can continue to dominate both battlefield and global strategic environment: increase military power, improve flexibility and agility, reduce spending, and strengthen global partnerships. Efforts over the last several decades have resulted in slow movement toward this end; however, despite its progress, the US has a long way to go if it is to meet the demands of the twenty-first century security environment while adjusting to ongoing challenges on the domestic front and further abroad.

Moving toward integrated operation and a force based on the idea of "jointness" means emphasizing the military's agility and flexibility and bringing together its current capabilities (including military powers of other nations). Command, initiative, tailoring force size, integrated operations through flexibility, globally integrated operations, merging military prowess across domains, developing and exercising light-footprint capabilities, and emphasis on intelligence are critical steps in building a truly joint US military force (Garamone, 2012). Adapting the current US military overall in these ways becomes even more critical in the new security environment with the rise of hybrid threats, which can be seen as "interest-based rational actors" (presenting both a qualitative and quantitative challenge) that are effectively revolutionizing the operational art and, in the words of Fleming (2011: ii), a "potential paradigm shift as a doctrinal and organizational Revolution in Military Affairs (RMA).

Notes

1 The concept of nontraditional security challenges notions of what constitutes a security threat. NTSs include, but are not limited to, transnational and organized crime, terrorism, disaster relief, cyber security, climate change, public health and welfare, food and water security, human trafficking, migration, and desertification.

2 We use the term *hybrid threats* to refer to regular, irregular, and criminal actors working independently or synchronically to achieve a single or set of social, political, economic, and religious goals.

3 "Security" and "threats" are abstract concepts. Indeed, the confluence of material provided in scholarly fields (e.g., political science, international relations, security studies, military studies, and war [strategic] studies) shows that adopting an "objective" lens for the study of both is problematic. Therefore, they must be treated subjectively. "Security"

and "threat" (as ideas) become manifest both ontologically and reiteratively. That is, anything can become a security concern or threat if they are labeled as such. The practice of security is a reactionary craft that is constantly broadening (or broadened) to include new and changing actions (i.e., threats). This is reflected in documents used in the following analysis. Broadly speaking, we use the term *security* non-"traditionally" to refer to action(s) and/or series of events within such fields as economic, political, social, religious, sustenance and nutrition, health, environmental, physical, and societal/community. Threats are as illusory as the concept of security. Threat is also taken as action(s) and/or a series of measures that either directly/indirectly or advertently/inadvertently impact a political, social, or economic element (tangible or intangible) in negatives ways. These two terms are closely linked, given that both are abstract and can become focal referent objects in micro or macro-securitization programs.

4 US military bases and installations are typically classified along four lines: (1) air force bases, (2) army or land bases, (3) navy bases, and (4) communication and spy bases (Dufour, 2013).

References

Avance, D. W., Ceplacha, C. S., Clay, R. E., Featherston, T. M., Grantham, D. S., Kelleher, P. A., Kelly, D., Pendleton, G. L., Rupp, J. and Yelder, C. E. (1996). "The Joint Strike Fighter," a development study presented to the Directorate of Research Air Command and Staff College. [Online]. Available at: <http://www.fas.org/man/dod-101/sys/ac/docs/96-207.pdf>.

Bedeski, R. E. (1992). "Unconventional Security Threats—An Overview," *Interdisciplinary Peace Research*, 4(2): 78–90. [Online]. Available at: <http://www.tandfonline.com/doi/abs/10.1080/14781159208412755#.U2ESta1dUug>.

Combined Arms Support Command (CASCOM) (Aug. 30, 2013). "Army 2020 and Beyond Sustainment White Paper." [Online].

Cordesman, A. and Hess, A. (2013). "The Evolving Military Balance on the Korean Peninsula and Northeast Asia," report of the CSIS Burke Chair in Strategy. Lanham, MD: Rowman & Littlefield.

Department of Defense (DoS) (January 2012). "Sustaining US Global Leadership: Priorities for 21st Century Defense," The White House, Washington, DC. [Online]. Available at: <http://www.defense.gov/news/defense_strategic_guidance.pdf>.

Dufour, J. (Dec. 14, 2013). "The Worldwide Network of US Military Bases—The Global Deployment of US Military Personnel," Global Research (Center for Research on Globalization), Montréal, Canada. [Online]. Available at: <http://www.globalresearch.ca/the-worldwide-network-of-us-military-bases/5564>.

Fleming, B. P. (2011). "The Hybrid Threat Concept: Contemporary War, Military Planning and the Advent of Unrestricted Operational Art," School of Advanced Military Studies, United State Army Command and General Staff College, Fort Leavenworth, Kansas. [Online]. Available at: <http://cgsc.contentdm.oclc.org/utils/getdownloaditem/collection/p4013coll3/id/2752/filename/2753.pdf/mapsto/pdf>.

Garamone, J. (Sept. 28, 2012). "Dempsey Releases Concept to Build Joint Force 2020," US Department of Defense, Washington, DC. [Online].

Gray, C. S. (April 2011). "Hard and Soft Power: The Utility of Military Force as an Instrument of National Power in the 21st Century," Carlisle Papers, Carlisle Barracks, PA: US Army War College Strategic Studies Institute. [Online].

Goldgeier, J. and Suri, J (2016). "Revitalizing the U.S. National Security Strategy," In *The Washington Quarterly*, 38:4, pp. 35–55.

Grothe, G. K. (2009). "Innovation Versus Adaptability: Seizing the Initiative Through Creative Thinking Versus Reacting to the Enemy," School of Advanced Military Studies, United States Army Command and General Staff College, Fort Leavenworth, Kansas. [Online]. Available at: <http://www.au.af.mil/au/awc/awcgate/sam/innov_vs_adapt_grothe.pdf>.

Hoffman, F. G. (2007). *Conflict in the 21st Century: The Rise of Hybrid Wars*. Arlington, VA: Potomac Institute for Policy Studies.

Hoffman, F. G. (2009). "Hybrid vs. Compound War," *Armed Forces Journal*. [Online]. Available at: <http://indianstrategicknowledgeonline.com/web/4198658.pdf>.

Hoffman, F. G. (2012). "Review Essay: History of Hybrid Warfare," *Small Wars Journal*. [Online]. Available at: <http://smallwarsjournal.com/jrnl/art/review-essay-history-and-hybrid-warfare>.

Johnson, K. D. (June 2009). "China's Strategic Culture: A Perspective for the United States," Carlisle Papers, Carlisle Barracks, PA: US Army War College Strategic Studies Institute. [Online].

Joint Chiefs of Staff (JCS) (1995). "Joint Publication 1: Joint Warfare of the Armed Forces of the United States," Washington, DC. [Online]. Available at: <http://www.dtic.mil/dtic/tr/fulltext/u2/a306543.pdf>.

Joint Chiefs of Staff (JCS) (2006/08). "Joint Publication 3-0: Joint Operations," Washington, DC. [Online]. Available at: <http://www.dtic.mil/doctrine/docnet/courses/operations/jfcon/jp3_0.pdf>.

Joint Chiefs of Staff (JCS) (2012). "Capstone Concept for Joint Operations: Joint Force 2020," Washington, DC. [Online].

Korb, L. (2009). "Defence Transformation," in Wheeler, W. and Korb, L. (eds.), *Military Reform: An Uneven History and an Uncertain Future*. Stanford, CA: Stanford Security Studies.

Krepinevich, A. F., Jr. (2002). "The Army and the Future of Land Warfare: Transforming the Legions," *Joint Forces Quarterly*: 76–82. [Online]. Available at: <http://www.dtic.mil/dtic/tr/fulltext/u2/a483750.pdf>.

Kumar, A. (2010). "The Evolving Doctrine of Joint Warfare in the USA and its Effects on the Military Modernisation of China: Implications for India," Paper presented at the 18th Biennial Conference of the Asian Studies Association of Australia, Adelaide, Australia. [Online].

Manea, O. (Feb. 2, 2016). "The Strategy of Hybrid Warfare," *Small Wars Journal*. [Online]. Available at: <http://smallwarsjournal.com/printpdf/38740>.

Monaghan, A. (2016). "Putin's Way of War: 'War' in Russia's 'Hybrid Warfare,'" *Parameters*, 45(4): 65–75. Strategic Studies Institute, Washington, DC. [Online].

Mullen, M. G. (February 2011). "The National Military Strategy of the United States of America," Washington, DC: The Pentagon. [Online].

Murray, W. and Mansoor, P. R. (eds.) (2012). *Hybrid Warfare: Fighting Complex Opponents from the Ancient World to the Present*. New York, NY and Cambridge, UK: Cambridge University Press.

National Security Strategy (NSS) (May 2010). The White House, Washington, DC. [Online]. Available at: <http://www.whitehouse.gov/sites/default/files/rss_viewer/national_security_strategy.pdf>.

National Security Strategy (NSS) February 2015, The White House, Washington, DC, available at <https://www.whitehouse.gov/sites/default/files/docs/2015_national_security_strategy.pdf>

National Strategy for Counterterrorism (June 2011). The White House, Washington, DC. [Online]. Available at: <https://www.whitehouse.gov/sites/default/files/counterterrorism_strategy.pdf>.

New York University (2014). "The New Global Security: Assessing the Threats of the 21st Century." [Online]. Available at: <http://www.nyu.edu/global/global-academic-centers/washington-dc/nyu-washington--dc-events/the-new-global-security--assessing-the-threats-of-the-21st-centu.html>.

Obama, B. (President of the United States) (June 2011). "National Strategy for Counterterrorism," Washington, DC: The White House. [Online].

Rand Corporation (2011). "Preparing for 'Hybrid' Opponents," Research Brief, Santa Monica, CA. [Online]. Available at: <http://www.rand.org/content/dam/rand/pubs/research_briefs/RB9600/RB9620/RAND_RB9620.pdf>.

Rees, W. (2016). "The United States and Transformation," in Eadie, P. and Rees, W. (eds.), *The Evolution of Military Power in the West and Asia: Security Policy in the Post-Cold War Era*. Abingdon, UK: Routledge.

Rember, B. (2000). "Tools for Transformation: The Military Requirements Process," Hoover Institution, Stanford University, Stanford, CA. [Online]. Available at: <http://oai.dtic.mil/oai/oai?verb=getRecord&metadataPrefix=html&identifier=ADA393962>.

Rubel, R. C. (2000–01). "Principles of Jointness," *Joint Force Quarterly*: 45–9. [Online]. Available at: <http://www.dtic.mil/dtic/tr/fulltext/u2/a525717.pdf>.

Skelton, I. (1993-4). "Taking Stock of the New Joint Era," *Joint Forces Quarterly*, 4: 15–21. [Online]. Available at: <http://www.dtic.mil/doctrine/jfq/jfq-13.pdf>.

Snider, D. M. (2003). "Jointness, Defence Transformation, and the Need for a New Joint Warfare Profession," *Parameters*. [Online]. Available at:<http://www.dtic.mil/cgi-bin/GetTRDoc?AD=ADA486430>.

Steele, D. (Nov. 2012). "Setting the Azimuth for Joint Force 2020: Globally Integrated Operations and Mission Command," *Army*: 27–29. [Online]. Available at: <http://www.ausa.org/publications/armymagazine/archive/2012/11/Documents/Steele_1112.pdf>.

Terrill, W. A. (2011). "The Saudi–Iranian Rivalry and the Future of Middle East Security," Carlisle Papers, Carlisle Barracks, PA: US Army War College Strategic Studies Institute. [Online].

United States Africa Command (AFRICOM) (Mar. 2011). "Statement: AFRICOM Commander on Commencement of Military Strikes in Libya," General Carter Ham, Commander, US Africa Command, Stuttgart, Germany. [Online]. Available at: <http://www.africom.mil/NEWSROOM/Article/8090/statement-africom-commander-on-commencement-of-mil>.

US AFRICOM Public Affairs (Mar. 20, 2011). "Overview of 1st Day of US Operations to Enforce UN Resolution 1973 Over Libya," US AFRICOM, Stuttgart, Germany. [Online]. Available at: <http://www.africom.mil/NEWSROOM/Article/8097/overview-of-1st-day-of-us-operations-to-enforce-un>.

US Department of Defense (DoD) (Aug. 1, 2012). Joint Publications (JP) 3-07.3 – "Peace Operations." [Online]. Available at: <https://fas.org/irp/doddir/dod/jp3-07-3.pdf>.

US Department of Defense (DoD) (Aug. 11, 2011). Joint Publication (JP) 3-0 – "Joint Operations." [Online]. Available at: <http://www.dtic.mil/doctrine/new_pubs/jp3_0.pdf>.

US Department of Defense (DoD) (Jan. 2012). "Sustaining US Global Leadership: Priorities for 21st Century Defense," Washington, DC. [Online]. Available at: <http://www.defense.gov/news/defense_strategic_guidance.pdf>.

US Department of Defense (DoD) (July 16, 2014). Joint Publication (JP) 3-05 – "Special Operations." [Online]. Available at: <http://www.dtic.mil/doctrine/new_pubs/jp3_05.pdf>.

US Department of Defense (DoD) (Mar. 23, 2013). Joint Publication (JP) 1 – "Doctrine for the Armed Forces of the United States." [Online]. Available at: <http://www.dtic.mil/doctrine/new_pubs/jp1.pdf>.

US Department of Defense (DoD) (May 30, 2000). "Joint Vision 2020."

US Department of State (DoS) (Mar. 27, 2011). "TRANSCRIPT: NATO Enforcing All Aspects of UNSCR 1973 in Libya," US AFRICOM, Brussels, Belgium. [Online]. Available at: <http://www.africom.mil/NEWSROOM/Article/8165/transcript-nato-enforcing-all-aspects-of-unscr-197>.

US Joint Forces Command (JFC), (2010). "Joint Operating Environment 2010." [Online]. Available at: <http://www.fas.org/man/eprint/joe2010.pdf>.

Vallor, S. (2015). "Beyond Originary Givenness? Postphenomenology, Digital Imaging, and Evolutionary Responsibility," in Friis, J. K. B. O. and Crease, R. P. (eds.), *Technoscience and Postphenomology—The Manhattan Papers*. Lanham, MD: Lexington Books.

Walsh, E. (Jun. 4, 2011). "Non-Traditional Security Threats in Asia: Finding a Regional War Forward," East Asia Forum. [Online]. Available at: <http://www.eastasiaforum.org/2011/06/04/non-traditional-security-threats-in-asia-finding-a-regional-way-forward/>.

Widnall, S. E. (Hon.) and Fogleman, R. R. (1995). *Global Presence 1995*. Washington, DC: Government Printing Office.

Williams, P. and Fellab-Brown, V. (2012). *Drug Trafficking, Violence, and Instability*. Carlisle Barracks, PA: U.S. Army War College Strategic Studies Institute. [Online]. Available at: <http://www.strategicstudiesinstitute.army.mil/pdffiles/pub1101.pdf>.

Part 2

Non-traditional warfare

4 Asymmetric wars and the future of US warfare

Frank Jacob

Introduction

Historians seldom concern themselves with the future, but understanding the past can provide valuable insights into what lies ahead of mankind—especially when it comes to war. The world saw 138 wars between 1945 and 1989, which, collectively, caused the deaths of 23 million people (Luttwak, 1995, p. 111); as such, the twentieth century has been referred to as the century of extremely violent societies (Gerlach, 2011). That being said, wars are a prominent part of human history—Anna Simons (1999), professor of Defense Analysis at the Naval Postgraduate School, stated that "the idea that we can now out-think, out-maneuver, outmode, and maybe even outlaw war is revolutionist in the most traditional sense" (p. 75). Her reasoning is that even if "we might wish war away, [...] the fact that states continue to prepare to fight other states" (Ibid.) is inevitable. Simons emphasized three factors that affect the future of warfare:

1 Causes are inherent (ergo, war will always be likely).
2 Conditions create situations that call forth war (change the conditions and we can limit war).
3 War emerged at a particular point in time for particular reasons (thus, it can also be made to disappear in time via reason).

(Ibid.)

Simons, along with many others, perceived war as inevitable. However, an important question remains—how is the United States (US) and its military supposed to prepare for such a future? As Luttwak (1995) asserted, the end of the Cold War yielded the dissolution of an "entire culture of disciplined restraint in the use of force" (p. 111). Hot wars were the consequence of this recklessness, and the US—now more than ever—must accept that such a reality will lead to other nations taking preventive military actions. American military planners have not only to "minimize the exposure of U.S. military personnel to the risks of combat" (Ibid., p. 112) but also to convincingly explain the necessity of war to the public.

Several factors determine the responsibilities of US military planners, and these factors require analysis. The present chapter will suggest possible scenarios for the future of US warfare based on historical perspectives. After conducting an initial analysis of existing future visions on warfare,[1] I will focus on the current concerns that current military planners have to incorporate into their strategies. My main topics are as follows: (1) the future of troops who have to fight in asymmetric wars, be knowledgeable of modern technology, and be trained to encounter civilians on the battlefield; (2) the future forms of war, particularly in the context of cyber warfare and economy; and (3) the US's potential enemies in future wars. Since common ideologies in warfare are directly influenced by previous ones, very few current military perspectives are particularly unique; this is why it makes sense to analyze wars from oldest to newest when speculating on scenarios that could occur in upcoming years or decades.

The past of the future war

Military luminaries and ordinary soldiers alike have devised countless war tactics throughout the ages by adapting better strategies, using new technologies (Dworok and Jacob, 2015), or taking advantage of geographical aids (Danielsson and Jacob, 2016). Sun Tzu (2002, p. 6), Machiavelli (1995, pp. 46–47), and Clausewitz (1873) have all expressed their thoughts on issues like geography in war, but they also emphasize that a successful general must rely on more than these intelligences to win battles. It is important to consider lessons from the past in order to improve future war strategies, and commanders must be aware of all possible determinants of a future war.

Hudson Maxim (1908) put this knowledge into an easy formula: "To-day it has become a recognized truism of military science that victory depends upon the concentration of attack upon the most vital points of an enemy's position, while offering to the enemy the minimum of vital exposure" (p. 821). Winning a war is simply beating the enemy without being beaten, and even in 1908, Maxim understood these very fundamentals that the US military continues to be challenged by: "With improvements in weapons, warfare becomes more and more a matter of exact science and the military man becomes more and more a civil and mechanical engineer. In the military land operations of the future, science will more than ever be supreme above mere brute force" (Ibid., p. 822). Maxim was also likely in favor of the weapon-producing industries, stating that

> The peace advocates are so short-sighted that they do not see that if we built but a few guns, we are obliged to slaughter with them, whereas if we were to build guns enough, we could then make war on war and put an end to slaughter. When we have only a few guns, and not enough to prevent war, then we must use them for killing. If we build guns enough, then we prevent war, and the gun is converted from a death-dealing implement into an instrument for saving life.
>
> (p. 831)

Regardless of such dubious statements, Maxim still had to consider the very things that today's military planners do. While much of the Western public believes that society has lived in peace for many years, commanders have continued to create military plans and devised many proxy war operations in the so-called "Third World" during the Cold War.

American scientist and inventor John Hays Hammond, Jr. (1916) described a military problem that the US continues to face: "[C]onsidering [...] the defense of the United States, there are two main propositions involved: First, the problem of what materials as weapons we need in our defenses; and second, what form of organization there shall be in the handling and development of these materials" (p. 373). Hammond also said, "[W]e are still battering each other with missiles, but with far more precision, greater range, greater rapidity of fire and greater destructive effect" (Ibid., p. 374). He also described the First World War as "a curtain of fire, an iron curtain of thousands of explosive shells, that [swept] the enemy's lines" (Ibid.); this turmoil was caused by technology that has since been developed to be even more destructive.

During the Second World War, Colonel Leon A. Fox (1942), an officer of the US Army Medical Corps, was concerned with the future use of biological weapons in war. He claimed that "bacterial warfare is one of the recent scare-heads that we are being served by the pseudo-scientists who contribute to the flaming pages of the Sunday annexes syndicated over the nation's press" (p. 5). Like Maxim, Fox was frustrated by the public's misunderstanding of war: "The space and thought that have been given to this question by feature writers have not been without effect and many people now believe that bacterial warfare represents a real threat and problem for future generations" (Ibid.). Fox criticized the ideologies of his era, stating that "the mere fact that this great body of peace workers considers bacterial warfare seriously enough to prohibit its use justifies military men in considering this agency of warfare. We know how little treaties protect, so we should study the question to see if the use of biologic weapons is a real problem for the military minds of the future" (Ibid.). It is thus apparent that in this era, there was a great conflict of interest between the public and the military. The former proclaimed that peace was the best possible way to secure a better future, while the latter asserted that preparing for war was of utmost importance. Since war is more likely to occur than not, one could argue that the military was right to think as they did, and Balmer's (2002) suggestion that military scientists—or scientists whose research could be used for military purposes—need to justify their work to a peace-oriented public is thus sensible.

What, then, can the past teach us about the future? One adage that rings true, particularly regarding war, is that some things never change. While technological races are often a problem, so are the continuous conflicts of interest between military planners and the public. While militaries constantly prepare for war, the public often denies the very possibility of it ever occurring in their lifetimes. This especially applies to the US; since no wars have occurred on US soil since the end of the Civil War in 1865, the American population generally disbelieves the possibility of future attacks—besides terrorist attacks. Although

I wish not to make an argument for more war efforts, I believe the public should be more aware of potential threats and thus better understand the US's decisions to fund new military technologies. Even so, although the development of these technologies is a large part of the issues at hand, it is not the only important facet of US warfare's future.

Current issues of planning for future wars

According to Hy S. Rothstein (2007), a senior lecturer at the Naval Postgraduate School in Monterey, CA, the US has a problem:

> The American strategic culture is not so flexible. As a result, irregular threats are handled quantitatively, based on the level of importance or threat posed to US interests, even if a quantitative approach undermines the accomplishment of strategic outcomes. Therefore it seems that, for irregular threats, particularly those where local government legitimacy is in question, U.S. success is inversely related to the priority senior U.S. officials (civilian and military) attach to the effort. Only when the issue at stake is of secondary importance can the U.S. response result in a thoughtful, tailored approach to a non-standard threat.
>
> (p. 276)

Instead of relying on ad-hoc, flexible solutions in response to potential problems, the US tends to overreact when threatened. For Rothstein, this suggested that "new policies, doctrines and technologies will be shaped by these characteristics," and that "[d]efence decisions will be made within the context of this unique way of war" (Ibid., p. 277). Instead of wishing complete annihilation upon its opponents, the US military could instead diversify its tactics. This issue of strategy raises questions about the future of troops participating in future military campaigns.

Troops

Troops have to deal not only with literal battlefields but also with cultural ones. They must make peace with their individual "mode-of-production differences, ideological and ideational differences, and/or differences in political organization and social type" (Simons, 1999, p. 84). Asymmetric wars and low-intensity conflicts (Münkler, 2004; van Creveld, 2008) have largely replaced conventional ones, and the soldiers in these conflicts are more than just fighters. It is challenging to fight an invisible enemy, especially since "[g]uerillas and terrorists typically engage in small-scale actions designed to undermine everything that large-scale conventional militaries and states stand for: protection and security" (Simons, 1999, p. 84). When it comes to fighting enemies on the battlefield, the structure of a soldier's own troop can be decisive when it comes to victory or defeat.

While T. E. Lawrence assumed that smaller units would outperform large ones, other authors—including Stalin—claimed that crushing victories could only be achieved by masses of soldiers (Rothstein, 2007, p. 275; see also Weigley, 1977). Today, the argument for the use of smaller units is more persuasive:

> Personal relationships create a commitment among people and differ significantly from faceless relationships. Consistent and visible commitment tends to bring about reciprocal behavior. This can generate an atmosphere of trust that is arguably a necessary attribute for a high-performance organization. Trust relationships and commitment to team-mates generate extraordinary result.
>
> (Rothstein, 2007, p. 275)

Conversely, Stalin emphasized the importance of a quality army, stating that "[n]umbers, or sheer mass, can make up for inferiorities in other important areas" (Ibid.), and that small groups are disadvantageous in asymmetric wars against enemies who generally avoid open forms of battle (i.e., guerilla warfare). However, the US campaign in Afghanistan following 9/11 is evidence against this claim. During this campaign, "[a] very small contingent of extraordinarily capable men defeated an enemy army and ousted a government" (Ibid, p. 287). This small force was successful in the early years of the war in Afghanistan, but the massive number of troops that were later deployed complicated the situation:

> The rapid, effective and efficient success of the US military in the initial phase of the war in Afghanistan has been followed by a build-up of more than 20,000 military personnel and a massive logistics infrastructure. Unfortunately, the build-up of forces marked the beginning of a less efficient, less effective operation and a countryside that has become less secure with each day that passes.
>
> (Ibid.)

The combination of indigenous troops, US special forces, and precise air strikes caused the Taliban to collapse within a few weeks, but after that period, the war became even more asymmetric—ultimately, the US never found an appropriate strategy to alter the course of the war back in their favor. Stephen Biddle (2003) said that some military analysts "see the 'Afghan model' as warfare's future and think it should become the new template for U.S. defense planning. Others, however, see Afghanistan as an anomaly—a non-repeatable product of local conditions" (p. 31). These ambivalent perspectives demand a closer look at this particular operation.

On October 7, 2001, the US Air Force started its air strikes against the Taliban. Eight days later, special operations forces (SOF) were sent in to support the northern indigenous allies. Three phases followed: the battles in the north to gain control of Mazar-i-Sharif; battles in the south around Kandahar; and a campaign against Taliban and al-Qaeda at Tora Bora, as well as Operation Anaconda, which was supposed to crush the enemy's forces in the Shah-i-Kot Valley. The US's victory over Mazar-i-Sharif on November 10, 2001, paved the way to Kabul's fall, and Kunduz surrendered not long after this, on November 26. The US's invasion of this region thus proved to be one of its "splendid little wars." However, the campaign also demonstrated the difficulty of battling troops who use civilians as camouflage. Biddle (2003) provided an example for this problem, outlining the crux of the issue in asymmetric warfare:

> Al Qaeda fighters wearing the flowing robes of local herdsmen and traveling in small parties among the mountains, however, are nearly impossible to distinguish at a distance from the noncombatants who tend goats or travel through such areas routinely. And defenders able to operate under radio silence by communicating using runners, landlines, or other non-broadcast means can reduce signal intercepts to a level that makes identifying specific fighting positions very difficult. Against such targets, it is far from clear that any surveillance technology coming anytime soon will ensure reliable targeting from standoff distances.
>
> (p. 37)

Technology alone could not solve this dilemma. Once the Taliban boasted such a powerful element of surprise, they also became more dangerous than the US could have predicted. Consequently, more US ground forces were required to launch successful counter-surprise attacks. Because of Operation Anaconda, US and Canadian forces—as well as their Afghan allies—"provided significant ground forces that ultimately shouldered an essential load of old-fashioned close combat against surviving, actively resisting opponents" (Ibid., 44). Events in Afghanistan thus indicated that the future of US warfare would have to focus on both precise firepower and skilled soldiers who could maneuver on the ground. However, it is doubtful that future campaigns could progress as the one in Afghanistan had, and the SOF's training would be an essential element of every future strategy. The campaign in the Middle East also suggested that shorter lines of command and immediate reactions are more efficient than long chains of command, since "the typical slow phase-by-phase model for the build-up of forces did not fit the reality on the ground nor the political need to strike back quickly and effectively" (Rothstein, 2007, p. 287). SOF was able to take action without direct orders, which was only possible though prior planning in Washington. Therefore, a short chain of command, collective trust in the operation, and tactical skills of ground soldiers in Afghanistan were what made the initial war phase in the nation a success.

Approximately 100 SOF soldiers, as well as tactical air force controllers, orchestrated the campaign from the ground while having access to the complete arsenal of US weapons. These capable soldiers demonstrated that well-trained military personnel with proper access to resources can achieve results.

Martin van Creveld (1985) emphasized that independent operations with short chains of command, however, are only one option to achieve a successful military command. A loose control strategy, which can only result from trust in the education and training of the officers, can accomplish even more than one that relies on close monitoring. Such "[b]ottom-up solutions, where US cadres are integrated with local counterparts, produce measurable and, more importantly, durable results" (Rothstein, 2007, p. 293) and consequently demand two things: (1) that the SOF officers are adequately educated, and (2) that commanders have trust in the abilities of those sent into battle. A good soldier, however, not only needs good training but also a reliable and efficient weapon. Technology and technological progress are therefore important to keep the soldiers' arsenals up to date.

Technology

Although "military technology has never stood still" (Cohen, 1996, p. 38), it progresses more slowly than one would assume; its development consists of an evolutionary process more than anything else (Jacob and Visoni-Alonzo 2016). Fox (1942) summed up the obstacles that new technologies must overcome to be integrated into existing arsenals:

> Every advance, every discovery of a new weapon by the fighting man, has had to overcome two groups of opponents: (1) The fixed and established military group that is always sure the new weapon "won't work," "Is not as good as older weapons," "Not practical," etc. (2) The pacifist groups—the shepherd group that considers each new weapon more terrible than the former and cries out against it.
>
> (p. 6)

Despite this resistance, several key technologies (e.g., clocks, engines, computers, and networks) have changed the nature of warfare on a whole (Bousquet, 2008, p. 917). Since "[n]o effective weapon once introduced has ever been abandoned until it was displaced by a more effective weapon or protection developed that rendered the instrument useless" (Fox, 1942, p. 8), military personnel must ensure that newly developed weapons surpass the abilities of their predecessors. This is all the more crucial if the enemy already possesses superior weaponry—this often occurred during the First World War, when new inventions were introduced to the battlefield on nearly a weekly basis.

It is also true that "some weapons have had a punctuated equilibrium effect" (Simons, 1999, p. 80), for example, the atomic bomb. In contrast to Maxim's assumption, no weapon can ever prevent war; instead, the "mutual fear of big weapons may produce, instead of peace, a spate of smaller wars" (Waltz, 1959, p. 236). In order to be successful, the military personnel of a nation state must therefore be in excellent condition and boast a steady supply of advanced technology. In addition, strong militaries must realize that ground forces cannot be replaced; neither air nor sea power can end wars on their own, so ground forces will undoubtedly remain an essential aspect of war in the future. President Bush, due to Operation Iraqi Freedom, had a "vision of military transformation that would rely on information technology and long-range, precision strike weapons systems" (Rothstein, 2007, p. 278). These elements alone, however, could not end the war—they could only support ground operations. Rothstein (2007) made it clear that "air power is neither a 'magic bullet' nor a way to win inexpensively" (Ibid.).

When at war, the US has often held that "the marshalling of muscle in terms of manpower, equipment and industrial capacity has always been of greater importance than the actual employment of force" (Ibid.), but to win future wars against enemies that behave asymmetrically, industry alone cannot suffice. Instead, the US will have to equip its soldiers with the best possible technology in preparation for ground operations, without which no war could be won. In fact, Eliot A. Cohen (1996) provided a scenario for such a future 20 years ago:

> Future warfare may be more a gigantic artillery duel fought with exceptionally sophisticated munitions than a chesslike game of maneuver and positioning. As all countries gain access to the new forms of air power (space-based reconnaissance and unmanned aerial vehicles), hiding large-scale armored movements or building up safe rear areas chock-a-block with ammunition dumps and truck convoys will gradually become impossible.
>
> (p. 45)

With regard to the military of the future, Cohen stated:

> The balance between quality and quantity has shifted in favor of quality, and it is no mere drive for economy (because professional militaries are not cheap) that has led countries either to give up conscription or to create two separate militaries, one conscript and one professional, with ever more attention and resources going to the latter. At long last, after a reign of almost two centuries, the age of the mass military manned by short-service conscripts and equipped with the products of high-volume military manufacturing is coming to an end.
>
> (Ibid., p. 47)

Modern technology has surely caused the onset of such changes, and the sophistication of this weaponry might eventually transform large armies into mere cannon fodder on the battlefield. Fewer and more adequately trained soldiers, better equipment, and a looser chain of command are ideal in comprising the armies of the future, although military forces "still need, and will always need, specialists in direct combat" (pp. 49–48), especially since ground victories are what end wars. One important factor has yet to be examined: military training in counterinsurgency.

Counterinsurgency

The Cold War transformed not only war itself but also common ideas about war. As James D. Atkinson (1965–1966) remarked, "[w]arfare has more and more become a battle of the spirit, of ideas, and of the human will" (p. 163). In order to win wars, propaganda must be used to sway the general population into approval of its nation's military activities. However, the Second World War demonstrated that propaganda required specific strategies in order to be effective, meaning that the target of the propaganda measures needed to be chosen wisely (McGranahan, 1946, p. 447). In the context of the Second World War,

> The thing to do, rather, was to concentrate on those non-ideological sore spots that intelligence exposed—war weariness, the feeling of hopelessness in a particular battle situation, the desire to escape the torture of front-line fighting against overwhelming enemy material, the knowledge of our industrial might, our reputation for humanitarianism, etc.
>
> (Ibid.)

Despite the valuable information brought to light by the World Wars, American psychological warfare never became sophisticated. Hans Speier (1948) criticizes this:

> In the two World Wars American psychological warfare was improvised after the outbreak of hostilities and conducted according to hit-or-miss methods. No peacetime agency existed which had preserved and developed pertinent skills and knowledge in this field, trained expert personnel, or engaged in appropriate research and planning.
>
> (p. 5)

In Speyer's view, "material superiority" prevented the US's interest in such measures (Ibid., p. 8). Since the US is concerned with combating anti-American radicalism from all over world, it is important for the state to obtain the general populace's trust.

It is also important to teach soldiers about counterinsurgency: "Embittered by their defeat in Vietnam, the military services practically abandoned teaching the subject of insurgency [...] [and] the army exorcised itself doctrinally of the fundamental problems that irregular threats presented" (Rothstein, 2007, p. 280). Defeat caused the US military to lose interest in these issues, but in reality, the instruction of counterinsurgency has become more important than ever before.

The US strategy is very often "ill-suited against an ideologically oriented, non-state enemy" (Ibid., p. 281) and has to develop strategies and training programs to make their soldiers understand how to persuade the population to support their future war efforts. Asymmetric and guerrilla warfare are enlivened by civil support, which, in many countries, is an extension of anti-American feelings. Nowadays, ideology is far more powerful than logistics or material superiority. The "total and uncompromising nature of modern ideologies" creates a strong "dehumanization of others outside their nationalistic, racial, and/or ideological boundaries" (Kaempf, 2009, p. 658). If the soldiers of the future are incapable of overcoming ideological obstacles, they will never be able to gain indigenous support in war. Events like the torching of the Quran in Afghanistan by American troops (Henderson, 2012) have highlighted the US military's recklessness while serving in foreign lands—to avoid such behavior, troops must be trained not only to efficiently fight enemies but also to incentivize innocent bystanders to trust the US military even more than the demagogues of their own kin. The nation state, in contrast, must prepare itself politically and economically for future wars, since they will be different from the major wars in US history.

Beyond modern wars

Economic warfare is more common than it was in the past, since it imposes both diplomatic pressure and financial harm on opponents. Economic warfare can even be carried out in small doses during times of peace, all the while acting as a small yet concrete part of a full-scale war; it is a hybrid military measure that will likely predominate in future conflicts (Førland, 1993, pp. 151–152). Such measures, however, "may shorten armed conflicts but not replace the use of armed force" (Ibid., p. 160). That said, economic warfare is not the only type of warfare with which future nations should be concerned.

Cyber warfare will likely prevail in the future; with it, atomic power plants can be hacked (Kelley, 2013) and government websites can be brought to a halt (Shekhar, 2016). Although the computer was developed because of the Second World War, its purpose during the war was related to "code-breaking and the computation of nuclear physics" (Bousquet, 2008, p. 922). It has since become one of the most influential technologies of the twentieth and twenty-first centuries. Combined with the Internet, computers can bring warfare to an unprecedented digital level. William Westmoreland (1969), during the Vietnam War, explained how such technology can alter the nature of war:

> On the battlefield of the future, enemy forces will be located, tracked, and targeted almost instantaneously through the use of data links, computer assisted intelligence evaluation, and automated fire control. With first round kill probabilities approaching certainty and with surveillance devices that can continually track the enemy, the need for large forces to fix the opponent becomes less important. I see battlefields that are under 24-hour real or near-real time surveillance of all types. I see battlefields on which we can destroy anything we can locate through instant communications and almost instantaneous application of highly lethal firepower

> [...] In summary, I see an Army built into and around an integrated area control system that exploits the advanced technology of communications, sensors, fire direction, and the required automatic data processing [...] With cooperative effort, no more than 10 years should separate us from the automated battlefield.

The US military used more than two million computers at the beginning of the twenty-first century, and this number has steadily grown since then (Kelsey, 2008, p. 1432). However, since the Internet provides the world with universal connectivity, every computer connected to the Internet is a potential victim—or a potential weapon.

Jeffrey T. G. Kelsey (2008) emphasized the importance of cyber warfare, stating that

> [It] is more than just an issue for websites and some forms of communication; indeed, some cyber attacks may cause serious physical, tangible effects [...]. Had the attack been more global in nature, millions of people could have been without communications for some time—a situation both life threatening and damaging to the global economy. Similarly, cyber attacks on purely military targets may be capable of effectively neutralizing a hostile threat without risking lives or equipment.
>
> (p. 1434)

Cyber warfare could bring down air defenses, infiltrate defense networks, and attack power plants or media stations; it is a particular threat to nation states that are unprepared for such attacks. The US government has already reacted by forming a "new command devoted to warfare in cyberspace" (Döge, 2010, p. 486). However, it is essential to provide sufficient personnel in this field, and training and education alike must be adequately provided to secure future national defenses. Human beings' dependency on computers and the Internet are vulnerability, and this reality must be taken into consideration to make solid military plans—especially when terrorists can use such a vulnerability to their advantage.

Proving the origin of cyber attacks is extremely challenging, especially when Internet protocol (IP) addresses can be faked (Ibid., pp. 487–489). The Internet makes it easier for terrorists to adversely affect others, but individual hackers are also capable of launching serious attacks, and the advent of asymmetric, universal cyber warfare will invariably blur the line between peace and war in the civilian sphere.

Many definitions highlight cyber warfare's relative infancy, as explained by Döge (2010):

> The terms computer network attack (CNA), [offensive] information operation (IO), and information warfare (IW) are frequently used interchangeably. The U.S. Department of Defense (DOD) defines CNA as "actions

> taken through the use of computer network to disrupt, deny, degrade, or destroy information resident in computers and computer networks, or the computers and networks themselves." By IO, the DOD means "the integrated employment of the core capabilities of electronic warfare, computer network operations, psychological operations, military deception and operations security, in concert with specified supporting and related capabilities, to influence, disrupt, corrupt or usurp adversarial human and automated decision making while protecting our own.
>
> (p. 488)

As "a byproduct of the information revolution" (Berkowitz, 2000, p. 8), computers and the Internet can be lethal weapons when in the wrong hands. Consequently, education and training are vastly important in preparing US soldiers for cyber warfare, and said education strategies must be of the utmost quality to ensure the US's efficacy within this "critical new dimension of military power" (Ibid.). Cyber warfare is dangerous because it "defies borders, challenges state boundaries, and most significantly, provides the military of a nation to realize certain political goals, allowing for a more precise form of propaganda" (Chansoria, 2012, p. 105), and the US military must prepare their technology and personnel alike to accommodate this aspect of warfare in the future.

Taking a closer look at the US's enemies can highlight the importance of this matter in potential wars. Thomas Donnelly *et al.* (2009) emphasized the US's problems in East Asia, where "the Chinese military has invested heavily [not only] in [...] ballistic missiles and attack submarines" (p. 147), but also in cyber warfare. The 12th Chinese Five Year Plan (2011–2015) already emphasized the importance of "net-based combat" (Chansoria, 2012, p. 113), and viruses are potent tools for weakening the US. When "Chinese cyber warfare advances, coupled with China's increasing skill at neutralizing information-transmitting satellites and other capabilities, it fulfills a part of its military objective of crippling potential foes, in the event of a crisis or confrontation" (Ibid., p. 118), and this form of advancement is both powerful and cost-effective for nation states whose hardware cannot match that of the US.

Like China, US foreign policies—as well as military planners in the air force, army, and navy—must consider the dangers of Russia. Despite the fact that the prevention of nuclear proliferation depends on US–Russian cooperation, Joseph Biden (1999) claimed that "both countries seem poised to accept an increased risk of war" (p. 58), an idea that the events in Ukraine and the antagonistic policies in Syria support. Russia, like China, wants to invest in affordable measures to weaken the US, and this is why cyber warfare attracts Moscow. However, proxy wars will also invariably continue, much like during the Cold War. It is essential for the US to consider multiple perspectives when planning for future wars, and they must acknowledge that potential proxy wars will require both an educated military and a saturation of specialists who can effectively partake in cyber warfare.

Conclusion

Thomas Donnelly *et al.* (2009) is correct in saying that "Washington faces an ever-changing threat environment; it does not have, nor will it ever have, the luxury of moving wholly from one clearly defined regimen of warfare to another" (p. 147). The best way for the US military to prepare for war is to develop a diverse strategy. It must prepare its troops for asymmetric wars in which proper training and a loose chain of commands are the determinants of victory. It has to provide counterinsurgency training to soldiers to prevent defeat against anti-American ideologies in the zones of war.

The US military must not only ensure that its weapons are of state-of-the-art quality, but also that they can manage and defend against cyber warfare and related strategies. One cannot prepare for every possible event or scenario, but the better the US is prepared for the aforementioned cases, the higher the nation's chance of remaining undefeated in future wars. In the past few decades, US foreign policy has lacked foresight; the military's actions have been uncoordinated, their results have been poor, and the solutions to their problems have seemed unattainable. These are signs of unpreparedness and are primarily the results of over-reliance on hasty ad-hoc decisions. The future will only be brighter for the US once it acknowledges the threats to the nation and begins to react appropriately. Rather than hoping that the sheer size and power of the US economy can guard the nation from potential threats, the US military must invest in better training and equipment in order to be truly effective in future conflicts.

Note

1 Future visions of war always existed, but they seldom exactly matched the real war. For some of these visions, see Förster (2016).

References

Atkinson, J. D. (1965–1966). Liddell Hart and Warfare of the Future. *Military Affairs*, 29:4, pp. 161–163.

Balmer, B. (2002). Killing 'Without the Distressing Preliminaries': Scientists' Defence of the British Biological Warfare Programme. *Minerva*, 40:1: *Ethics and Reason in Chemical and Biological Weapons Research*, pp. 57–75.

Berkowitz, B. D. (2000). War Logs on: Girding America for Computer Combat. *Foreign Affairs*, 79:3, pp. 8–12.

Biddle, S. (2003) Afghanistan and the Future of Warfare. *Foreign Affairs*, 82:2, pp. 31–46.

Biden, J. (1999). A Strategic Vision: Mapping a Future for US–Russian Relations. *Harvard International Review*, 21:3, pp. 58–63.

Bousquet, A. (2008) Chaoplexic Warfare or the Future of Military Organization. *International Affairs*, 84:5, pp. 915–929.

Chansoria, M. (2012). Defying Borders in Future Conflict in East Asia: Chinese Capabilities in the Realm of Information Warfare and Cyber Space. *The Journal of East Asian Affairs*, 26:1, pp. 105–127.

Clausewitz, C. von (1873). *On War*, trans. by Col. James John Graham. London: N. Trübner.

Cohen, E. A. (1996). A Revolution in Warfare. *Foreign Affairs*, 75:2, pp. 37–54.

Danielsson, S. K. and Jacob, F. (2016). *War and Geography: The Spatiality of Organized Mass Violence*. Paderborn: Schöningh.

Döge, J. (2010). Cyber Warfare: Challenges for the Applicability of the Traditional Laws of War Regime. *Archiv des Völkerrechts*, 48:4, pp. 486–501.

Donnelly, T., Dur, P. and Krepinevich, A. F., Jr. (2009). The Future of U.S. Military Power: Debating How to Address China, Iran, and Others. *Foreign Affairs*, 88:6, pp. 146–151.

Dworok, G. and Jacob, F. (Eds.) (2015). *The Means to Kill*. Jefferson, NC: McFarland.

Førland, T. E. (1993), The History of Economic Warfare: International Law, Effectiveness, Strategies. *Journal of Peace Research*, 30:2, pp. 151–162.

Förster, S. (2016), *Vor dem Sprung ins Dunkle: Die militärische Debatte über den Krieg der Zukunft 1880-1914*. Paderborn: Schöningh.

Fox, L. A. (1942). Bacterial Warfare. *The Scientific Monthly*, 55:1, pp. 5–18.

Gerlach, C. (2011). *Extrem gewalttätige Gesellschaften: Massengewalt im 20. Jahrhundert*. Berlin: DVA.

Hammond, J. H., Jr. (1916). The Future Mechanism of Warfare. *The North American Review*, 204:730, pp. 373–380.

Henderson, B. (2012). US troops 'burned 100 copies of the Koran in Afghanistan'. http://www.telegraph.co.uk/news/worldnews/asia/afghanistan/9502439/US-troops-burned-100-copies-of-the-Koran-in-Afghanistan.html (Last access, April 28, 2016).

Jacob. F. and Visoni-Alonzo, G. (2016). *The Military Revolution in Early Modern Europe: A Revision*. London: Palgrave, 2016.

Kaempf, S. (2009). Double Standards in US Warfare: Exploring the Historical Legacy of Civilian Protection and the Complex Nature of the Moral-Legal Nexus. *Review of International Studies*, 35:3, pp. 651–674.

Kelley, Michael B. (2013). The Stuxnet Attack on Iran's Nuclear Plant was 'Far More Dangerous' Than Previously Thought. http://www.businessinsider.com/stuxnet-was-far-more-dangerous-than-previous-thought-2013-11 (Last access, April 28, 2016).

Kelsey, J. T. G. (2008). Hacking into International Humanitarian Law: The Principles of Distinction and Neutrality in the Age of Cyber Warfare. *Michigan Law Review*, 106:7, pp. 1427–1451.

Luttwak, E. N. (1995). Toward Post-Heroic Warfare. *Foreign Affairs*, 74:3, pp. 109–122.

Machiavelli, N. (1995). *The Prince*, ed. by George Bull. London: Penguin.

Maxim, H. (1908). The Warfare of the Future. *Science*, New Series 28:728, pp. 820–831.

McGranahan, D. V. (1946). U.S. Psychological Warfare Policy. *The Public Opinion Quarterly*, 10:3, pp. 446–450.

Münkler, H. (2004). *Die neuen Kriege*. Berlin: Rowohlt.

Rothstein, H. S. (2007). Less Is More: The Problematic Future of Irregular Warfare in an Era of Collapsing States. *Third World Quarterly*, 28:2: *The Long War: Insurgency, Counterinsurgency* and *Collapsing* States, pp. 275–294.

Shekhar, S. (2016). Indian Hackers Target Pakistan Government Websites to Avenge Pathanok Attack, http://www.dailymail.co.uk/indiahome/indianews/article-3385888/Indian-hackers-target-Pakistan-government-websites-avenge-Pathankot-attack.html (Last access, April 28, 2016).

Simons, A. (1999) War: Back to the Future. *Annual Review of Anthropology*, 28, pp. 73–108.

Speier, H. (1948). The Future of Psychological Warfare. *The Public Opinion Quarterly*, 12:1, pp. 5–18.

Sun Tzu and Minford, J. (2002). The Art of War. *New England Review*, 23:3, pp. 5–28.

Van Creveld, M. (1985). *Command in War*. Cambridge, MA: Harvard University Press.

Van Creveld, M. (2008). *The Changing Face of War: Combat from the Marne to Iraq*. New York: Ballantine Books.

Waltz, K. N. (1959). *Man, the State, and War: A Theoretical Analysis*. New York: Columbia University Press.

Weigley, R. F. (1977). *The American Way of War: A History of United States Military Strategy and Policy*. Bloomington: Indiana University Press.

Westmoreland, W. (1969). Address to the Association of the US Army, 14 Oct. Cited in Dickson, P. (1976). *The Electronic Battlefield*. Bloomington, IN: Indiana University Press, pp. 215–223.

5 The moral, physical, and technological

Communication and the wars of the 21st century

John M. Callahan

Introduction

The United States (US) finds itself in conflict with a wide variety of adversaries, many of whom challenge the core tenets of the American "Way of War." That way of war emphasizes the black and white: One is either at war or not at war. War is a struggle of compulsion between nations, to be fought on battlefields, and only includes cities and noncombatant populations when absolutely necessary. Victory in war is brought about by overwhelming conventional military force, even if no longer by massive superiorities in manpower and material. It is, to paraphrase a line from the movie *The Avengers*, about the application of a "boot" to an "ant." Unfortunately for strategists, particularly since the end of the Cold War, the non-Western world sees war quite differently and far more holistically, to the point that America, in spite of retaining overwhelming power and capability, finds itself outmaneuvered on those fields of conflict, which are not kinetic. Key competitors, such as Russia and the Islamic State, have developed considerable expertise at exactly those types of efforts.

There is further dispute over a critical area of operations, the information environment. Differing motivations for the use and dissemination of information, combined with the rise of technologies that have served to break down the traditional roles of the government, and the media in doing so, have generated a situation where the US government finds itself continually behind the curve in a conflict of rhetoric that in many ways eclipses the importance of kinetic conflicts that the US is very good at prosecuting. That conflict not only includes the time of war, but the period before and, perhaps more importantly, after kinetic action has taken place.

The gaps generated by these two issues lead to noteworthy losses of face and actual failure for US policies around the world, a trend that must be arrested. This article will discuss the nature of war and the primacy of the information environment in it. Historical cases will show the impact of technology on the information environment and the interplay between decision-making styles and conflict in the realm of rhetoric. The role of framing and the narrative space, both in justifying America's wars at home and winning them abroad, will be examined. It is critical that military and government participants in the rhetorical struggle overcome the tyranny of their own bureaucratic methods in order to succeed in conflict with highly flexible, decentralized, tech-enabled adversaries.

War has not changed, but America's understanding of its tenets must do so. Never has the role of public affairs and strategic communication in conflict been more prominent, and never has the need for a re-examination of both the information environment and the role of public affairs in it been more pertinent.

On war, power, technology, and opposing capabilities

> *In the 21st century, we have seen a tendency toward blurring the lines between the states of war and peace.*
>
> (Gerasimov, 2013)

War consists of actions designed to compel an opponent to do one's will. Every act performed with coercion in mind, using any aspect, active or passive, of national power, is a part of the struggle to dominate and should rightly be considered to lie within the definition of *war*. It centers on the servicing of one's (individual, national, or organizational) interests. It therefore follows that every act of national power should be bent toward compulsion. At its very root, the essence of warfare has not changed since man first raised a club, or a voice, at his fellow man. Perhaps more importantly to the joint warfighter, a second level of war, that of armed struggle—the basic tenets of which were broadly laid out by Sun Tzu, Machiavelli, and von Clausewitz—has not changed significantly in the modern era. Use of terms such as *hybrid* or *fourth-generation* warfare, or conflict in the *Gray Zone*, primarily reflect US efforts to come to grips with methods and styles or warfare which, while timeless in a global sense, do not fit with America's current military and national security mindset. The continual generation of such new terms, while typically American, actually clouds the basic point, captured by von Clausewitz most succinctly, that coercion is what warfare is all about (Paret, 1989). Similarly, terms such as *hard* and *soft power* have sprung from an academic debate about the nature of the world, but they also obscure a holistic approach to the compulsive nature of power.

Operations is another concept that has become focused to the point of obscuring a broader utility. A truer definition of the term would be all activities using power that lead to coercion. Over the past decade, intelligence and information categories have been included under the umbrella of operations, and this is a timely and useful development. The global information environment is more active than ever, with actors implementing communication strategies underpinned by narratives based on terror and/or perception management or manipulation in order to create a desired end state of their choosing. While the use of information as a primary enabler to achieve an objective is not new in any way, recent and evolving technology has provided new levels of access to audiences, bypassing traditional fourth-estate channels and associated filters. Technology is the great equalizer and force multiplier, imputing the same global access and audience reach to the lone individual in a cave as that of a corporation or government. While they have not changed the nature of warfare, technological advances have brought about massively shortened decision cycles. These often prove to be anathema to the top-down hierarchical military system of issue assessment and decision-making processes of the past, processes with which we are still burdened today, to our loss. This is particularly true in battles of the narrative in which the US is often crippled by overburdensome and time-consuming command review processes for the release of public information.

From the use of nationalism, counterterrorism, and protection-of-minorities narratives to justify actions in the Crimean Peninsula and Ukraine to the Middle East and Africa, where fabricated ideologies are published and promoted based on manipulated and weaponized religious narratives in order to recruit and resource their campaign, information is the weapon of choice and the information domain the chosen battlefield of modern conflict. Recent examples of successful use of the narrative include the Russian response to the Turkish shoot down of one of its aircraft in December 2015 (Schapiro, 2015). Whether inspired by a strong understanding of Turkish military capability or by an unwillingness to run the risk of expanding the Syrian conflict further, the fact is that Moscow's response was masterful, up to and including demands to return the Hagia Sofia in Istanbul to its former status as a Christian church (*Thema News*, 2015). Seeking ways to develop capabilities that allow it to compete against the kinetic power of the US, Russia has placed considerable emphasis on the use of the narrative to weaken its opponents and achieve its political and military objectives. In the narrative realm, including significant use of social media fueled by external and internal events, and with its own version of foreign-focused, government-controlled news media in the form of the Russia Today news network (Shuster, 2015), Russia is capable of "holding its own" in information warfare (Thomas, 2014).

Dae'sh's use of a hyperflexible social media presence and its cunning use of public brutality, as well as Iran's propaganda usage of American sailors that it briefly held in custody in January 2016, also show a cunning understanding of how technology has impacted the information environment, operations, and the methods of coercion. "What has made ISIL particularly notable is their use of social media platforms in messaging broadly and targeting selectively among potential recruits, supporters, and sympathizers," stated Lesperance (2016), author of "The Origins of the Islamic State" in *The Islamic State*, in a personal interview.

Married up with technologically advanced messaging systems, Dae'sh also practices some of the principles that the US armed forces often preach but do not implement. For example, Dae'sh communicators generate simple, broad, enduring messages but then not only empower their tactical leaders to carry out operations, but also, critically, to flexibly deliver key messages to their audiences. This is something that US allies such as Israel excel at but which

US military communicators simply cannot replicate, for the reasons discussed above. This is not a technological but a psychological hurdle generated by risk-adverse senior leaders and micromanaging national security staffs enabled, oddly enough, by the very technologies that enable Dae'sh and others to be flexible.

Engaging, countering, and neutralizing such information campaigns requires the coordinated and synchronized use of the full spectrum of information disciplines and a whole-of-government approach. Whether this process is called *strategic communication* or *communication*, synchronization, the most publically visible communication discipline, is often that of public affairs. The Department of Defense (DoD) defines "those public information, command information, and community engagement activities directed toward both the external and internal publics with interest in the Department of Defense, as Public Affairs (PA)." PA (Joint Staff, 2010) is the main capability used to engage audiences in the narrative battlespace of today and tomorrow, and the department's leading tool to accomplish that mission is the Joint Public Affairs Support Element (JPASE). JPASE and other DoD PA capabilities join DoD Information Operations (IO) forces, when needed, in coordination with the Department of State's Public Diplomacy community to assist joint and interagency commanders with operations across the information environment.

The information environment: The past as prologue

> *Words are bullets and the first casualty of war is truth. But war is not the only entity that is weaponized. The militarization and weaponization of information has global roots and branches.*
>
> (Huhtinen, 2015)

The information environment is a vital, indeed integral, component of the operating environment as a place in which rivals clash. Historical examples of critical events in the information realm will be presented here as demonstrations of the inherent nesting of the information environment into operations and the impact of clashes of rhetoric that sometimes outweighed the importance of kinetics.

We cannot know how the information environment existed before the written word. We can guess that appearances and kinetic action were central to warfare, if for no other reason than that we can observe how social animals behave. Symbols mattered, physical size and physical prowess were important, and the loser knew that he had lost based on the bruising of his body and the blood he had shed, with the winner's proverbial boot at his neck. Therefore, the symbols of power took on a communicative aspect. Wealth and military strength came to mean that all conflicts need not become kinetic; that all disputes, even existential ones, need not result in the exercise of arms.

As mankind developed enhanced social institutions, the power of words and symbols became ever more important. The ability to write down words, to ensure the long distance communication of thoughts and ideas, was a central moment in human history and the history of struggle. The written word was combined with another critical element, the advent of fast transportation in the form of paved roads and powered ships, which allowed communication over vast distances in a timely manner. In the classical world, the Romans were by far the best information environment combatants. Over a 250-year period from the Punic Wars to the reign of the stoic Emperor Marcus Aurelius, Rome, as characterized by Luttwak (1976) and others, was concerned not only with the organization of the territory within its borders but also with the Romanization of as much of the neighboring areas as possible, in a way not unlike the efforts of the US to secure a free market space over the course of the twentieth century. These efforts fueled one another. A relatively small but incredibly proficient army maintained order and expanded the borders when needed. Meanwhile, the idea of Rome and what it offered morphed neighboring areas into proto-Roman areas. In both cases, the ease of communication offered by the road network was the enabling technology that allowed for the dissemination of information and influence in a coherent and ordered fashion.

For seventeen centuries after Rome fell, there was one significant change in the information environment, the development of the Gutenberg printing press, which made communication and the spread of thoughts and ideas incrementally faster than it had been under the Romans. By the nineteenth century, technological changes, which would rapidly decrease the decision cycle of leaders, had begun occurring in rapid succession. Steam-powered ships, the railroad, the telegraph, telephone, and finally, the radio, all served to compress the operational decision cycles of foreign policy leaders. But they also sent tremors through the information environment. The rise of the media was fueled by the ease of communication, as was the ability of contrarians to spread their ideas rapidly and relatively securely (in the sense that communication became so easy that the government no longer controlled its means and mechanisms). The need for the government to explain its actions to its stakeholders was not limited to the transparency requirements of democratic societies. Through this most rapid of time compressions, one thing did not change.

The example of Germany provides a pair of interesting contrasts. In 1918, the primacy of the narrative struggle over the kinetic was demonstrated by the allies' mismanagement of the defeated Germany, a mistake that was not repeated in 1945. By 1918, the Imperial German Army was exhausted from years of largely victorious warfare, and when Germany surrendered that fall, the problem was that the language, symbolism, and actions of the conquering powers were not synchronized, allowing the pervasive and perfidious feeling to arise in Germany that the war was not lost but rather that the defeat was the result of a betrayal. While Germany's armies had been beaten, those actions took place in occupied territory, not in the homeland, leading to a myth that corrupt politicians had stabbed them in the back. The allies needed to win the peace, but they failed to rally the symbolic tools needed to demonstrate why they had won and failed to utilize new technologies to try and shape the situation on the ground. Germany was not occupied, except for brief periods and in small regions, such as the occupation of the Ruhr by the French and Belgians. Such incursions were sold as hostile acts or invasions by growing radical German political movements. Furthermore, forcing Germany to take the blame for the start of the war directly controverted the publicly known actions of the summer of 1914 that had started the conflict. A failure to coordinate messaging and actions caused this dissonance and set the stage for renewed conflict.

On the other hand, the Second World War demonstrates a successful integration of the kinetic and rhetorical realms of conflict. In 1945, the goals were the same as they had been in 1918: to win a peace through demonstrations of power and the ability to provide stability. The technology was largely the same, but the methods used were vastly different and ultimately successful. The openly recognized fault of the Nazi regime in initiating the Second World War in Europe was augmented in the minds of the German people by the years of aerial bombardment they had suffered, followed by the utter destruction of the German armed forces and the complete and semi-permanent occupation of the country and dismantling of its government by the victorious allies. There was absolutely no question in anyone's mind as to who won and who lost the war. As a result, the victorious allies also won the war of the narrative with the German people on both sides of the East–West divide to such an extent that German political thought is still dominated by the events of that era.

A final, cautionary example is provided by the failure of coalition forces to change the political landscape in Iraq after the invasion of 2003. While Secretary of Defense Rumsfeld and his team were proven right, in that only a relatively small force was required to defeat Iraq's regular armed forces, take Baghdad, and depose Saddam Hussein, that force was wholly inadequate to provide the symbolism that might have made a difference in securing a meaningful, transformative peace in the region. One missing piece was the visible rhetoric of the end of tyranny and the return of stability that a soldier on every street corner would have provided.

In each of these cases, there is a kinetic and a rhetorical segment, with a key lesson being that the rhetorical effort does not stop just because kinetic operations have ceased. For Rome, the desire was for stability, trade, and occasional growth, facilitated by cultural power and military power backed by a robust infrastructure. In 1918, the need was to provide the symbolic proof of allied victory coupled with a restoration of stability. Technology was irrelevant, and nothing could overcome the appearance of a stab in the back, undoing the fruits of the military victory. 1945 saw a perfect convergence of kinetics and rhetoric, and the narrative failure of 1918 was avoided. But 2003–5 saw a failure much like that of 1918, one in which it was not more and better types of media and communication tools that enabled the rhetoric but rather a symbolic failure to provide the imagery and reality of improved security through a robust occupation.

Public affairs in the information environment

> *The ends are clear (information dominance), and the means are known (producing and promulgating popular and effective messages); the ways are through loosely guiding the individual voices of defense patriots.*
>
> (David, 2015)

The nature of the world has not changed, but technology has impacted decision cycles and planning. Military commanders and their staffs must understand both the kinetic and information sides of their operations and operating environments. Public Affairs Officers fully understand the information environment, its evolution, and its implications, as well as the strategic communication process and subordinated concepts such as the rules of engagement. That level of expertise is the number one asset that the public affairs community provides to a variety of joint, interagency, and coalition commanders.

History has shown the necessity of powerful spokespersons in raising the profile and integration of their military skill sets. Artillery only became the equal of the infantry and cavalry when an artillerist, Napoleon Bonaparte, became Emperor of the French. Before then, the artillerist was often a lieutenant or captain in a world of colonels and generals, and the guns were a sideshow rather than an integrated maneuver element. The commander would often seek the advice of his senior, professional warriors who were adeptly skilled in their kinetic or nonkinetic arts, seasoned by years of experience. Most were experts and specialists in the weapons or skills of their field and they had the respect, and ear, of the commander for those very reasons. The public affairs advisor had to be no less a skilled and seasoned warrior with his or her skill set, the art of communication, which transcends both rhetoric and actions.

Understanding the mechanics of rhetoric, of how and why perception can become reality, and a solid knowledge of communication practices and policies are just some of the required traits of a professional communicator. Communication design, audience identification and selection, message construction, cultural framing, and knowing how to analyze and assess communication activities and products for effectiveness are just some of the skill sets required. Knowing the adversary's communication design and intent is key to being able to anticipate, neutralize, and overcome opposing information strategies, particularly in a world in which most adversaries have actively studied our communication models and are using them against us.

Military public-affairs teams are trained to liaise with relevant intergovernmental agencies and nongovernmental organizations (NGOs) in furtherance of communication objectives and can act to provide defense support to public diplomacy through direct cooperation with embassy staffs and Department of State regional bureaus. This becomes particularly important in successfully winning the battle of the narrative in post-conflict environments, a critical mission (Fortune, 2008) led by the State Department and supported by DoD. They can, in cooperation with intelligence and operational staffs and other collection sources, provide an assessment of the overall information environment. This provides the commander with a better understanding of the information environment in his or her area of responsibility, with insights on how the adversary may be structured to communicate, which communication model they may be using, their communication network capacity and capabilities, and how they may be shaping their information strategy and tactics. The resulting information environment assessment (IEA) will contribute directly to countering an opponent's narrative. It can also lead to kinetic and nonkinetic courses of action in neutralizing an opponent's capacity and capability to communicate in the first place. An example of this would be using kinetic force, such as a bombing run, to eliminate a known center of enemy messaging.

Going to war: The information environment and justifications for intervention

A communications strategy cannot work if it cuts against the grain of policy.
(Nye, 2004)

The future of US warfare requires communicators to win two narrative battles, one abroad and one at home. The former includes all of the phases of warfighting, from steady state operations to the need to ensure that post-conflict narratives prevent the positive results of conflict to be undone, as we have seen in the case of Germany in 1918, discussed above. The latter is in the justification of interventions to the American people. This is a problem of marketing (Schuessler, 2004), because military interventions, when not following some kind of obvious threat, are in conflict with core American values. Any intervention or conflict must be framed to the American public as a war of necessity, forced on the nation by dire circumstances or even by armed attack. According to Schuessler (2004), when America acts, it should do so not for low goals but with the goal of improving mankind's lot.

There is a significant burden on the government to prove the benefits of its actions, particularly when American lives are at risk (Haass, 2009). This adds to the burden of both deciding on and explaining armed interventions, which are, in the end, wars of choice rather than necessity. For Haass, the benefits of such an intervention must outweigh the costs, while the net result must be positive; "If this test cannot be met, the choice will appear to be ill advised and most likely is" (Haass, 2009). Thus, the "bang for the buck" test joins the other tests, motivators, and detractors for intervention that must be considered when examining each one and that serve as intervening variables for the cases. The inference here is that framing acts of choice, such as military interventions, is a more difficult and time-consuming act, less certain to succeed than is the case in wars of necessity.

The US government uses an interagency process to decide on when and how the nation will intervene abroad. At some point in this process, and not necessarily concurrent to it, leaders frame and communicate their themes and messages concerning their opponents and world events. Those communications are then executed by spokespersons across the whole of government, but most notably by the White House, Congress, and the Departments of State and Defense. Paul (2011) noted that successful framing or communication strategy calls for synchronizing messages with deeds. Even without the interference of personalities, however, a significant source of friction and conflict in the decision-making and communications processes is the radically different institutional mindsets of the various interagency players, most specifically the Departments of State and Defense. A key point here is the ongoing gap in funding, crisis response methodologies, training, outlook, and expectations between these two most significant stakeholders in the decision-making process, making such a process difficult to implement (Carter & Deutsch, 2000).

Since the government remains the primary source for information, this creates a system of asymmetric interdependence between the information holders and the media as well as the public, but this can only be taken advantage of if there is unanimity in the message, particularly in a media environment that is prepared to conflate news with entertainment (Bennett & Entman, 2001). Manuel Castells drew upon and added to the concept in *Communication Power* (2009), most particularly in his writing concerning what he called "Networks of Mind and Power." For this concept, Castells (2009) draws significantly on the models of framing developed by Entman (2004), who focuses attention on the efforts of the government to frame key events and the media's countervailing efforts to "counter-frame" them. Technology impacts this too, in that social media allows for individual actors to generate information and news stories. It also allows for spokespersons to skip the media and communicate directly with their core audiences in America and abroad. The implications of this for the future role of the media are poorly understood, but a world in which the majority of information is held by the government and communicated directly to the populace, without challenge, is Orwellian at best. According to Butler (2012), for frames and influence efforts to work, they must be credible, salient, and dynamic, and they must be delivered by trustworthy spokespersons.

Retired Admiral James Stavridis (2007), a noted advocate of and expert on strategic communication (SC), adopts a military approach, noting its power as an enabler for operational forces. He acknowledged the role of the national security apparatus and entreated leaders to "take an early and persistent role in deciding how ideas and decisions are shaped and delivered, and that this need required a shared understanding of what SC is and is not, especially in an international context" (Stavridis, 2007). This is re-enforced by Joseph Nye, who, in *Soft Power* (quoted at the beginning of this section), warned that "reiterating the need for coordination of messaging at the highest levels" was a critical piece of the communications puzzle. Paul (2011) reinforces this idea by noting that a successful framing or communication strategy calls for synchronizing messages with deeds. This must be done at all levels, from the national to the tactical, for the messages to have their intended impact.

This section, and the one before it, not only laid out the public-affairs role in the militarized information environment but also discussed the higher, policy-level implications of communications. According to Colonel Tad Sholtis (2005), the public affairs chief for Air Combat Command, "The American military must maintain public trust, and leaders must take action whenever communication efforts push or exceed the bounds of credibility." To be successful in warfare in the 21st century, against opponents who are empowered to deliver their messages at the lowest of levels, successful military operations require professionals who can understand all of these factors, and utilize them, whether working at the cabinet level, or at the spearhead of operations on the ground.

Conclusion

> *Because terrorism requires an open flow of information, Western-style republics need to be able to respond offensively—even preemptively—with their own convincing ideas.*
>
> (David, 2015)

Symbolism, technology, speed of information, compressed information cycles, and the continued blurring of the narrative and kinetic realms will be the trends of the next decade. This includes awareness of how the information environment is evolving, how adversaries or opponents use it, and how it will be used by potential adversaries or opponents in the future. The never-ending quest for power and control has not changed over the course of time. History has demonstrated most conflicts or crises are born of design and with intent by an actor with an agenda. Setting the stage is the first step in introducing chaos or building to a crisis. Doing so in today's Internet-connected, border-free, and stateless information environment has never been easier.

Building audience awareness and constructing a road to crisis are the early stages of crisis or conflict development. There is a period in which indicators of development can provide enough granularity to clearly identify if the activity is accidental or careless in nature or if there is design and intent underpinning the chaos. Historically, these indicators have included published manifestos directed at a sympathetic population who feel, or can be made to feel, a sense of oppression and the potential rise of charismatic actors, culminating in a decisive act that demonstrates the seriousness of the actor. With today's technology, including new and emerging social-media tools, audiences can be quickly and easily targeted with unerring precision, right down to any type of desired demographic, all aimed at reducing decision times—dancing with increasingly nimble, technology enabled adversaries.

However, this is a two-way road, and the contribution of the community of information practitioners assists commanders in understanding and reacting to trends in the information environment. Due to the nature of social media and the Internet, early rhetoric can be heard and actions seen, intents and objectives can be telegraphed, and end states can be previewed, all before a particular conflict can gain traction. This could, with early identification and analysis, be used to shut down a crisis or conflict before it reaches a tipping point. Further development is needed in how to better analyze the current, and future, information environment; how social media trends can indicate troubles spot or even predict troubles; and real-time crisis communication impact assessment and efficacy in achieving objectives.

Social media is increasingly impacting the roles of both government spokespersons and the media. This goes beyond the compressing of decision cycles and ultimately questions the very role of the media in the modern world. When spokespersons can reach their true audiences, the people of the US and the world, by releasing statements and even conducting interviews and media events over Twitter and other social media, the traditional media are no longer necessary to serve as interlocutors. What will this do to their watchdog function and their role as the unelected fourth estate of government? More to the point of this discussion, what will that mean for the role of the public affairs and information practitioner?

Communication challenges are ever present and evolving. On January 27, 2016, former Defense Secretary Robert Gates (2016) noted the atrophy of the strategic communication process. Technology in the form of hardware and new capabilities and platforms, especially in the social media realm, not only provide opportunities for successful engagement but also pose challenges in terms of an ever-shrinking decision and response cycle. Amidst all of this change, the absolute need for public-affairs officers to provide advice to commanders and to implement communication strategies in a timely and effective manner is critical to waging the battle of the narrative with adversaries and providing transparency to the American people. This is especially critical given that nearly all of America's conflicts are not existential ones, whereas they very much are for our adversaries. Moral authority, or America's self-perception of its moral authority, is a critical variable in securing support for operations and for their success.

In order to make this all work, commanders must embrace the principal of decentralized operations. While technology has served enemies such as Dae'sh well in the efforts to dominate the information space, the same factors that enable them often have the opposite impact on US efforts to defeat them. American military leaders, particularly in the army, are so risk averse that they use communication and information technologies not to enable their subordinates but rather to exert control over them, down to the very lowest levels. Senior leaders must learn to trust the individual service member or civilian employee to act in accordance with their oaths. This is the case both in the kinetic and informational realms, leading to the all-too-often-heard allegation that America's General Officer Corps are the best set of battalion commanders in the world (Ricks, 2012). The State Department, that other critical leg of diplomacy and communication, lost the struggle decades ago. Working in the public-affairs section of the American Embassy in Baghdad from 2004 to 2005, this author well remembers the smothering level of control that the National Security Council and the Near East Asia Bureau at main state exerted over the high and mid-levels of public affairs and public diplomacy operations. In fact, it is only the comparatively tiny budget of the department, particularly in the public-affairs arena, that keeps that control from being pushed down to the manipulation of the actions of individual diplomats, such as regularly happens with tactical units in the armed forces. Our best agents in the ongoing battle of the narrative are not always the high-level spokespersons but are often the soldiers, sailors, airmen, Marines, and civilians that are in constant contact with our audiences, both friendly and hostile. This is no more than falling into line with the implicit core value that they represent America. The bottom line in all of this is that social media is most functional when it relies on crowdsourcing versus the typical top-down review processes that the government in general, and the military in particular, employ (David, 2015).

While politically expedient in terms of protecting promotion cycles, the increasing centralization of decision making and communicating must be overcome or the US will lose not just the narrative struggle but also the overall struggle against its ever-increasing coterie of foes. This chapter has shown that this is more of a problem of organizational mindset than one of technology.

References

Bennett, L., & Entman, R. (2001). *Mediated Politics: Communication in the Future of Democracy*. Cambridge: Cambridge University Press.

Butler, M. (2012). *Selling a "Just" War: Framing, Legitimacy, and US Military Intervention*. New York: Palgrave Macmillan.

Carter, A., & Deutsch, J. (2000). "Strengthening the National Security Interagency Process." In *Keeping the Edge*. Harvard: Kennedy School.

Castells, M. (2009). *Communication Power*. Oxford: Oxford University Press.

von Clausewitz, Carl. *On War*. (1873) http://www.clausewitz.com/readings/OnWar1873/BK1ch01.html#a.

David, G. (2015). "Taking the OFFENSIVE." *US Naval Institute Proceedings*, Oct. 2015, Vol. 141, Issue 10.

Entman, R. (2004). *Projections of Power: Framing News, Public Opinion, and US Foreign Policy*. Chicago: University of Chicago Press.

Eder, M. (2011). *Leading the Narrative, the Case for Strategic Communication*. Annapolis: Naval Institute Press.

Fortune, F. (2008). "Strategic Communication: The Heart of Post-Conflict Processes." *Conflict Trends*.

Gates, R. (2016). "Comments on Strategic Communication," January 23, 2016, http://www.publicdiplomacycouncil.org/commentaries/01-23-16/quotable-robert-gates-%E2%80%9Cstarved%E2%80%9D-%E2%80%9Cdisarmed%E2%80%9D-and-%E2%80%9Cdismantled%E2%80%9D-strategic-communic.

Gerasimov, V. (2013). "The Gerasimov Doctrine and Russian non-Linear War" *Voenno-promyshlennyy Kuryer* (Feb. 27–Mar. 5, 2013), http://vpknews.ru/sites/default/files/pdf/VPK_08_476.pdf. An English translation by Robert Coalson and commentary by Mark Galeotti, https://inmoscowsshadows.wordpress.com/2014/07/06/the-gerasimov-doctrine-and-russian-non-linear-war/.

Haass, R. (2009). *War of Necessity, War of Choice: A Memoir of Two Iraq Wars*. New York: Simon and Schuster.

Huhtinen, A. (2015). "The Double Edge of the Information Sword." *The International Journal of Cyber Warfare and Terrorism*.

Joint Staff, (2010). "Joint Publication 3-6.1 *Public Affairs*." Washington DC: The Joint Staff.

Joint Staff (2012). "Joint Publication 3-13 *Information Operations*." Washington DC: The Joint Staff.

Kupchan, C. (2010). *How Enemies Become Friends: The Sources of Stable Peace*. Princeton: Princeton University Press.

Lesperance, W. (Apr. 27, 2016). Personal communication.

Luttwak, E. (1976). *The Grand Strategy of the Roman Empire*. Baltimore: the John's Hopkins University Press.

Nye, J. (2004). *Soft Power*. New York: Public Affairs.

Paret, P. (1989). *On War, by Carl von Clausewitz*. Princeton: Princeton University Press.

Paul, C. (2011). *Strategic Communication: Origins, Concepts, and Current Debates*. Santa Barbara, CA: Praeger.

Ricks, T. (2012). "General Failure." *The Atlantic*, November 2012. http://www.theatlantic.com/magazine/archive/2012/11/general-failure/309148/

Schapiro, R. (2015). "Russia cuts travel to Turkey after fighter jet shot down." *The New York Daily News*. http://www.nydailynews.com/news/world/russia-cuts-travel-turkey-fighter-jet-shootdown-article-1.2448529.

Schuessler, J. (2004). "When Wars of Necessity Become Contested as Wars of Choice." In *2004 Annual Meeting of the American Political Science Association*.

Sholtis, E. (2005). "Public Affairs and Information Operations, Strategy for Success." *Air and Space Power Journal*, Fall.

Shuster, Simon (2015). "Putin's On-Air Army." *Time*, March 16, 2015.

Stavridis, J. (2007). "Strategic Communication and National Security." *Joint Force Quarterly*, July.

Thema News, (2015). "Russian Lawmakers call for a return of Hagia Sofia to orthodoxy." *Thema News*, November 27, 2015. http://en.protothema.gr/russian-lawmakers-call-for-a-return-of-hagia-sophia-to-orthodoxy/.

Thomas, Timothy. (2014). "Russia's Information Warfare Strategy: Can the Nation Cope in Future Conflicts?" *Journal of Slavic Military Studies*.

6 A proposal for the future of US counterinsurgency

Military peace with economic incentives and political devolution

Glen M. E. Duerr

Introduction[1]

This chapter examines the history of United States (US) counterinsurgency operations over the last 25 years, with specific reference to the cases of Yugoslavia, Afghanistan, Iraq, and Libya. The overarching goal is to respond to American General David Petraeus' call to learn from previous experiences and make improvements (see Petraeus 2006). These divergent cases, in particular, can provide useful directives as to how to utilize counterinsurgency in the future. In all likelihood, future US military action will take place in similar situations, in a complex, heterogeneous state, with some commonalities to the aforementioned scenarios. (The possible exception to this scenario is a counterinsurgency operation in one of the most homogenous countries in the world: North Korea.) This conclusion has been reached, quite simply, because many countries in the world are heterogeneous (Ker-Lindsay 2012, 5). Complex ethnic, religious, societal, and linguistic states can be difficult to govern, but they can be even more difficult to put back together in the aftermath of war. This, therefore, is the great challenge of US counterinsurgency policy in the future: to better deal with complexity and heterogeneity within fragile states. Often, counterinsurgency fails when countries like the US do not properly adapt to changes in the environment (Marston and Malkasian 2008, 17), which often occurs because of misunderstood complexities.

These four major cases can provide guidance on what types of situations will likely arise in the future and provide insight into how best to conduct a counterinsurgency. Any given real-life scenario will have to be specifically tailored, but overarching general principles can be described. Lessons from Yugoslavia, Afghanistan, Iraq, and Libya are all useful in that there are not only overarching similarities in the cases but also significant differences, which can help policy makers tailor future counterinsurgency operations. These cases connect to the three-pronged argument laid out in this chapter as recent examples of what to expect in future scenarios.

Although there is a constant danger of fighting the last war, this analysis will take information from previous conflicts and assess what lessons can be drawn in a counterinsurgency operation in the future. After all, generalities exist. Blanket information on a potential scenario can be found and analysis made.

For example, if the US is fighting a counterinsurgency battle, the particular country has an insurgency; in essence, the baseline assessment is that a civil war is ongoing (Galula 2006, 2–3). The civil war is simply internationalized with the presence of American and allied operations. Additionally, there will be an asymmetry of power between the groups. The world's largest and most powerful military force—the US—will be trying to contain one or more insurgency groups that have far fewer resources but much more intimate knowledge of the terrain.

For example, in countries that have recently faced an insurgency, several factors are prominent. Some scholars argue that these states are poor, are politically unstable, have rough terrain, and have large populations (Fearon and Laitin 2003). Others argue that civil wars and insurgencies are correlated with civil wars and insurgencies in neighboring states (Sambanis 2001). These lessons can be utilized as part of a wider knowhow on what to expect in future theatres of operation.

Ultimately, the optimum situation in a post-conflict situation is to provide a solution that will provide fair and just governance structures to a population so that they can move on with their lives, raise their families, engage in productive commerce, worship freely, and enjoy a level of peace and stability. This is utopian in some senses but can be accomplished by good governance buttressed by a military solution and economic incentives.

The major failing of recent US counterinsurgency policy has been an inability to best comprehend the subtleties of ethnic divisions within several countries. These ethnic divisions are then sometimes buttressed, or sometimes complicated, by language, religious affiliation, history, economic circumstances, and a range of other different variables. In order to better reduce this gap, a three-pronged strategy is proposed. By combining military, economic, and political policy options, the goal is to provide a window of opportunity for positive change to occur and for an insurgency to diminish. This could have the net effect of saving lives, treasure, and time.

Perhaps the biggest challenge to counterterrorism is to balance creating the most necessary changes to a society whilst also allowing someone with enough power to accomplish the task. If one strongman is replaced with another, people tend to rebel. But it is difficult to accomplish systemic change to the society, and improvements to infrastructure, without someone that has the power to achieve such changes. If change takes too long, people are also generally unsatisfied, which can lead to rebellion and insurgency.

The decentralization of power may seem like an unlikely solution to solving the aforementioned puzzle. However, as a means of maintaining territorial integrity, this strategy relies on a level of national level coordination, even if the largest decision-making powers are devolved. There is little incentive to purposefully not work with others in order to further secessionist claims because little will be accomplished without national-level coordination, and little money will flow into the region as a result. Instead, with coordination amongst regions, all can theoretically prosper in a shorter time frame.

US counterinsurgency and counterterrorism policy

This chapter argues that creating a military peace in combination with economic incentives (such as openness to trade and association agreements with key international organizations) and the political devolution of power (the creation of some form of confederal or consociational governance structure) represent the best tactic to counter a nascent insurgency in heterogeneous states. This combination provides an overarching strategy and blueprint to move forward in a productive manner in the aftermath of US, or US-led, military action; that is, to turn the former enemy into a productive and stable international partner and, if possible, a reliable American ally in the future—much like Germany and Japan in the aftermath of World War II.

An examination of US counterinsurgency policy and strategy over the last 25 years is important because during this time, the world transitioned from the Cold War into a unipolar era. The US also engaged in a limited war with Iraq, the Gulf War, which was very successful if a few, measured objectives are used: to rid Kuwait of Saddam Hussein, restore the balance of power in the Middle East, and showcase US military superiority. Since then, the challenges of Yugoslavia, Afghanistan, Iraq (again), and Libya provide examples of what existing counterinsurgency strategies have attempted to accomplish in a (largely) unipolar era—obviously, strategy differs based on the president and even based on the changes within each presidential administration. The cases diverge in terms of strategy, but all cases have some element of military, economic, and political policy. Guiding these areas in a calculated way is the key to an optimal post-conflict resolution.

With the strongest and most technologically adept conventional military in the world, the US has the ability to project power across the world in multiple arenas. Other countries, like China and Russia, also have major military capabilities with significant spending, but neither, at of the time of writing (July 2016), matches the US. The gap is clearly closing with China, but the above assessment currently holds. An American military force, then, has the capability to win virtually any war in the short term. Creating a brief military peace is possible in virtually any scenario, but holding that peace has long been difficult (Vietnam, Afghanistan, and Iraq are all testament to this challenge).

Once a military peace has opened a short-term window, political and economic solutions need to be implemented. The central mistake of policymakers in recent cases has been to utilize a complex parliamentary system with too many options, as is the case with Iraq. Only a mature democracy like Belgium or Italy can handle such division (Iraq does not work), but, even then, it has a limited ability to pacify all groups or to maintain territorial integrity (see Duerr 2015).

Ultimately, US counterterrorism policy should enact a policy of loosening the political structures of the state in the short term. This would allow some of the fissures in the society to be dealt with as they come to the surface in the aftermath of removing a longstanding dictator. If consolidated, national political structures can be put back in place; then the country should move forward with a transition government. If not, secession may be the best path to reducing zones of conflict throughout the state. Essentially, the country will be pacified in sections, rather than efforts being made to maintain what are often very arbitrary boundaries. Confederal structures can work in the short term—the US, Switzerland, and Canada are all cases where this type of system was enacted historically. But, with political maturity and with existential threats, the best option is to move to a more federal system, which will not only unify the country but also allow for more regional expression. Consociational structures can also work, as shown in South Africa's transition to multiethnic democracy (1990–1994), Lebanon post-1991, and Nigeria post-1999, but all retain a level of instability without federal structures overlaid.

Finally, economic incentives are also useful in terms of rebuilding economies that have been shattered by conflict. Invariably, war is the biggest reason for economic contraction, so opportunities need to be speedily implemented in order to pacify complaints on the part of the populace, which often lead to violence. This situation needs to be approached in a delicate manner given the complexities of trying to rebuild a society post-conflict. By utilizing several different actors in the rebuilding process, this strategy can ameliorate the impact and responsibility on the US alone. The task is too great to accomplish without delegating some of the work to other entities that seek to assist.

The next section of this chapter investigates what happened in the post-1991 wars: Yugoslavia, Afghanistan, Iraq, and Libya. Much of the information is descriptive, but some brief historical notes will be mentioned so as to provide context. Then, each of the cases will be connected to the three-pronged argument laid out in this chapter to showcase the shortcomings of the counterinsurgency policy in each of the four cases.

Yugoslavia

The Yugoslav conflicts in the 1990s took Europe and the world by surprise. As the largest conflict on the European conflict since World War II, many people were rightly concerned about the potential for spillover and looked for outside intervention to stop further contagion. In total, this fissiparous ethnic conflict cost the lives of approximately 140,000 people and splintered a country formed in 1918 into six (possibly seven) new pieces (for a full overview, see Ramet 2005). A strategically important part of the world, as well as an ethnically diverse region, the Balkans is the crossroads of Western Europe, Russia, and Turkey. Cauterizing the conflict at the border of Yugoslavia was an important position for the international community.

Yugoslavia was a complex mix of different ethnic, religious, and social groups formed together in one state. The Kingdom of Serbs, Croats, and Slovenes, created in 1918 out of the old Austro-Hungarian and Ottoman empires, was from the outset a heterogeneous state by design. But, in 1929, the country was recast as the Kingdom of Yugoslavia in attempt to unify the more disparate groups within the state. In 1943, the Socialist Federal Republic of Yugoslavia was initiated to move toward a republican and communist system as World War II was moving toward a close. The rule of Josep Tito (1944–1980), the longstanding dictatorial leader of Yugoslavia, loomed over the society and kept a lid on ethnic tensions within the society. Most notable in the history of Yugoslavia was the decentralization of power in 1974 in the waning years of Tito's rule. Yugoslavia was, essentially, a unitary state under Tito but decentralized into eight sectors: Serbia, Croatia, Slovenia, Montenegro, Macedonia, Kosovo, and Vojvodina.

Even President George H. W. Bush, who came to the White House with very strong foreign-policy credentials, had to conduct formal research to understand the complexities of the heterogeneity of Yugoslavia. Even though the US did not become overtly involved in the early stages of the Yugoslav conflict in 1991 and early 1992, the government was involved in a United Nations humanitarian airlift to the besieged city of Sarajevo. From June 1992 to January 1996, Operation Provide Promise supplied 160,536 metric tons of food, medicine, and other supplies to the population of 380,000 people (Air Force Museum). In order to best ensure the safety of the C-130 (mainly French and American) to Sarajevo, a "no-fly" zone was established.

In Yugoslavia, military victory was achieved by a NATO air campaign, which lasted through the month of August 1995. Ultimately, this led to a ceasefire in October 1995, culminating in peace talks at Wright Patterson Air Force Base in southwest Ohio (Air Force Museum). The ensuing peace talks resulted in the Dayton Accords. Alongside the United Nations' early recognition of Slovenia and Croatia as independent countries, much of the former Yugoslavia transitioned more peacefully into separate entities. The Former Yugoslav Republic of Macedonia (FYROM) became independent without much confusion, and a rump Serbia was created in the aftermath of the fighting. The obvious challenge was to construct Bosnia–Herzegovina in an organized and peaceful manner, which, given the ethnic heterogeneity of the territory with a large Bosniak, Croat, and Serb population, was a very difficult task.

Later, Kosovo also proved to be a point of conflict when Serb and Kosovar forces became embroiled in a war over the status of Kosovo. The initial eruption of violence in 1998 was quite quickly stifled with US Air Force intervention supplemented by NATO ground troops under the K-FOR mission.

Even though there have been violent episodes in Kosovo in particular, as well as interethnic conflict in FYROM, the former Yugoslavia is much more peaceful in the 2010s than it was in the 1990s. Serbia has made major strides toward creating a thriving democratic system replete with respect for human rights and the rule of law. For example, in 2006, Montenegro even peacefully voted for independence. The referendum narrowly passed with 55.5 percent of the vote (the European Union [EU] mandated a threshold of 55 percent) (Darmanovic 2007).

Moreover, former Yugoslav republics such as Slovenia (2004) and Croatia (2013) have successfully joined the EU (as well as NATO); others, like Serbia, Montenegro, and FYROM are also in line for membership. In many respects, despite some of the major problems in stopping the conflict and stopping crimes against humanity, there were some noteworthy successes in the former Yugoslavia. For the US, the goal of counterinsurgency was never to solve all of the problems in Yugoslavia. This comes with a moral cost, but, in terms of creating a more peaceful region, there is significant reason for optimism in the Balkans, at least in the short and medium terms.

The major challenge, however, is that ethnic borders are still in flux. It remains possible that a return to violence could occur without overt and continued help from the EU as well as internal and external actors. But a useful combination of a military peace, along with strong governance structures and incentives to trade (and join) the EU, have been paramount in creating stability in the region.

Although the bloodshed in the region was severe, the Balkans is now a much more stable region. Each of the three prongs, as argued in this chapter, was initiated to some degree, which provides a window of possibility into what a successful counterinsurgency strategy might provide. The major issue is that the three prongs were applied unevenly. If EU membership or association, for example, was used more readily, it could have brought stability to the region more quickly. It also shows that a political agreement, like the Dayton Accord, has provided, thus far, over 20 years of stability. Finally, quicker military action through an immediate no-fly zone or earlier aerial assistance might have also shortened the conflict. But, in sum, this case provides a strong example of what can be accomplished with a three-pronged approach.

Afghanistan

The country of Afghanistan has long been at the crossroads of many different cultures and national groupings. It has also served as hosts to power struggles involving great powers from the United Kingdom to Russia/Soviet Union to the US, with both India and China also highly influential.

Unlike Yugoslavia, the US played a much more direct role in military action in Afghanistan. Operation Enduring Freedom utilized all branches of the US military to defeat Taliban and al-Qaeda forces (Air Force Museum). In response to the 9/11 terrorist attacks on New York, Washington DC, and Pennsylvania, a US-led coalition invaded Afghanistan in order to try to find the mastermind of the attack, Osama bin Laden, and to replace the Taliban government that allowed bin Laden's terrorist organization, al-Qaeda, a place to conduct training operations. The US obtained a United Nations Security Council Resolution 1378 authorizing military action and also utilized a NATO force to jointly engage in military operations.

One of the historic challenges of governing Afghanistan has been the ability to adequately control the territory. Afghanistan is made up of a patchwork of different major tribes, which have differing levels of allegiance to the central state (CFR). In some respects, governing Afghanistan requires much more than control of the capital city, Kabul, and the second most populous city,

Kandahar. The population of Afghanistan has transitioned in these two cities, but full progress has not been shown outside of these two population centers.

Unlike Yugoslavia, the goal in Afghanistan was much more encompassing. Ridding Afghanistan of al-Qaeda, and its supporting Taliban government, meant replacing the government of the country. At some level, the loftiness of the task may have been too challenging to accomplish without sending much larger numbers of troops, which would also have cost a lot more money as well.

The initial military victory did not take long to accomplish, but given the mountainous and disparately governed population of Afghanistan, establishing control proved to be difficult. Moreover, Afghanistan exists in a very different geographic region to Yugoslavia, meaning that providing economic incentives was much more difficult and more of an afterthought for policymakers. The most pertinent lesson here could have been to utilize other actors in Afghanistan's neighborhood to build economic opportunities for the people. China, India, and Russia, along with the US, could have built an organization designed to increase trade in tandem with the political rebuilding of Afghanistan. Establishing economic opportunities could have reduced the continued appeal of the Taliban.

In some respects, a level of informal power has been devolved down to the regions of Afghanistan, at least as measured by the ongoing power of tribal leaders in the country (CFR). The problem, however, is that this power was not formally passed to these leaders under the constitutional framework of the society. Given some vetting, this could have been a means of bringing more people into the process.

In some ways, the case of Afghanistan has always been difficult—even a three-pronged approach would likely have some limitations. Nonetheless, more options could have been applied. In tandem with the military prong that was utilized in late 2001, very little was done to apply an economic prong; the above examples could have helped here. Finally, devolving power might have brought some of the historic groups (tribal leaders, ethnic and linguistic groups, etc.) in the society together under a larger umbrella of governance.

Iraq

Iraq was founded in 1920 as a British mandate in the aftermath of the dissolution of the Ottoman Empire at the end of World War I. The Hashemite Kingdom was granted formal independence in 1932 and lasted until 1958, when a military coup d'etat ended the years of rule. The Ba'ath party came to power in 1968 after their own military coup, led by Ahmed al-Bakr. In 1979, Saddam Hussein, al-Bakr's second-in-command, ascended to the presidency (see Dawisha 2009).

Under Hussein, Iraq was, in theory, divided into 18 provinces, and, at some level, decisions were made at local and regional levels. However, as with many dictators, ultimate power in Iraq resided with Hussein alone. The constitution worked only on paper, and practice with the devolution of power was not a reality.

The new constitution of 2005 shifted the structure of Iraq from a centrally controlled dictatorship toward a more federal model, utilizing the existing 18-province structure of the state. For example, Article 1 states, "The Republic of Iraq is a single federal, independent and fully sovereign state in which the system of government is republican, representative, parliamentary, and democratic, and this Constitution is a guarantor of the unity of Iraq" (Iraqi Constitution). Article 115 of the constitution, for example, notes that all powers not mandated to the federal government will be devolved down to the regions. Some groups, like the Kurds, have also been granted much greater autonomy (under articles 117 and 141) (Iraqi Constitution).

Of course, in any real-world case, a written constitution may not be applied under a given leader. Under Nouri al-Maliki, prime minister of Iraq from 2006 to 2014, his fellow Shia Muslims were, by many accounts, given advantages in the state. This was done to the detriment of other minorities, especially Sunni Muslims. Allowing al-Maliki the latitude to make these changes was a mistake on the part of US policymakers, who could have used a range of tools to rein in his power.

As was the case in Afghanistan, once Saddam Hussein was removed from power, the US-led coalition had a much greater task of state building in Iraq (see Hashim 2005). This was much more comprehensive than the counterinsurgency-type operations in Yugoslavia and Libya. An initial military peace was won, but the seeds of insurgency developed as expectations for the future were not realized.

Iraq's neighborhood, again unlike Yugoslavia with the EU at its doorstep, proved to be more difficult. Economic opportunities were sparse, and with the empowering of the Shia community, the historic powerholders, the Sunnis, were left in a more difficult situation. If power had been devolved, the Sunni population might well have had greater opportunities in the society, rather than being wooed by the Islamic State of Iraq and Syria (ISIS). In order to provide for their families, many young Sunni men have worked with and for ISIS as a means to put bread on the table. Moreover, by encouraging trade with other wealthier actors in the region such as Saudi Arabia, the United Arab Emirates (UAE), and Israel, along with the US, a nascent economic grouping could have been utilized to create some opportunities to provide for the families of Iraq.

In terms of a three-pronged approach, there were two windows of opportunity wherein military success could have coincided with the other two prongs: after the initial military victory in 2003, and during the "surge" in 2007. If power had been devolved to larger groups in the society, such as the Sunni, Shia, and Kurds, during one of these two occasions, many of the preconditions for the rise of ISIS could have been undercut. Moreover, stronger trading links to the US and other coalition countries could have allowed Iraqi goods to better penetrate foreign markets—in this way, Iraq would have been better enriched economically and in a better situation to fend off radicals seeking to disrupt life in the country.

Libya

Historically part of the Ottoman Empire, Libya was colonized by Italy in 1911 as the Ottoman "sick man" of Europe continued to decline in health (McMeekin 2015). Italian suzerainty remained through the latter stage of World War II, when Libya fell under joint British and French rule. This lasted until 1951, when Libya gained full independence. King Idris al-Sanusi ruled the country until a bloodless coup in 1969 brought Muammar al-Gadhafi to power (Carpenter, Witherbee, and Lee 2014).

Unlike in Afghanistan and Iraq, US intervention in Libya was relatively brief and much more limited in scope. The overarching goal of NATO was to support rebels against former dictator Muammar al-Gadhafi and to help the transition toward democratic governance. Unfortunately, the immediate aftermath has not led to the grounding of representative democratic institutions, and similar policy failures have resulted in a fragile state much like Afghanistan and Iraq.

Under UN Security Council Resolution 1973, NATO forces engaged in military confrontation in Libya from March to October 2011 (UNSC Resolution 1970 laid the groundwork in February 2011). As noted above, the goals were much more limited than in Afghanistan and Iraq and revolved around missions such as enforcing a no-fly zone and naval blockades of the Mediterranean coast.

In comparison with the other cases, the counterinsurgency strategy in Libya resembles the case of Yugoslavia much more than those of Iraq and Afghanistan, mostly in terms of commitment level. Without obvious incentives to move toward normalization, though, Libya could remain a very weak state for the foreseeable future. In some ways, Libya has already been divided into at least two main regions and is beginning to devolve back into its previous Ottoman era provinces of Tripolitania, Cyrenaica, and Fezzan.

A three-pronged counterinsurgency approach had an opportunity for success after the death of Gadhafi. Further security support would have been helpful, but also utilizing the oil resources in the various regions of the country could have helped each region move forward whilst the central government was in transition. Better economic connections with EU countries, especially those in proximity to Libya, could have undercut the conditions for extremism that moved into the society in the years following Gadhafi's removal.

A three-pronged approach

As noted earlier, the major argument in this chapter is that creating a military peace in combination with economic incentives (such as openness to trade and association agreements with key international organizations) and political devolution of power (the creation of some form of confederal or consociational governance structure) is the best tactic to counter a nascent insurgency in heterogeneous states.

The first prong is to create a military peace, that is, for the formal military conflict to come to an end in an expedient timeframe. Of course, this is difficult in general, but the US military has proven to be remarkably capable in this situation—for example, in all of the cases noted in this chapter, the US, or coalition forces, won the military peace very quickly; it was the aftermath in insurgency that proved particularly troubling. Foreseeably, there are future scenarios wherein larger challenges may arise—for example, if China and/or Russia or any other outside force is also embroiled in the conflict, or when facing the North Korean one million person army (with 10 million plus reservists)—but these are unlikely at the present (unless North Korea is the conflict).

The initial military victory, as noted in the above cases, tends to happen quickly when the full force of the US military is enabled, but, as an important note, full resources must be provided (see Egnell 2011). Victory over an insurgency force, however, is much more difficult. Therefore, when the initial peace is created, the second and third prongs need to be initiated immediately in order to take advantage of the small window of time between military victory and large-scale insurrection.

The future of the military prong of the counterinsurgency strategy requires creating—and then maintaining—stability in parts of the country for long enough that the second and third prongs of the strategy can take hold. In essence, the goal for the military is to provide enough space for the other components of the strategy to be simultaneously implemented and for the norm of peaceful transition to develop.

The second prong is economic incentives, which, in the modern era, are a key component to increasing prosperity in a country. As noted in the introduction of this chapter, it is most likely that future US counterinsurgency operations will be conducted in a country that is quite poor. Providing economic opportunities could be a central part in alleviating the underlying conditions for insurgency. As a preliminary step, it could mean cajoling the government to incentivize private enterprise and move away from reliance on natural resources. This would help the country to move away from the "resource curse" (Ross 1999). Recent scholarship argues that the resource curse tends to hold in countries that are not democratic, whereas countries that democratized and then realized their natural resources, like Norway, become wealthy yet still remain democratic (Haber and Menaldo 2011).

The economic prong requires the creation of a market system with varying level of government input. Depending on the situation, policymakers can use some flexibility. Once a rudimentary free market has been established, possibly with some government assistance akin to the South Korean model, further incentives can be applied. For example, if the US (US-led) counterinsurgency takes place in North Korea, the South Korean government can more naturally assist with the restructuring of Pyongyang's economy. Flexibility is useful because, depending on the case, one option may be better than another, provided that the end outcome of economic growth is achieved. Where an economy falters in the short and medium terms, counterinsurgency becomes much more challenging if there is no emphasis on improving the economic lives of the citizens. As an example, a US-style economic system was advocated in Russia at the beginning of the Cold War. Although desirous of a good outcome, the policy did not take into account more localized challenges of the Russian economy. The economy then faltered in the midst of shock therapy, and the average Russian became disillusioned with the system.

As an example, a country that is geographically close to the US or allied countries could be given some form of preferential trading partnership. For example, if US intervention occurs in a country in Central and South America, an incentive could be to include the country within the North American Free Trade Agreement (NAFTA) or the Dominican Republican-Central American Free Trade Agreement (DR-CAFTA). Or, as another example, Libya in 2011 could have been given a partnership with the EU. This could mimic the association agreement that countries like Ukraine, Georgia, and Moldova signed in 2014. It might have averted some of the problems that Libya is facing today.

The final prong is the political devolution of power. Policymakers should evaluate two main options here. First, the option is to create a consociational government wherein members of different groups within the society share governing responsibilities (see Lijphart 1979). It is a way of maintaining territorial integrity whilst simultaneously trying to undercut the incentives for war. Second, the option is to temporarily create a confederal state wherein the existing major divisions become miniature countries, at least until peace and stability is restored. People within each unit could then be given the option to vote as to whether or not the central government is once again strengthened, whether the region should become independent, or whether the new status quo of confederalism should remain.

Consociation has been used by a variety of deeply divided societies in the world (see Lijphart 1979; McGarry and O'Leary 2004). Perhaps the best known contemporary case is Lebanon, where the major ethnic and religious groups share power. In Lebanon's consociational system, a Maronite Christian serves as President; a Sunni Muslim as Prime Minister; a Shia Muslim as Speaker of Parliament; and cabinet positions for a range of different minority groups. This system has, at least thus far, managed to avoid widespread bloodshed that plagued the country during the Lebanese civil war from 1975 to 1991. Another example is Nigeria, which uses a somewhat more simple formula. In effect, the presidency rotates between a Christian and a Muslim, with the respective vice-presidents being of the opposite religious affiliation as the president.

Confederalism is a different model of state structure. Although this was once fairly popular, virtually every country in the world now uses an alternate model. Many small states in the world utilize a unitary style of governance wherein the parliament in the capital city exercises tight control over state policy. In many large (in terms of land mass) countries of the world, a federal model is instead used. This structure grants powers to subnational units in order to facilitate more local governance. This should be taken on a case-by-case basis but should involve a strategy of slowly, and then peacefully, putting the country back together. Ultimately, the people will decide one way or another, but in order to avoid bloodshed, localized plebiscites could be utilized. This should be conducted very differently to the way in which the League of Nations led plebiscites after World War I, which, in many respects, fueled the conditions for the rise of Hitler in Germany and ongoing hostility in parts of Europe.

In the short term, this might involve heavy decentralization to govern the country at the provincial or subnational level before moving to put the parts of the country together again. Once a province/subnational unit is safe, then the national government can come in and restore governance at the national level. This approach is piecemeal. It has risks if one is opposed to maintaining territorial integrity. But it provides flexibility as a counterinsurgency strategy to allow the US military and its allies an ability to create safe zones. Once two or more contiguous safe zones are created, then comprehensive national governance structures can be put back in place.

In Iraq, the consociational government formed as a mechanism of power sharing between major groups in the society did not stop Prime Minister al-Maliki from effectively turning Iraq into a Shia-dominated state, much to the consternation of other groups. Had Iraq also adopted a confederal system on top of the consociation and limited the power of al-Maliki, or of any prime minister for that matter, the early seeds of ISIS could have been averted. Although the international media has focused on overseas recruits for ISIS, the majority of its support is found amongst disenfranchised Sunni Iraqis.

Conclusion

One of the central debates in the literature on partitioning or decentralizing states is whether the action serves as glue or solvent (Bird, Villancourt, and Roy-Cesar 2010). Under this proposal, there is a very real possibility that states could partition. This has its challenges because partition is a rarity in the post-World War II international arena, and, when it does occur, the outcomes are not known. However, if the goal is better security and then perhaps the implementation of democracy, then a solution based on decentralization (and possible partition) should be more palatable.

Some scholars openly argue that either secession or some form of confederacy would best help to solve the problem of intractable civil conflict in the midst of US military involvement. In Syria, for example, professors Steven Simon and Jonathan Stevenson argue that one of the options "would be to partition the country to some degree and establish a confederacy of sorts to replace central rule from Damascus" (Simon and Stevenson 2015, 9). In the same volume of *Foreign Affairs*, a former special assistant to various US ambassadors in Iraq argues for "openly encouraging confederal decentralization across Iraq and Syria—devolving powers from Baghdad and Damascus to the provinces while maintaining the two countries' territorial integrity" (Khedery 2015, 39).

On the other hand, Professor Stephen Walt argues that in order "to undercut the Islamic State's local support, Washington should continue to press the Shiite-dominated government in Baghdad to adopt more inclusive policies toward Sunnis" (Walt 2015, 50). This is a useful component to help Iraqis, but if power were formally devolved down to majority Shia, Sunni, and Kurd regions (with requisite protections for minorities), then greater autonomy could be reached for all of these respective major groups. In tandem with the realization of autonomy along with the development of a political sense of security, economic opportunities must also be found. By partnering with major economies in the region, who also have an incentive to play a positive role in the rebuilding of these states or risk stability concerns themselves, economic development could flourish and detract from activities that are counterproductive to economic growth.

In the forthcoming years, the US may yet play a role in the conflicts in Syria and Yemen (Cordesman 2015). Depending on the size and scope of any US role, policymakers would be wise to consider a policy of decentralization as a means of rebuilding the country. Syria could be linked to the EU through existing agreements with Turkey as well as states in the North Caucasus like Georgia. Yemen is more challenging geographically, but it is in the interests of regional actors like Saudi Arabia, Oman, Bahrain, and UAE to avoid instability in the neighborhood. By reaching out to Yemen through the Gulf Cooperation Council, a level of stability could be accomplished in the Persian Gulf region.

There is a reticence amongst the American population to become further embroiled in conflicts in the Middle East, but the arbitrary nature of border creation with the 1878 San Stefano, 1878 Berlin, 1916 Sykes-Picot, 1920 Sèvres, and 1923 Lausanne treaties means that the current national lines are, at best, placeholders in history. Conflict will likely continue in the Middle East, and, if the US remains a great power/superpower in the world, Washington will play some role. Counterinsurgency strategy should be limited, but where a greater commitment is required, a combination of economic, military, and political components can best move the particular case forward toward peace.

Note

1 The author would like to thank his research assistant for the 2015–2016 university calendar year, Spencer Woody, for his exceptional help in compiling the necessary research for this chapter.

References

Air Force Museum. December 2015. Riverside, OH, USA.

Bird, Richard M., François Villancourt, and Edison Roy-Cesar. "Is Decentralization 'Glue' or 'Solvent' for National Unity?" *International Center for Public Policy Working Paper Series* (2010).

Carpenter, Michael, Amy Witherbee, and M. Lee. "Libya." Salem Press Encyclopedia (January 2014): Research Starters, EBSCOhost (accessed December 16, 2015).

Cordesman, Anthony. "Rethinking the Wars against ISIS and the U.S. Strategy for Counter-Terrorism and Counter-Insurgency," *CSIS*, September 28, 2015.

Council on Foreign Relations. "Afghanistan." http://www.cfr.org/afghanistan/taliban-afghanistan/p10551. Accessed on October 29, 2015.

Darmanovic, Srdjan. "Montenegro: A Miracle in the Balkans?" *Journal of Democracy* 18, no. 2 (2007): 152–159.

Dawisha, A. I. *Iraq: A Political History from Independence to Occupation*. Princeton, NJ: Princeton University Press, 2009.

Duerr, Glen. *Secessionism and the European Union: The Future of Flanders, Scotland, and Catalonia*. Lanham, MD: Lexington Books, 2015.

Egnell, Robert. "Lessons from Helmand, Afghanistan: What Now For British Counter-insurgency?" *International Affairs* 87, no. 2 (2011): 297–315.

Fearon, James D. and David D. Laitin. "Ethnicity, Insurgency, and Civil War." *American Political Science Review* 97, no. 1 (2003): 75–90.

Galula, David. *Counterinsurgency Warfare: Theory and Practice*. Westport, CT: Greenwood Publishing Group, 2006.

Haber, Stephen and Victor Menaldo. "Do Natural Resources Fuel Authoritarianism? A Reappraisal of the Resource Curse." *American Political Science Review* 105, no. 1 (2011): 1–26.

Hashim, Ahmed. *Insurgency and Counter-insurgency in Iraq*. Ithaca, NY: Cornell University Press, 2005.

Iraqi Constitution. http://iraqinationality.gov.iq/attach/iraqi_constitution.pdf. Accessed on January 22, 2016.

Ker-Lindsay, James. *The foreign Policy of Counter Secession: Preventing the Recognition of Contested States*. Oxford: Oxford University Press, 2012.

Khedery, Ali. "Iraq in Pieces: Breaking Up to Stay Together." *Foreign Affairs*. 94, no. 6 (2015): 33–41.

Lijphart, Arend. "Consociation and Federation: Conceptual and Empirical Links." *Canadian Journal of Political Science* 12, no. 3 (1979): 499–516.

Lijphart, Arend. *Patterns of Democracy: Government Forms and Performance in Thirty-Six Countries*. New Haven, CT: Yale University Press, 1979.

Marston, Daniel, and Carter Malkasian, eds. *Counterinsurgency in Modern Warfare*. Oxford: Osprey Publishing, 2008.

McGarry, John, and Brendan O'Leary. *The Northern Ireland Conflict: Consociational Engagements*. Oxford: Oxford University Press, 2004.

McMeekin, Sean. *The Ottoman Endgame: War, Revolution, and the Making of the Modern Middle East, 1908–1923*. New York: Penguin, 2015.

Petraeus, David H. *Learning Counterinsurgency: Observations from Soldiering in Iraq*. Army Combined Arms Center: Fort Leavenworth, KS, 2006.

Ramet, Sabrina P. *Thinking about Yugoslavia: Scholarly Debates about the Yugoslav Breakup and the Wars in Bosnia And Kosovo*. Cambridge: Cambridge University Press, 2005.

Ross, Michael L. "The political economy of the resource curse.' *World Politics* 51, no. 2 (1999): 297–322.

Sambanis, Nicholas. "Do Ethnic and Nonethnic Civil Wars Have The Same Causes? A Theoretical and Empirical Inquiry (Part 1)." *Journal of Conflict Resolution* 45, no. 3 (2001): 259–282.

Simon, Steven and Jonathan Stevenson. "The End of Pax Americana: Why Washington's Middle East Pullback Makes Sense." *Foreign Affairs* 94, no. 6 (2015): 2–10.

Walt, Stephen. "ISIS as Revolutionary State: New Twist on an Old Story." *Foreign Affairs* 94, no. 6 (2015): 42–51.

7 US special warfare units overseas

Roger Chapman

Introduction

"It may well be that the era of U.S. military expeditions, so much a feature of our history in the late nineteenth and entirety of the twentieth centuries, is drawing to a close," stated John D. Negroponte (2012, p. 2). What Negroponte, the former United States Ambassador to Iraq, was in fact stating is that the wars in Iraq and Afghanistan—indeed the entire global war on terrorism (GWOT)—led to a tipping point in which Americans in the foreseeable future will be less patient with large deployments for overseas combat. Negroponte, however, was not suggesting that the United States is becoming isolationist. There are, and will be in the foreseeable future, military deployments, but they will be characterized by a light footprint and with largely indirect combat engagements. As further explained by Negroponte (2012):

> So, if we are not going to deploy abroad in large numbers in the future, the question presented is just what kind of overseas interventions by U.S. forces can we reasonably expect. I think we are already beginning to see the shape of our response to that question. Take, for example, the increased use of drones and other air assets against terrorist targets in countries such as Pakistan or Somalia. There has also been a concomitant increase in the use of Special Operations Forces (SOF). As time goes on and as the withdrawals of ground forces from Iraq and Afghanistan are completed, we are likely to witness a growing reliance on technology such as drones and on SOF.
>
> (p. 2)

According to Thomas and Dougherty (2013, p. 30), prior to the terrorist attacks of September 11, 2001, only about 2,800 SOF personnel were deployed overseas. But due to the subsequent GWOT, including the wars in Afghanistan and Iraq, SOF presence overseas peaked at 12,000 and then remained constant. From 2001 to 2012, the US Special Operations Command (USSOCOM) expanded by 68 percent, increasing its personnel size from 25,000 to 38,000. The USSOCOM budget rose from $2.3 billion in 2001 to about $10.4 billion in 2013 (Thomas & Dougherty 2013, p. 31). Despite what may seem like a staggering cost, SOF in fact constitutes only about 5 percent of the entire US military budget (Robinson 2013a, p. 6). SOF had the starring role in the fight against terrorism, but this overuse with numerous deployments and missions—a situation officially referred to as high operations tempo (OPTEMPO)—came at a price of fraying (Thomas & Dougherty 2013, pp. 39–41). With the official combat missions having wound down, SOF looks to an immediate future of regrouping by renewing training and education that had to be postponed due to frequent deployments. Still, there will be SOF units and detachments spending time overseas, as the world remains unstable with extremist groups carrying out violent attacks as well as nation states like Russia and China flexing muscle in their respective regions.

SOF consists of elite military units that carry out missions that are often classified in nature. In fiscal year 2013, the special operators of these elite units were 33,000 strong, but the entire uniformed and civilian personnel under USSOCOM jurisdiction stood at 66,594 (Feickert 2013). All branches of the military, including active duty, National Guard, and reserve, have SOF personnel. About 45 percent of SOF is part of the US Army and includes the Special Forces groups, Delta Force, the 75th Ranger Regiment, the 160th Special Operations Aviation Regiment (Airborne), and the 95th Civil Affairs Brigade (Airborne). The US Air Force, at 28 percent, is the second largest SOF component, including the 1st Special Operations Wing and the 720th Special Tactics Group, both located at Hurlburt Field, Florida, as well as the 352nd and 353rd Special Operations Groups in England and Japan, respectively. The Sea, Air and Land teams (SEALs), numbering around 2,400 active-duty personnel, are part of the Naval Special Warfare Command. The US Marines constitute the smallest portion of SOF with a special operations regiment founded in 2006.

Indirect missions

SOF missions are usually clandestine, typically carried out in hostile or politically denied or official "off limits" territory, and can involve raids, intelligence gathering, rescue, insurgency, counterinsurgency, psychological operations (psyops), counternarcotics, and counterterrorism for the purpose of achieving military, diplomatic, and/or economic objectives (Chapman 2011, p. 729). Direct missions, such as raids, are typically carried out at night for the purposes of stealth and surprise. This represents the "commando" aspect of SOF (Fitzsimmons 2003, p. 206).

Indirect missions (sometimes termed *special warfare*), which is the present emphasis, involves aiding host nations that are combatting forces that the US regards as the enemy. Specifically, indirect missions may involve the training and advising of friendly host nations' armies and police forces, as well as friendly militias and tribes; providing airlift to the military of host nations; sharing appropriate intelligence; accompanying the partner force on raids; and conducting civil affairs projects (e.g., building schools, offering medical services, and other humanitarian assistance) (Robinson 2013a, pp. 11–12). Of the SOF personnel deployed to an estimated 139 countries in 2015, a majority were engaged in indirect mission, primarily the training of the defense forces of partner nations (Gallagher 2016, 6). One success of this effort was the December 2015 Iraqi recapture of Ramadi from the Islamic State (ISIS); the unit that spearheaded this successful operation consisted of 500 elite commandos who had been trained by Americans (Bradley & Nabhan 2016).

The so-called "Afghan model" has been a synonym for indirect missions—based on the US counterinsurgency strategy that was applied in Afghanistan—but there is spirited disagreement over what in that war zone was achieved and how it was achieved (Biddle 2005/2006). Until the 2012 publication of the Army Doctrine Reference Publication (ADRP) 3-05, *Special Operations*, the concept of "special warfare" was largely absent from American military doctrine, despite the fact that for preceding decades it was an important part of SOF thinking (Harkins 2015, p. 112). In the new ADRP, however, special warfare has been expansively addressed. Thus, in addition to unconventional warfare, counterinsurgency, and psychological warfare, there is a new focus on foreign internal defense, stability operations, special reconnaissance, and security force assistance (Harkins 2015, p.114). All of these are what US special warfare units will be doing in the foreseeable future. This is in harmony with an epiphany experienced by General Stanley McChrystal, when he served as commander of US Forces in Afghanistan: "it became clear to me and to many others that to defeat a networked enemy we had to become a network ourselves" (Thomas & Dougherty 2013, p. 13).

The trend of the network approach often involves a blending of the surgical strike and special warfare ("Special Operations Research Topics" 2016, p. 12–13). The special warfare component is a crucial aspect of the SOF mission when partnering with friendly foreign forces. An understanding of culture is critical for allowing SOF to train allies and work and fight beside them. The surgical strike, a strictly lethal approach, will be carried out more often than not by indigenous fighters with American SOF acting as observers, guides, and, when necessary, as periphery fighters. The delivery of the strike team to the objective, however, will invariably require SOF aviation assistance. This hybridization of tactics and mixed fighting force will have its political advantages when the United States wishes to maintain a veil of secrecy about the extent of its overseas operations. The Trump administration will have to decide whether it is prudent to continue the Obama administration's approach of announcing when an SOF operator dies in the course of a surgical strike, such as was the case in October 2015 after Delta Force operator Master Sgt. Joshua L. Wheeler was killed during a hostage rescue officially carried out by friendly Kurdish commandos (Pérez-Peña, Rogers, & Phillips 2016).

For some time, USSOCOM has been developing a global SOF network that has involved the training of foreign SOF counterparts, the conducting of joint exercises, and advising (Szayna & Welser IV 2013). This trend toward partnering is designed to not only reduce America's footprint but also to expand its capability to address regional security concerns as well as respond rapidly to terrorist incidents, hostage situations, and outbreaks of violence in general (Davis 2012).

Such partnering, of course, is fraught with risk. First, there is no guarantee that allies will remain allies. Even if these countries that are part of the SOF network remain on good terms with the United States, it cannot be expected that there will be unanimity on how to respond to specific crises when they occur. The long-term viability of the SOF network will largely be dependent upon American funding, which is subject to the vagaries of Washington politics. The extreme partisanship and populism that have been plaguing American society can only increase such budget uncertainties. What will become of the global SOF network should the United States significantly scale back or eliminate entirely its expenditures for foreign militaries? As Szayna and Welser IV (2013, p. 5) have explained, the budgetary process is already convoluted, and there are disagreements on who pays for what. Near the end of his tenure, President Obama complained about the "free riders," those allies he felt were overly relying on the blood and treasure of the United States to provide their security (Landler 2016).

An inherent risk of partnering is the diffusion of American special warfare tactics to enemy forces. There is no guarantee that some of today's foreign SOF operators will not later become terrorists or insurgents themselves. How to gauge another person's allegiance is easier said than done and is all the more complex in a cross-cultural setting. When specially trained personnel switch to the other side, they bring their expertise with them. In some cases, such as happened in Libya, the American-supplied weapons for the training of elite forces end up in the hands of the enemy (Schmitt 2014). There is also no guarantee that these specially trained units will not be misused by the host nation for the purpose of squelching political dissent—the past history of the US Army's School of the Americas, now known as the Western Hemisphere Institute for Security Cooperation, has been criticized over the years for training Latin American military officers who afterwards committed human rights violations against their own people (Chapman 2014).

Global SOF network

Currently, the American SOF is divided into seven Theater Special Operations Commands (TSOCs):

- Special Operations Command South (SOCSOUTH), Homestead Air Force Base, Florida
- Special Operations Command Africa (SOCAAFRICA), Stuttgart, Germany
- Special Operations Command Europe (SOCEUR), Stuttgart, Germany
- Special Operations Command Central (SOCENT), MacDill Air Force Base, Florida
- Special Operations Command Pacific (SOCPAC), Camp Smith, Hawaii
- Special Operations Command Korea (SOCKOR), Yongsang, South Korea
- Special Operations Command U.S. Northern Command (SOCNORTH), Peterson Air Force Base, Colorado

In February 2013, the TSOCs, though they continued to be operationally controlled by the Geographic Combatant Commanders (GCCs), were reassigned to USSOCOM for organizing, training, and equipping, as it has always done for all SOF units (Feickert 2014). According to the ADRP 3-05 (2012), the US Army doctrine manual for Special Operations, SOF theater operations focuses on four major areas: security, deterrence, limited intervention, and peacetime engagement with allies.

There have also been plans for Regional SOF Coordination Centers (RSCCs) modeled on the NATO SOF Headquarters, which was created in 2007 and offers analysis and disseminates lessons learned from multiple theaters. One plans calls for RSCCs in the Western hemisphere, Asia-Pacific, and Africa. According to Szayna and Welser IV (2013), "RSCCs will likely range in ambition and scope to ensure effectiveness without overreach" in order to "increase the capabilities of partner SOF, improve their training and education, increase their interoperability, enhance their information-sharing, and enable more effective cooperation and coordination with their U.S. counterparts" (p. 4). The same authors recommended up to ten RSCCs to cover the Caribbean, Central America, and South America; Southeast Asia and possibly South Asia (the latter if there can be a reduction of Afghan–Pakistani and Pakistani–Indian tensions); and northern Africa, including the Sahel and the Horn, western Africa, and south and eastern Africa.

The hopeful outcome of the bureaucratization of SOF is that it will optimize overall capability by enhancing partnering (Svendsen 2014). The challenge will be integrating all players, including conventional forces. Also, it remains to be seen if the partnering nations that are not a part of NATO will, in the global SOF network, have as close ties with the United States as the United States has had with its NATO allies. No matter how strong the bonds, there will invariably be challenges because of the hegemonic realities of the United States acting as more than a mere facilitator. The global SOF network will greatly depend on liaison personnel stationed at pertinent US embassies to assure and cajole host nations (Szayna & Welser IV 2013, p. 2).

As SOF looks to the future and its overseas role, there will be an emphasis on fighting against Islamist violent extremist networks (VENS), countering weapons of mass destruction (WMD), and the directing of counterinsurgency in proxy wars. The Russian military interference in Ukraine has led to SOF placing an emphasis on developing "counter-unconventional warfare," which, at its core, is a hybridization of unconventional warfare (UW), foreign internal defense (FID), counterinsurgency (COIN), stability operations (SO), and counterterrorism (CT) (United States Army Special Operations Command 2014, pp. 22–23). In addition to Russia, counter-unconventional warfare will be applied against Iran and other state actors responsible for "causing chaos and disrupting civil order, while seeking to provoke excessive responses by … security organs" (United States Army Special Operations Command 2014, p. 4). In numerous countries across several continents, SOF will continue to partner with friendly counterparts, collect intelligence, pinpoint high-value targets, and conduct covert counterterrorism.

The question is whether the trend of SOF bureaucratization will prevent the mistakes of the past, such as uncovered by Vandenbroucke (1993) in his meta-analysis of the Son Tay raid, the Bay of Pigs invasion, the *Mayaguez* operation, and the Iran rescue mission:

> Faulty intelligence, poor interagency and interservice cooperation and coordination, provision of inadequate advice to decision makers, wishful thinking, and overcontrol of mission execution by officials far removed from the theater of operations have repeatedly jeopardized the ability of the United States to conduct such missions successfully.
>
> (p. 152)

The "wishful thinking" mentality can sometimes lead to situations in which operators and analysts are pressured to produce the most positive spin on the situation, such as in 2015 when the Central Command allegedly "rewrote classified assessments for intelligence officials and policy makers that played down the setbacks in Iraq" (Cooper 2016).

It is safe to say SOF, in the next few years, will continue to pivot from direct action missions, primarily in the Middle East, to indirect missions on a global scale pertaining to the ongoing GWOT (Nelson & Wise 2012). The goal is for SOF to maintain a "light footprint," but the Pentagon (perhaps due to political considerations) has not defined exactly what that means (Luján 2013, p. 8). What can be noted, however, is that the combat troops being deployed are largely SOF personnel, and even those numbers are about half of what they were just a few years earlier. Considering the rise of ISIS, it is prudent to predict that a substantial part of the deployed SOF force will be focused on the Middle East and Africa. During the wars in Iraq and Afghanistan, the majority of the 13,000 SOF personnel that were deployed overseas were sent to those two countries. By the end of 2015, only 7,500 SOF were deployed overseas with half of that number sent to 85 countries outside of the Middle East or South Asia (Mazzetti & Schmitt 2015).[1]

As explained by Cooper and Schmitt (2015), the threat of ISIS in the Middle East prompted President Obama to send SOF groups back into Iraq, with the number reaching 3,500 by the end of 2015. Some of these forces included Delta Force operators who used Iraq as a base from which to attack ISIS in Syria, including the May 2015 raid involving Black Hawk helicopters and V-22 Ospreys that killed Abu Sayyaf, the ISIS leader who generated revenue for the Islamic State through the sale of oil and gas. That raid resulted in a major intelligence breakthrough with the seizure of laptop computers, cell phones, and other materials. The success of that raid prompted American defense officials to expand to 200 the number of SOF commandos, primarily Delta Force, operating in Iraq (Cooper, Schmitt, and Schmidt 2016). Another success was Delta Force's capture of a top ISIS chemical specialist, who, during interrogation, explained his group's production of sulfur mustard; this information was used to carry out bombings of production plants (Cooper & Schmitt 2016). Prior to the pivot back toward Iraq, Obama cancelled his plans to withdraw by October the remaining 5,500 troops in Afghanistan, and this was followed by additional conventional troops being sent, expanding the thin American force there to 9,800 (Jolly 2016). And some nine months later, it was announced that 8,400 troops would remain in Afghanistan indefinitely. Additional SOF personnel were sent to Syria to organize anti-Islamist Arabs and Kurds (Baker, Cooper, & Sanger 2015). Beginning in 2013, Special Forces and Delta Force have been training indigenous forces in Libya, Niger, Mauritania, and Mali for the purpose of creating elite antiterrorism units for opposing al Qaeda, Boko Haram, ISIS, and other Islamist fighters (Schmitt 2014). In Senegal, US Special Forces and Dutch operators teamed up to provide riverine training to Senegalese naval commandos in defense preparation against the spreading reach of Boko Haram (Schmitt 2016). And in Somalia, about 50 American SOF, acting as both advisers and operatives, have been conducting an ongoing campaign against al-Shabaab (Ibrahim 2016).

Meanwhile, Russia and China, perhaps taking advantage of the global preoccupation with ISIS and other Islamist extremists, have embarked on provocative actions in their regions. Russia has backed a breakaway remnant in eastern Ukraine and, prior to that, annexed Crimea. The United States has responded to the Russian provocations by conducting SOF exercises in the Baltic region with other nations, namely Estonia, Latvia, Lithuania, Finland, Sweden, Norway, and Poland. China has been engaged in "island building" of island chains (the Paracels and the Spratlys) in the South China Sea, territory that is under dispute by neighboring nations. Both Russia and China clearly wish to exert regional hegemony and at the same time would like to diminish America's global reach. "The ambivalent future of U.S.–Chinese relations is an open question and the role of SOF facing a potential Chinese advisory is not yet fully formulated" (Johnson 2006, p. 291). Potentially, SOF will have a role in counterproliferation, especially with respect to Iran and North Korea; and always worrisome is the possibility of a coup in Pakistan, where there are easily portable tactile nuclear weapons (Johnson 2006, p. 291). Although it is less than mainstream to equate climate change with terrorism, "Water is becoming the new oil" ("Special Operations Research Topics" 2016, p. 5) and the shortage of this resource in various parts of the world will invariably lead to conflict that draws in SOF.

Technology and Special Ops

In its overseas roles, SOF will continue to focus on technology for mission enhancement. Invariably, there will be continual experimentation with robotics, which may include driverless vehicles and situations in which a truck with a human driver is followed by a convoy of drone transports. In general, robotic systems tend to be too noisy, making them unsuitable for SOF's stealth requirements (Magnuson 2016). Perhaps a greater concern is the vulnerability of autonomous technologies, especially those involving weapons, which are susceptible to systems failure or being hacked and taken over by adversaries (Scharre 2016). The fear of systems failure is making some SOF commanders reconsider the overreliance on GPS for navigation purposes; it will be imperative, should global-positioning technology be disabled, that skills are proficient enough to revert to using old-fashioned protractors and grid maps.

For intelligence, there will be the ongoing challenge of handling and evaluating big data in order to note enemy trends and patterns for anticipating future action. Social media could be used for crowdsourcing or for implementing psyops. The monitoring of the online presence of terrorist groups will be important for intelligence gathering (Cohen 2015); ISIS threats have already been exposed due to monitoring of social media (Schmitt and Eddy 2016). It has even been suggested that SOF may in the future play a role in creating "cyber rebellions" ("Special Operations Research Topics" 2016, p. 19). Technology will allow "virtual interactions," enabling at reduced costs smoother operational autonomy of SOF personnel deployed in isolated parts of the globe ("Special Operations Research Topics" 2016, p. 25). This can be extended to the operator's personal life, allowing them to keep up with family even though deployed. The cyber world, especially social media, raises serious OPSEC concerns for SOF, however, as carelessness could even put SOF family members at risk.

Technology will need to be bolstered to offset the technology utilized by the enemy forces, including the use of miniature drones that will be the next generation of IEDs (improvised explosive devices). A counterdrone system, called *Venom*, is in the testing stage and is envisioned to identify, track, and target the hobbyist, copter-like drones that are likely to be more and more employed by insurgents, especially those operating in urban areas (Versprille 2016b). SOF will be deploying micro-sized unmanned aerial vehicles, some perhaps weighing under 20 pounds, which will be less expensive systems, operate quieter, not be tied to a runway, and allow units in forward-operating bases to have more sensors on the battlefield (Versprille 2016a).

By 2020, SOF units may be provided with fire support by AC-130J Ghostrider gunships outfitted with laser weapons (Tadjdeh 2016; Harper 2016). Laser weaponry would not only provide a cost saving (perhaps as low as $1 per shot), but they could also be used with precision accuracy and thus prevent less collateral damage. In a 3–6-second burst, the laser would give off a burst of directed energy the equivalent of 150 kilowatts, capable of burning a hole in the target. Lasers are expected to be a feature of the next generation of drones as well as army trucks. Eventually, one can imagine a SEAL team desert patrol vehicle (DPV) armed with a miniature laser, range finder, and weapon all in one.

The "Iron Man suit"—otherwise known as *TALOS* (tactical assault light operator suit)—is anticipated to make its debut in a couple of years (Votel 2015). TALOS represents a new generation of body armor ("liquid armor," in fact), but it is designed to do more than stop bullets. As touted, the battery-operated computerized suit, weighing approximately 13 pounds, will enable the special operator to heft heavy objects easier, manage blood flow, and control heating and cooling. The exoskeleton design of the suit is also supposed to help relieve strain on the wearer's knees. The TALOS, which could end up costing $80 million, was inspired by the death of an American SOF who was killed by a Taliban fighter upon entering a room during a raid (Diamond & Starr 2015).

Conclusion

In the near future, the temptation will be for military planners to view SOF as the solution for every kind of emerging problem. The last years of the Obama administration were a challenge for the US military's fight against ISIS while following the rule of "no boots on the ground" (other than SOF boots). The Trump administration may or may not deploy more American personnel to address the terrorist threat, but the emphasis on airpower and SOF will undoubtedly be a major approach to such situations. The tendency to rely on SOF so much may lead to a deterioration of the units, which would be stretched too thin and would not be at their operational best when worse crises occur. "In the process, a naturally-constrained resource is squandered as opposed to optimized" (Yarger 2013, p. 69). Retention of SOF operators, despite reenlistment bonuses up to $150,000, will remain a challenge, especially since there are higher-paying civilian jobs with military contractors and even with the CIA's paramilitary branch (Johnson 2006, p. 288).

Despite the status and prestige SOF has been enjoying, there are critics. The hidden nature of SOF missions has drawn the most criticism. Gallagher (2016, p. 7), for instance, describes SOF as "a postmodern Praetorian Guard" and warns, "The secrecy surrounding Special Ops keeps the heavy human costs of the war off the front pages. But in doing so, it also keeps the nonmilitary public wholly disconnected from the armed violence carried out in our name." This in turn leads to "our state of perpetual warfare" and "ensures a recurring cycle of armed conflict." The nature of the GWOT will see a continual blurring of the clandestine and the covert; SOF "over the next several decades" will be operating "in and against countries with which the United States is neither allied or at war, but which have not granted permission to U.S. troops to operate on their soil" to go after nonstate actors (Johnson 2006, pp. 288–289).

In remarks before the Senate Armed Services Committee in March 2015, General Joseph L. Votel, the USSOCOM commander, detailed the sacrifice of SOF in the GWOT: 2,500 wounded or killed and 7,500 enrolled in the SOF Wounded Warrior program. Of those in the Wounded Warrior program, "many" were suffering from "the 'invisible wounds' of traumatic stress" (Votel 2015). According to the *New York Times*, about 50 American SOF operators committed suicide in a 2.5-year period between 2011 and 2014, which exceeded the number for the previous 5 years for the SOF community (Shanker & Oppel, Jr. 2014). Moreover, that suicide spike occurred when the rate for the overall military was on the decline. Assessing risks factors for SOF suicides will continue to be a vital part of mental health maintenance, involving education and counseling. The question is how the SOF community, which has a tradition of not displaying weakness, can overcome the stigma of seeking help for depression and other mental health–related problems. Related to this issue is the challenge of extended deployments (the multiyear campaigns) and OPTEMPO in general, which frays the important bonds of home and hearth, family and friends. According to Votel (2015), over a 14-year period, the average operator deployed four to ten times overseas, usually spending less than 12 months at home between deployments. Shorter deployments, however, will tend to erode the capability of operators to master the language and culture of partnering nations. A larger SOF force may be necessary, but with any expansion comes the question of cost and the challenge of maintaining elite standards.

The real question is whether SOF has been asked to do more than it can deliver. The world is large; by comparison, the SOF network is small. If SOF is to no longer be the tip of the spear, then what role do conventional forces play? Lt. Col. James Jay Carafano (2014) suggests that SOF is at "an evolutionary dead end" (p. 45) and part of the problem is that it has been "tasked to do virtually everything" (p. 47). In his last year of office, President Obama faced mounting pressure from Pentagon officials who were warning that ISIS could not be defeated by SOF alone (Schmidt & Cooper 2016). The overuse of SOF is largely an acknowledgment that seldom are wars won by airpower alone, but the overuse of SOF has the same flaw as the attempt to defeat an enemy solely by bombing. In other words, there is a tactical imbalance when a sizeable force is not sent to destroy a ruthless enemy. Since the intransigent jihadist movements threatening world stability have not stirred a corresponding political commitment for a large-scale offensive response, SOF will, for the unforeseeable future, be deployed to carry out indirect missions. The consequences of this "dead end" (what could be termed *SOF overstretch*) is that countries such as Iran, Russia, and China may interpret such American strategy as a sign of weakness, a confirmation of post-American realities, which in turn could embolden them to embark on adventures that put the world at further risk.

As pointed out by Steven P. Bucci (2016), any overreliance on SOF that is structured to make easier the drawdown of conventional forces is in the end self-defeating. In order to accomplish its overall mission, SOF depends upon regular military forces, but those forces are being downsized just as a heavier burden is being placed on SOF. Furthermore, the conventional units serve as a recruitment pool for future operators and support staff, meaning that any cuts in the regular forces automatically reduces the number of prospective SOF warriors. Even if military downsizing was not occurring, there would be the challenge of growing SOF at a fast enough pace. SOF expansion, due to the special training requirements of operators, is limited to 3–5 percent annually. So the dilemma facing the future of SOF overseas, at least in the near term, is how to deploy more operators to more areas of the globe with less support and without eroding operational capability. This is why some are cautioning that SOF is at an evolutionary dead end.

Note

1 Sources skeptical of official US military reporting—that is, Turse (2015)—believe SOF have been deployed to a far larger number of countries.

References

Arango, T. (2016, February 24). Kurds rescue Swedish girl in territory held by ISIS. *The New York Times*, p. A6.

Baker, P., Cooper, H., & Sanger D. (2015, October 31). U.S. troops sent to Syria to aid forces fighting ISIS. *The New York Times*, p. A1.

Biddle, S. D. (2005/2006). Allies, airpower, and modern warfare: The Afghan model in Afghanistan and Iraq. *International Security* 30(3), 161–176.

Bradley, M., & Nabhan, A. A. (2016, January 9). In Ramadi, small force carried burden. *The Wall Street Journal*, p. A12.

Bucci, S. P. (2016). The importance of special operations forces today and going forward. *2016 Index of U.S. Military Strength*. Washington, DC: Heritage Foundation. Retrieved from http://index.heritage.org/military/2015/important-essays-analysis/importance-special-operations-forces-today-going-forward/.

Carafano, J. J. (2014, December). Special operations for the 21st century: Starting over. *Army*, 45–48.

Chapman, R. (2011). Special operations forces. In G. P. Hastedt (Ed.), *Spies, wiretaps, and secret operations: An encyclopedia of American espionage*: Vol. 2 (pp. 729–731). Santa Barbara, CA: ABC-CLIO.

Chapman, R. (2014). School of the Americas. In R. Chapman & J. Ciment (Eds.), *Culture wars in America: An encyclopedia of issues, viewpoints, and voices*: Vol. 2 (pp. 579–580). Armonk, NY: M. E. Sharpe.

Cohen, J. (2015). Digital counterinsurgency: How to marginalize the Islamic State online. *Foreign Affairs* 94(6), 52–58.

Cooper, H. (2016, January 7). Obama picks a commander for Mideast. *The New York Times*, p. A7.

Cooper, H., & Schmitt, E. (2015, December 3). U.S. sees elite military unit as key to bigger push against ISIS. *The New York Times*, p. A14.

Cooper, H., & Schmitt E. (2016, March 10). Detainee said to direct U.S. to ISIS caches. *The New York Times*, p. A1.

Cooper, H., Schmitt, E., & Schmidt, M. S. (2016, March 2). U.S. captures ISIS operative, ushering in tricky phase. *The New York Times*, p. A10.

Davis, A. D. (2012). The regional special operations headquarters: Franchising the NATO model as a hedge in lean times (Unpublished paper), Air Force Fellows Air University, Maxwell AFB, AL. Retrieved from http://www.dtic.mil/cgi-bin/GetTRDoc?AD=ADA571850.

Diamond, J., & Starr, B. (2015, October 6). U.S. military is on its way to getting its Iron Man. CNN. Retrieved from http://www.cnn.com/2015/10/06/politics/special-operations-iron-man-talos-suit/.

Feickert, A. (2013, February 6). *U.S. special operations forces (SOF): Background and issues for Congress*. Washington, DC: Congressional Research Service. Retrieved from http://www.au.af.mil/au/awc/awcgate/crs/rs21048.pdf.

Feickert, A. (2014, May 8). *U.S. special operations forces (SOF): Background and issues for Congress*. Washington, DC: Congressional Research Service. Retrieved from https://www.fas.org/sgp/crs/natsec/RS21048.pdf.

Fitzsimmons, M. (2003). The importance of being special: Planning for the future of US special operations forces. *Defense & Security Analysis*, 19(3), 203–218, doi/10.1080/1475179032000118795.

Gallagher, M. (2016, January 31). Welcome to age of commando. *The New York Times*, Sunday Review, p. 6.

Harkins, H. W. (2015). What is old is new again: The reemergence of special warfare. *Special Operations Journal*, 1(2), 112–118, doi/10.1080/23296151.2015.1096687.

Harper, J. (2016, January 20). U.S. SOCOM officials lay out technology challenges. *National Defense Magazine*. Retrieved from http://www.nationaldefensemagazine.org/blog/Lists/Posts/Post.aspx?ID=2065.

Ibrahim, M. (2016, March 10). Somalia raid aided by U.S. kills fighters from Shabab. *The New York Times*, p. A4.

Johnson, M. (2006). The growing relevance of special operations forces in U.S. military strategy. *Comparative Strategy*, 5(4), 273–296, doi/10.1080/01495930601028622.

Jolly, D. (2016, February 10). U.S. will send troops to assist Afghan forces pressed by Taliban. *The New York Times*, p. A6.

Landler, M. (2016, March 10). Obama criticizes the 'free riders' among America's allies. *The New York Times*, p. A9.

Luján, F. M. (2013). *Light footprints: The future of American military intervention*. Washington, DC: Center for a New American Security. Retrieved from http://s3.amazonaws.com/files.cnas.org/documents/CNAS_LightFootprint_VoicesFromTheField_Lujan.pdf.

Magnuson, S. (2016, January). Army to integrate ground robots into forces. *National Defense Magazine*. Retrieved from http://www.nationaldefensemagazine.org/archive/2016/january/Pages/ArmytoIntegrateGroundRobotsIntoForces.aspx.

Mazzetti, M., & Schmitt, E. (2015, December 28). Obama's 'boots on the ground': Special forces. *The New York Times*, p. A1.

Negroponte, J. D. (2012). United States engagement with the world in the wake of conflicts in Iraq and Afghanistan. *American Foreign Policy Interests*, 34, 1–4. doi/pdf/10.1080/10803920.2012.653292.

Nelson, R., & Wise, R. (2012). The future of special forces. *Global Forecast*, pp. 75–77. Retrieved from http://csis.org/files/publication/120412_gf_nelson_wise.pdf.

Pérez-Peña, R., Rogers, K., & Phillips, D. (2016, October 24). Life recreated in Army ends in combat. *The New York Times*, p. A13.

Robinson, L. (2013a). *The future of U.S. special operations forces*. New York: Council on Foreign Relations. Retrieved from http://www.cfr.org/special-operations/future-us-special-operations-forces/p30323.

Scharre, P. (2016). *Autonomous weapons and operational risk*. Washington, DC: Center for a New American Security. Retrieved from http://s3.amazonaws.com/files.cnas.org/documents/CNAS_Autonomous-weapons-operational-risk.pdf.

Schmidt, M. S., & Cooper H. (2016, January 29). More is needed to beat ISIS, military concludes. *The New York Times*, p. Al.

Schmitt, E. (2014, May 27). Elite U.S. troops helping Africans to fight terror. *The New York Times*, p. A1.

Schmitt, E. (2016, February 16). U.S. and Dutch train Senegal force as terror fears grow. *The New York Times*, p. A4.

Schmitt, E., & Eddy, M. (2016, January 1). Monitoring off threats has risen, official says. *The New York Times*, p. A8.

Shanker, T. & Oppel, Jr., R. A. (2014, June 6). War's elite tough guys, hesitant to seek healing. *The New York Times*, p. A1.

Special Operations Research Topics (2016). MacDill AFB, FL: Joint Special Operations University. Retrieved from http://www.soc.mil/swcs/SWEG/_pdf/GRAD/JSOU%20Special%20Operations%20Research%20Topics%20AY2016.pdf.

Szayna, T. S., & Welser IV, W. (2013). *Developing and assessing options for the global SOF network*. Santa Monica, CA: RAND International Security and Defense Policy Center. Retrieved from http://www.rand.org/pubs/research_reports/RR340.html.

Tadjdeh, Y. (2016, March). Special operations aircraft to be outfitted with laser weapon. *National Defense Magazine*. Retrieved from http://www.nationaldefensemagazine.org/archive/2016/March/Pages/SpecialOperationsAircrafttobeOutfittedwithLaserWeapon.aspx.

Turse, N. (2015, January 20). US special forces are operating in more countries than you can imagine. *The Nation*. Retrieved from http://www.thenation.com/article/us-special-forces-are-operating-more-countries-you-can-imagine/.

Thomas, J., & Dougherty C. (2013). *Beyond the ramparts: The future of U.S. special operations forces*. Washington, DC: Center for Strategic and Budgetary Assessments. Retrieved from http://csbaonline.org/publications/2013/05/beyond-the-ramparts-the-future-of-u-s-special-operations-forces/.

United States Army Special Operations Command (2014). *Counter-unconventional warfare*. Fort Bragg, NC: USASOC. Retrieved from https://info.publicintelligence.net/USASOC-CounterUnconventionalWarfare.pdf.

Vandenbroucke, L. S. (1993). *Perilous options: Special operations as an instrument of U.S. foreign policy*. New York: Oxford University Press.

Versprille, A. (2016a, January). Affordable surveillance a priority for special operations. *National Defense Magazine*. Retrieved from http://www.nationaldefensemagazine.org/archive/2016/january/Pages/AffordableSurveillanceaPriorityforSpecialOperations.aspx.

Versprille, A. (2016b, February). Army tests counter-drone technology. *National Defense Magazine*. Retrieved from http://www.nationaldefensemagazine.org/archive/2016/February/Pages/ArmyTestsCounterDroneTechnology.aspx.

Votel, J. L. (2015, March 18). Statement before the Senate Armed Services Committee. Retrieved from https://fas.org/irp/congress/2015_hr/031815votel.pdf.

Yarger, H. R. (2013). *21st century SOF: Toward an American theory of special operations*. JSOU Report 13:1. Tampa, FL: Joint Special Operations University. Retrieved from http://www.au.af.mil/au/awc/awcgate/jsou/21st_century_sof_yarger_apr2013.pdf.

8 The US home front

Drug cartels and security along the US–Mexico border

Martin S. Catino

Introduction

The 262 miles that stretch across the rugged and parched terrain of southern Arizona from the New Mexico state line to Yuma County is known as the Tucson Sector—key terrain designated by the United States Customs and Border Protection. This boulder-strewn and jagged landscape has had among the highest levels of drug trafficking and illegal immigration in the United States and currently leads all other sectors in marijuana smuggling and assaults (United States Customs and Border Protection, 2015, pp. 1–4). The Tucson Sector and its residents, primarily the ranchers who reside near or along the US–Mexico border, have been in the center of drug trafficking and illegal immigration activities, encountering violence, threats, home invasions, property damage, and theft, which occur frequently and over the course of decades.

These ranchers of southern Arizona comprise a unique group of citizens who have observed firsthand the historical trends, levels of activity, key characteristics, sociopolitical context, and security measures of this critical area associated with the interaction of illegal immigration and drug cartel activities. In essence, the ranching community of southern Arizona serves as a watchman witnessing the changing security environment of not only the US–Mexico border but also the United States as the very cartels that cross these lands continue to deepen, lengthen, and strengthen their operational lines throughout the country and thereby create the opportunities to harm local communities with violence, crime, drug peddling to minors and others, and recruitment into gangs (National Gang Intelligence Center, 2013, pp. 3–5). The assessments of these Arizona ranchers, particularly their encounters with these drug and human smugglers, are therefore essential for determining the strategic nature of drug cartel activity and thus security along the border.

Relying on observations, extensive interviews, local history, and document analysis, this chapter asserts that current perceptions of ranchers in Arizona have emerged from violent and hostile encounters with cartels, ineffective border-security operations, and the political marginalization of the ranching community. The evidences and circumstances of these ranchers demonstrate that the federal government has not deployed the necessary resources, security forces, and mission requirements requisite for interdicting the high volume and adaptive cartel movements operating on these ranch lands.

This perception of insecurity and imminent danger of the Mexican drug trafficking organizations (MDTOs) emerged from the volatile period of approximately 2000–2011, when high volumes of illegal immigration and drug-smuggling activities created frequent encounters with local ranchers. During this period threats, intimidation, confrontations, robberies—and, on occasion, murder—occurred as a routine part of the lives of these cattlemen and their families as they collided with cartels overrunning the southern border. The decrease in volume of illegal immigration and episodes of violence and confrontation, shifting patterns of illegal movements away from Arizona ranches and into federal and tribal lands, the thinning of cartel movement lines and operations, and an increase in border-security manpower and resources since that time have done little to assuage these fears of imminent threat from the cartels, which continue to operate with freedom and with increasingly advanced capabilities (Dannels, 2015, pp. 5–6).

The testimonies, documentation, and perceptions of ranchers also vividly reveal the growing capabilities of drug cartels, the multilayered domains of operations (financial, economic, political, social, informational, and military), and the increasing presence of special interest aliens from the Middle East (Ibid). This strategic posture of the MDTOs is juxtaposed with a failed domestic security apparatus that Arizona ranchers attribute to political interests, lack of public and government will, ineffective rules of engagement, and security forces postured in rear rather than front areas, leading to slow response times and regular movements of cartels across ranch lands.

Violent, threatening, and criminal encounters

While the US Border Patrol debates the very definition of border security with terms like *operational control* versus *monitoring*, (Longmire, 2014, p.1) the ranchers of southern Arizona have no ambiguity when it comes to understanding their own safety. They assert that border security is achieved when cartels and illegals can no longer enter or move freely, an outcome that would assure the safety of the families, friends, workers, neighbors, and other Americans who live in and frequent this area where the most industrialized and prosperous nation of the world stands next to one of the poorest and most threatened members of the international community, Mexico.

"We'll be the ones to determine if security is achieved on the border," rancher Ed Ashurst remarked, "not the officials in Washington." (Ed Ashurst, personal communication, August 25, 2014). If security is certainly understood in this community, so is insecurity. Every rancher interviewed in this study provided detailed stories of violence, threats, theft, trespasses, and intimidations that occur as cartels move their human and drug cargo across the border and into the United States. The agreement and consistency of their testimonies are striking.

Indeed, death is a common sight under the sweltering Arizona sun that shines on the path of cartel footprints eerily etched in these ranch lands. The dehydrated bodies often found by ranchers and border patrol are more the victims of cartel activities than the misfortunes of illegals failing to seize the American dream. The cartel's rapid operational tempo, evasive movements through difficult terrain to avoid local security, and abandonment of illegals tricked into paying without delivery to their destination all drive the weak, aged, infirm, slow, and unlucky border crossers to an early death, making Arizona the deadliest path of illegal entry into the United States (Velzer, 2014; Roger Barnett, personal communication, November 1, 2014).

After witnessing the rapid movement of illegals given "mule juice" – a cocktail of drugs to increase their endurance – a local rancher noted, "You have never seen anything like it. It would be difficult to catch them on a horse as they could even outrun a horse they are moving so quickly." (Arizona Rancher, personal communication, January 9, 2015). Another rancher recalled the time when a young girl crossing his ranch from Mexico had been bitten by a poisonous snake and left by her "coyotes" (illegal guides) to die on the very spot (Barnett). Yet another rancher remembers her heart pounding in despair when returning home to see a dead body in the driveway leading to her ranch house. "I thought it was my son because he had been threatened during that time by the cartels; it really looked just like him but we found out it wasn't him but an illegal" (Arizona rancher, January 29, 2015)—another disposable asset of cartel business. Therefore, the price of crossing the border with the cartels is far more than the $1,500.00–3,500.00 USD equivalent commonly paid to coyotes and the burden of carrying 50-lb packs of marijuana.

The death toll among illegals entering the Tucson Sector could be substantially higher had not many in the ranching community routinely given them emergency aid, including notifying the border patrol so it can provide relief (Susan Krentz, personal communication, October 22, 2014). The problem is particularly evident in the abandonment of illegals by cartel coyotes, who receive payment but fail to guide their customers to their destination. Arriving on ranches exhausted, dehydrated, ragged, and desperate, these victims of the backdoor path to America are often lost and hundreds of miles from their destination. Rancher Don Kimble recalled this common Occurrence. "They (illegals) are dropped off at my ranch and told by their coyotes they are only within eye-shot of Phoenix, which is over 300 miles away. When I told one illegal this truth he broke down and cried. After giving him water I asked if he wanted the border patrol to come and rescue him, and he answered 'yes.'" (Don Kimble, personal communication, April 3, 2015).

The ranchers of southern Arizona are not easily shaken by death, violence, and despair—life in this Wild West. In addition to the aforesaid problems, "rip crews," the criminal and often heavily armed cartel groups that rob and terrorize illegals and other cartels operating on the border, routinely commit violence, including the killing of border patrol agent Brian Terry in December 2010. Large groups of heavily armed individuals dressed in camouflaged clothes and camping in remote areas of ranch or federal lands are not an uncommon occurrence, adding to the insecurity of the southern border (Ashurst). But earlier that year, the death of cattleman Rob Krentz shook the ranchers to their dusty boots. "I thought it would happen to me before Rob," Roger Barnett explained. "Rob was one of the nicest ranchers, caring, and always trying to help illegals who were hungry and thirsty, in [dire] need." (Barnett).

Krentz was most likely killed by a "South-bounder," a cartel agent heading south to Mexico after delivering their human and drug cargo. "These are the most dangerous," Ashurst noted while pointing to a picture of South-bounders his remote cameras took in the deep hours of night as these dark figures operated in 2014 not far from where Krentz was killed [See picture above]. "These illegals rob, damage property, and even threaten ranchers as these drug mules return home," he explained further" (Ashurst). These cartel operatives are often hungry, desperate, and willing to die before being caught by border patrol agents, who can easily retrieve their criminal record, or before finding a worse fate at the hands of their "capos," who ruthlessly punish failure.

The death of Rob Krentz remains a disturbing event in the lives of Arizona ranchers because it calls attention to the fragility of security along the border and the tenuous relationship between ranchers and cartels. Even the most accepting coping strategies of the ranchers, illustrated in Krentz's humanitarianism, can meet with violence. Likewise, commonly accepted notions that drug dealers keep their violence among their own kind and no significant "spillover" occurs prove false in such tragedies a mere 30 miles from the border. The failure of border patrol to actively pursue the killer, a fact witnessed by Ashurst and others, adds to the frustration of this community (Ibid). Also, the rumor that "El Chapo" himself (Joaquín Guzmán Loera, former head of the Sinaloa Cartel) ordered his subordinates to avoid Arizona ranchers after the incident, and possibly ordered the execution of Krentz's killer, is little comfort to residents of southeastern Arizona, who witness the continued activities of cartels moving with near impunity.

Moreover, rapes are another common problem occurring in this inhuman terrain. The smugglers guiding illegals to their northern destinations often rape the women and children along the way and leave their stretched or torn undergarments and clothes displayed openly in "rape trees" that testify to this violent journey—a message left to intimidate both locals and other illegals passing along the way north [See picture above]. A Texas rancher remarked that the howling cries of coyotes in the moonlight were in fact the sounds of crying victims raped by the cartels. Arizona is no different. The pharmacies in Altar, Sonora, Mexico sell birth control pills to women headed across the border to Sasabe, Arizona, a practice of preventive medicine used to overcome a fate that awaits as high as 80 percent of female border crossers on their illegal journey (*Vice News*, 2014). In the high volume period of illegal crossings from 2001 to 2011, rancher Fred Davis noted that a well-dressed local woman had to flee in her vehicle as illegals gave pursuit threatening to rape her—another indication of spillover (Fred and Peggy Davis, personal communication, February 8, 2015).

Assaults in the Tucson Sector are also a major problem. The ranchers interviewed in this study testified of assaults occurring frequently in the high volume period and lessening afterward only because of two factors: ranchers learning to cope with cartel operations by not directly interfering or pursuing, and cartel operatives moving deeper into remote areas while increasing the capabilities of their intelligence operations, tactical movements, and infiltration of border security communities and security apparatus (Ashurst).

Yet the memories of past assaults remain vivid in rancher testimonies today for no other reason than they can occur again given the freedom of movement of cartels and the low interdiction rates by border patrol. Roger Barnett noted an incident that occurred as "illegals tried to throw one of the ranch hands off of his horse," most likely to steal it. Assaults and attempted assaults were indeed a common practice of cartel operations seeking transportation northward. One anonymous rancher noted the problem and placed it in context:

> All three of our grandchildren have been either threatened or attacked by these people who enter our country illegally. Our land...which has been in the family for six generations, is trashed. Our fences and gates are torn down, the water lines cut, piles of garbage including clothes, backpacks, blankets, plastic jugs, medications, toiletries, dirty diapers, soiled female pads, black runny human feces and toilet paper litter our countryside... We never leave the house without cell phones and radio to call for help. We carry firearms to protect the ones we love; praying we will never be forced to take another human beings life. Our home is locked; the gates chained...The illegals have poisoned our dog, attempted to steal our horses and truck; broken into our neighbor's homes and out-buildings, formed human chains and carried boulders to lay across the road in an attempt to stop us and steal our vehicles. They killed, mutilated, and cooked and ate a 1 day old calf in our neighbor's corral. The stories go on and on.
>
> (Arizona Rancher, unpublished journal, 2010)

The same ranch family cited above further testified of the violence that so frequently occurs on the border. The mother and daughter were driving on the remote roads of their ranch in the early spring of 2000 when they encountered an oncoming vehicle with several illegals. Pulling over to let it pass, the mother stated: "The driver stuck his head out of the window and made some extremely obscene hand gestures...He laughed the most evil laugh I have ever heard....Then he said in Mexican, 'We want you and we will get you.' At that point in time, he immediately began to turn the truck around." (Ibid).

The truck gave pursuit, but one flat tire and the difficulty of turning in the narrow road allowed the mother and daughter to escape along a hidden road. The border patrol responded but failed to catch the individuals. "If those dirty no-good wackos had been able to catch you, they would have definitely carried out their threat" (Ibid), the border patrol added. Another fatal incident had been averted, but just barely.

Yet among the most disturbing stories among Arizona cattlemen involved a young rancher named Charlie, who was carrying out his normal duties in 2002 when he found his drive to a nearby ranch turning into another violent encounter with illegals. The darkness of the rainy day, sharp turn in the road, and flooding provided the cover and opportunity necessary for the illegals to use boulders to block the road. Charlie arrived just as the boulders were set in place. His family later recalled that "the two trespassers stood firm behind the road block…not making any attempt to move out of the road. Charlie swerved to miss the aliens and the rocks and turned the pickup on its side in the running water. He finally crawled out of the window on the driver's side…then the illegals were running, and there was no way to pursue them in the rain" (Ibid). The illegals were not apprehended.

On January 29, 2015, Charlie's family took me to the very location of this event on their ranch. The troubling fact was not only the failure to apprehend these criminals but also the sophistication of the tactic—evidence that the illegals were not merely unskilled migrants. The boulders were set during favorable cover provided by the weather, during a time when a rancher used this remote road, and at a location difficult to see because of the sharp turn in the road (Arizona rancher, personal communication, January 29, 2015). The boldness that the illegals demonstrated during the incident was also noteworthy as an indicator of skilled operators. Insurgents use the same TTPs (tactics, techniques, and procedures) in Iraq and Afghanistan against the United States and its allies and partners, and the comparison is noteworthy for demonstrating the increasing capabilities demonstrated by cartels operating along the border.

These assaults on ranchers indicate that another incident of the type involving Rob Krentz is certainly possible. The fact there are not many more incidents like it is due to the vigilance of ranchers and not the success of border patrol—an outcome that adds to the insecurity of the ranching community. These "close calls" with illegals certainly could have ended in a fatality had not the ranchers taken evasive action (Ashurst; Daniel, 2012, pp. 54–62). These cases stand out as a different category than assaults occurring over the theft of transportation vehicles, food, and other provisions, as the objective is not only to procure material items but also to physically harm ranchers.

But violent, threatening, and criminal encounters are not merely local problems caused by the cartels and their human and drug smuggling. The death of Kathryn Steinle serves as an example of a victim living (and dying) far from the border. On July 1, 2015, the 32-year-old Steinle was walking with her parents along a waterfront in the Embarcadero district of San Francisco when an illegal alien fired three shots, one piercing the young lady's heart, causing her to cry out to her father, Jim, as she collapsed on the ground. Despite the loving efforts of this father to save his daughter, she died soon after at San Francisco General Hospital (Mai-Due, 2015).

Although Juan Francisco Lopez-Sanchez, the illegal alien who fired the shots, was on probation in Texas at the time of the shooting, had seven prior felony convictions, and had been deported five times, the San Francisco

Sheriff's department had refused earlier the request of federal immigration authorities to turn him over for deportation. Lopez-Sanchez later admitted that the sanctuary policy of San Francisco, which shelters illegals from prosecution and removal, attracted him to the city. Even if the shooting was an accident, as the illegal asserts, the accident never would have happened if the laws were enforced (Ibid).

Home on the illegal range

Rancher John Ladd likewise finds life on the border to be a perilous existence filled with encounters with cartels. He and his family have staked their claim along the US–Mexico line, ten miles of ranch that hugs the border fence in Naco, Arizona and dates back 118 years to the time when his great grandfather came to nearby Bisbee and became its first justice of the peace. Ladd takes little comfort in the decrease in overall presence of illegals on his land. He recalls the shift of illegal trespassing in the last 5–10 years from poor farm workers to drug traffickers and gang members looking "lean and mean." He further remarks, "They are usually wearing hoodies, dark clothes, or military uniforms." (John Ladd, personal communication, September 25, 2014). But his most pressing concern is the three-truck drug convoys that cut through the border fence and race across his ranch to a highway only three miles away. Ladd described one such crossing he witnessed, with the driver making a desperate run while the passenger waved an Uzi automatic weapon at the onlooking cattleman.

Ladd's testimonies of life on the border also demonstrate another common problem: home invasions. During my second visit to his ranch in April 2015, while a documentary was being filmed, two illegals walked over the border and directly along his driveway past his house. Ladd called border security and immediately the trespassers were apprehended [See picture above]. But not all uninvited guests merely pass his ranch house. He recalled that in 2010 he watched an illegal entering the property and approaching his house. When seeing the trespasser arriving to open the front door, Ladd maneuvered to the side of the door and pointed a pistol at the intruder as he entered and stuck his head in. This unwanted visitor immediately departed the premises (Ibid).

Among the more tragic cases of home invasions among ranchers involves an elderly woman living in a remote area of Cochise County, Arizona. The cartels regularly invade the elderly woman's home, robbing her food and using her phone. They threaten to kill her if she does not comply. This home invasion occurs so frequently that she uses a ladder to climb to the top of her roof whenever these narco-terrorists arrive, doing so to evade the abusive treatment (Arizona rancher, personal communication, January 29, 2015). The ability of MDTOs to use their scouts effectively to case homes like this one and the slow response time of border patrol embolden such activities and create the frustration, anger, and outrage that are so commonly experienced by ranchers and other residents of the border.

Encountering cartel messages

Among the many types of encounters between ranchers and cartels are direct messages from these nocturnal predators. Numerous testimonies by Arizona ranchers indicate that the cartels crossing these sage-brush lands seek to strengthen their operational lines by chasing ranchers away from their lands, cowing them, gaining their open or tacit cooperation, or gaining a tactical advantage including immediate access to food, water, communications (telephone), transportation, directions, or shelter. These strategic and tactical objectives are often achieved by directly approaching ranchers and talking to them, a fact that increases rancher apprehensions (Ashurst; Kimble).

Arizona ranchers are keenly aware that drug cartels control both the human and drug smuggling, and thus illegal immigration and drug smuggling are intertwined organizationally and operationally. The point is salient as drug smugglers disguise their operations as mere migrations and seek aid from locals as well as using the illegals as decoys while drug lines move evasively in coordination and in opposite directions. Kimble noted that cartel members possibly hide their weapons before approaching his ranch house and present themselves as desperate sojourners merely seeking immediate help (Kimble).

But not all messaging is so clear and direct. In addition to direct messaging verbally, and the aforesaid discussion of rape trees as visible messages, cartels seek to mark territory in order to create confusion, uncertainty, and fear—necessary attitudes for shaping their operational environment. Rancher and businessman Glenn Spencer demonstrated this fact as he pointed to an unusual sight on his property that hugs the border fence. Strewn across a barbed-wire fence on his property are male pants and a shirt, in addition to what appears to be a torn blanket (Glenn Spencer, personal communication, September 3, 2015). [See picture above.] Similar cloth or garbage bag markings appear on the Ladd ranch but are more likely communications to cartel members for pick up and drop areas for drug smuggling (Ladd). The inhuman terrain of southern Arizona is thus littered with the messages of cartels seeking to increase their operational effectiveness as their drugs and human victims move northward.

The Arizona ranchers of the Tucson Sector have encountered drug and human smugglers for decades and thus serve as extraordinary witnesses to the nature of drug and human smuggling affecting not only the border but also the entire United States, which continues to witness the growth of drug, criminal, and gang activities all related to the organizational expansion of Mexican drug trafficking organizations. The violent, threatening, invasive, and brutal cartel activities imprinted on these ranch lands have created the inhuman terrain layering the landscapes of the southern border. The US Border Patrol has failed to check the increasing capabilities and adaptability of cartel activities operating in the remote areas of southern Arizona, thus forcing the ranching community to rely on coping strategies rather than on the border patrol. In the early 1870s, when Brewster M. Higley wrote the lyrics of the classic Western song *Home on the Range*, he never envisioned this present time when more than deer and antelope play. The players today are highly skilled and determined drug cartels that will continue to make the border their home as long as the will, resources, manpower, and strategies are lacking to create the security so urgently needed.

References

Daniel, William R. (2012). *One if by land: What every American needs to know about our border.* Tucson: Wheatmark.

Longmire, Sylvia. (2014). *Border insecurity: Why big money, fences, and drones aren't making us safer.* London: Palgrave.

Mai-Duc, Christine. (2015, September 1). Deportee accused of killing Kathryn Steinle to stand trial on murder charge. Los Angeles Times online. Retrieved September 4, 2015, from http://www.latimes.com/local/lanow/la-me-ln-kathryn-steinle-sf-shooting-murder-trial-20150904-story.html

Vice News. (2014, June 25). Mexico's immigrant oasis: Last stop before the border. [YouTube]. Retrieved June 25, 2014, from https://www.youtube.com/watch?v=hB_66Z7G7_Y

National Gang Intelligence Center. (2013). *2013 National gang report.* Retrieved February 13, 2015.

United States Customs and Border Protection. (2014). *United States Border Patrol Sector Profile—Fiscal Year 2014 (Oct 1st through Sept. 30th).* Retrieved May 3, 2015, from http://www.cbp.gov/sites/default/files/documents/USBP%20Stats%20FY2014%20sector%20profile.pdf

U.S. Senate Committee on Homeland Security and Governmental Affairs. (2015). *Testimony of Mark Dannels. Securing the Southwest Border: Perspectives from Beyond the Beltway, Homeland Security Brief 2015*, March 17, 2015. Retrieved May 21, 2015, from http://www.hsgac.senate.gov/hearings/securing-the-southwest-border-perspectives-from-beyond-the-beltway

Velzer, Ryan. (2014, August 8). Arizona's border deaths remain highest in the U.S. *The Arizona Republic.* Retrieved August 8, 2014, from http://www.azcentral.com/story/news/local/arizona/2014/08/08/arizona-undocumented-border-deaths-highest-in-the-country/13738253/

9 Psychological warfare

Us paramilitary operations in indonesia and iraq

Suzie Sri S. Sudarman and Senia Febrica

Introduction

This chapter analyzes the use of psychological warfare by the United States (US) to win the hearts and minds of the people of Indonesia during the 1950s and of Iraq and Muslim majority countries in the aftermath of the 9/11 attacks. This chapter explains both the change and historical development of US psychological warfare, the agencies involved, and the policy options at its disposal. It argues that the US's uses of psychological warfare without clear understanding of the domestic politics of the targeted country produced results that were not only harmful to the people of the country targeted but also to US interests. Drawing from the past and present practices of US psychological warfare, this chapter seeks to identify the changes in US psychological warfare and the mistakes that the US has made, as well as the type of challenges that it will face in the future from state and nonstate actors.

US foreign policy and psychological warfare

Communications and information issues were somewhat peripheral to the concerns of US diplomacy, but World War II changed this. For the first time, a global US military presence highlighted the need for dependable overseas communications. After the war, American mass media came into its own as an international force, as filmmakers, television broadcasters, and print publishers expanded abroad (Dizard Jr., 2001: 76). James Schwoch illustrated the technological transformation in global television and other new electronic communications technologies of the era. There was an effort in the 1950s to build international and intercontinental TV networks and to bridge the Atlantic Ocean with live television with microwave relay towers. This constituted the early linkage of television, telecommunications, and information technology to strategic security issues (Schwoch, 2009: 11).

There was also the increasing importance of computer technology and even early visions of global information technology "wideband linkage" (now known as *broadband convergence*) via the Internet. The interface between computers and public relations or information diplomacy allowed for discussion of the importance of the United States Information Agency (USIA) exhibitions (Schwoch, 2009: 12). Its impact set the tone and pace of American digital diplomacy in recent decades.

Information issues became increasingly important in American Cold War strategy, but there were no guidelines for dealing with these issues. The rising tensions of the Cold War demonstrated to American policymakers that issues such as propaganda, information dissemination, public opinion, and global attitudes were significant, but they were neither sure about how to conceptualize nor operationalize statecraft around such issues (Schwoch, 2002: 3).

A quantum leap in connection technologies has significantly impacted the way nations conduct their foreign policy. American diplomacy in particular thereby evolved and is continuously being reshaped by information-intensive forces. Traditional diplomacy consists of government-to-government relations (G2G) and has been about governments talking to global publics (G2P) that include efforts to inform, influence, and engage the public in support of foreign policies (Snow, 2009). More recently, with the rise of user-friendly communications technologies, public diplomacy involves practices or attempts by both government and private individuals and groups to directly and indirectly influence the public attitudes and opinions that bear directly on another government's foreign policy decisions (P2P) (Snow, 2009: 6).

The changes brought about by the global information age touched on every aspect of American national interests. The technical breakthroughs have far outpaced American political, economic, and social attempts to deal with them effectively (Dizard Jr., 2001: ix). Electronic communications and information resources are influencing American foreign policy, and their effects are not merely confined to strategic issues, as they have also resulted in a different type of relations between nations. For example, economically, the US has to deal with the cyber security of the economy. Its economic future depends on an array of cyber-security policies, particularly in the financial industry, which has high profit margins. There need to be regulations on electronically controlled utilities such as electrical transmission grid, oil and natural gas, water, waste water, chemicals, telecommunications, transportation, banking, and finance.

The problem of cyber inferiority can also lead to problems in national security and military affairs, particularly in an era replete with utopian ideas articulated by radical technophiles who believe that transparency and personal privacy are the foundations of a free society. They view secrecy and surveillance as gateways to tyranny (Scherer, 2013: 22–9).

Walter Lafeber noted how historians of US foreign relations have done very little to relate policymakers' views of technology to their foreign policies at home and abroad (Lafeber, 2000: 1–19). He pointed out that it was Secretary of State George Shultz who concluded that technology had "outstripped the traditional methods" of diplomacy (Lafeber, 2000: 13). Secretary Shultz declared further that "Walter Wriston says an 'information standard has replaced the gold standard.'" Secretary Shultz had also decided that "over centralized plans and decisions increasingly will be inept" (Lafeber, 2000: 14).

From the above discussion, it can be seen that the central questions that we intend to broach as a topic of this chapter are as follows: How do the changes in the development of information technologies influence the conduct of US psychological warfare? What are the weaknesses in US psychological warfare? What type of challenges will the US face in its future psychological warfare both from state and nonstate actors?

Psychological warfare and the US covert operations in Indonesia: 1957–1958

As explained earlier, there was an effort in the 1950s to link communication technologies to US strategic security interests across the Atlantic. However, during the same time, the US did not make such efforts in Indonesia. The US efforts to win the war against communism in Indonesia in the late 1950s were limited primarily to the use of traditional measures. In order to undermine the morale of the Soekarno administration, the US focused on the use of covert operations to assist regional rebellions. The 1950s was marked by the victory of the Chinese Communist Party over the Kuomintang government, the invasion of South Korea by Soviet-backed North Korea, and Chinese support for Vietnamese communists (Gardner, 1997). The US carried out the "largest covert operations since World War II involving CIA, the US Navy and camouflaged American Air Force" to make a psychological advance against the Soekarno administration to undermine, if not replace, the Indonesian government that had been deemed as growing closer to communist influence (Kahin & Kahin, 1995; see Jones, 1999: 1180–1).

US psychological warfare in Indonesia provoked a major rebellion in West Sumatra (Revolutionary Government of the Republic of Indonesia/PRRI) and North Sulawesi (Universal Struggle Charter/Permesta) and triggered civil conflict (Kahin & Kahin, 1995). The US intention was to damage the morale of the Soekarno administration through the support of rebel groups. Despite the fact that the main bases of the rebel groups were the small areas of the Bukit Barisan highlands in West Sumatra and North Sulawesi, the Indonesian central government sincerely believed that other provinces in the archipelago would support the rebel cause or follow a similar path (Kroef, 1958: 72). The US support of the rebel groups worked against Jakarta's immediate aim in postcolonial Indonesia, maintaining the country's integrity. This had been a key national aim that Soekarno and other Indonesian prominent officials had pursued with fierce intensity since the country gained its independence in 1945.

As the crisis in Indonesia continued to unfold, in July 1957, the Indonesian Communist Party's (PKI) success in winning 27.4 per cent of the vote in provincial elections held in Java had made the PKI the largest party on the island (Jones, 1999: 1183). The US Interdepartmental Committee's "Special Report on Indonesia," issued on September 3, 1957, pointed out that US security interests in Southeast Asia, Western Pacific, and Australia "would be seriously affected" if the PKI gained political control in Jakarta or Java (Kahin & Kahin, 1995: 93).

Since October 1957, the US had been providing the PRRI and Permesta rebels with funds, arms, and munitions, as well as equipment and personnel training at US Pacific bases in the Philippines, Okinawa, and the Marianas (Kahin & Kahin, 1995: 132). An offshore task of the US Seventh Fleet accompanied the CIA efforts to infiltrate arms and personnel to assist the regional rebellions against Jakarta (Scott, 1985: 246). As part of the covert operations, the US used its diplomatic prominence to gain cooperation from the Philippines, the United Kingdom, Taiwan, and South Korea (Kroef, 1958: 78). The CIA delivered American-made arms and other aid to the rebels through the agency's Taiwan-based subsidiary company, Civil Air Transport (CAT) (Jones, 1999: 1180, Doeppers, 1972: 191). British facilities such as Labuan airfield in Borneo and Singapore had been crucial to support bombers coming from Taiwan to Manado for the refuelling and support of CAT missions originating in Bangkok, Taiwan and the Philippines (Doeppers, 1972: 192; Jones, 1999: 1186).

Despite the US extensive military aid to the dissidents, the Indonesian government managed to put down the regional rebellion in 1958. The failure of the US covert operations in Indonesia had significant implications on Jakarta–Washington relations. Rather than winning the hearts and mind of the Indonesian people to be pro-West, US psychological warfare paved the way for strengthening the communist influence and anti-Americanism in Indonesia. During the PRRI/Permesta rebellion, both the Soviet Union and China had offered unqualified support—including military supplies—to the Indonesian government (Kroef, 1958: 78). In Indonesia, the communist party, the PKI, became more influential following the rebellion. The growing evidence of the US involvement in the rebellion, especially following the capture on May 18, 1958 of an American pilot, Allan Lawrence Pope, who flew a bomber plane for the rebels, served as an effective propaganda tool for the PKI and contributed to generating anti-Americanism in Indonesia (Jones, 1999: 1209). The regional rebellion precipitated the introduction in 1959 of President Soekarno's Guided Democracy, which ended the country's parliamentary democracy system (Evans III, 1989: 26). This led to a condition that put the PKI at the center of Indonesia's power politics along with Soekarno and the army (Evans III, 1989: 26).

In the aftermath of the PRRI/Permesta rebellion, the Eisenhower government sought desperately to repair its relations with Jakarta, including initiating a military assistance program (McMahon, 1981: 324). Yet, this move met with only limited success. The failure of the US psychological warfare in Indonesia stemmed from Washington's oversimplification of Indonesia's complex domestic politics into the Cold War equation of Communist vs. West. By mid-1957, President Eisenhower, Secretary of State John Foster Dulles and CIA Director Allen Dulles were deeply concerned that the Indonesian government, particularly President Soekarno, was under the growing influence of the PKI (Kahin & Kahin, 1995, 16). The anxiety led them to deal with the situation in Indonesia as an "adjunct of the Cold War" (Doeppers, 1972: 195). By September 1957, senior officials in the US administration decided that discontent in the outer islands served as an instrument that they could exploit to prevent communist takeover in Indonesia (Kahin & Kahin, 1995: 17). The planning and implementation of the covert operations were done with little knowledge of internal political development in Indonesia, and there was no consultation with the US Embassy in Jakarta. As a consequence of taking the uninformed psychological warfare policies in Indonesia, by mid-1958, the US found itself supporting the losing side. Having explained the practice of US psychological warfare in Indonesia during the Cold War, the next section of this chapter will proceed by elaborating the strategies that the US adopted following the 9/11 attacks in Iraq, the Middle East, and the Horn of Africa. It will provide details of the US policy options and challenges that they have encountered.

US psychological warfare in the post 9/11: Iraq and beyond

After the 9/11 attacks, President Bush refocused US foreign and military policy on the war on terror. There was no initial attempt by the media or the government to transform the image of the 9/11 horror into responsible discourses of reflection and action (Der Derian, 2001). In the rush to judgment, there was little time for deliberation, for understanding the motivations of the attackers, or for assessing the consequences of a military response (Der Derian, 2001). US political warfare is obviously harnessed to a new national security of operations strategy, one in which dominance had to be established through the clear projection of a preponderance of power as stated in the US Defense Planning Guidance, which is, according to Barnett, a rather "full-throated" policy version of the "reconstitution" pillar already in vogue in force structure planning circles" (Barnett, 2009: 3).

During the Global War on Terrorism and the implementation of Secretary of State Rice's Transformational Diplomacy, there was a growing realization that there had been a huge void of guidance in foreign policy from those that ought to be formulating and executing it. Or, to be more specific, some truly believed that it was time to realign the US political–military boundaries and political–military thinking on addressing the boundary problems. The boundaries the US set for conducting foreign policy have little or no relationship to the boundaries for US combatant commanders (Wilcox LTC USA (Ret.), 2006: 1–7). Some career foreign service officers, by contrast, felt that the diplomatic corps had become too militarized (State Department, 2008). Some others underscored the fact that the war on terrorism was as much a battle of ideas as it was a military struggle, and that public diplomacy or information diplomacy has been the most effective instrument.

The source of the threat was and is the ideology that promoted the use of terrorism. Although it is still true that the organizations that used terrorist tactics to mount attacks and encouraged others could be dismantled by military means, if necessary, the effectiveness of military means should not be exaggerated at all. This ideological nature of the threat seemed to necessitate US involvement in the "battle of ideas."

The intelligence agencies, the Department of Defense (DOD), the Department of State, and the United States Agency for International Development (USAID) have been implementing various efforts to identify, monitor, and counter the support for and funding of the global propagation of Islamic extremism (US General Accounting Office (GAO, September 2005: 11)). In December 2002, the US Department of State announced the establishment of the Middle East Partnership Initiatives (MEPI) as a Presidential Initiative to support the administration's policy of promoting democracy in the Middle East and North Africa. Other US agencies' efforts seemed to have emanated from various strategies, including the National Security Strategy, National Defense Strategy, National Strategy for Combating Terrorism, and State Department and USAID Strategic Plan for Fiscal Years 2004–2009 (GAO, September 2005: 11).

DOD has implemented several efforts to track facilitators and terrorist financing as part of its counterterrorism force protection efforts. One example is the Air Force Office of Special Investigations (AFOSI) Collections and All Source Analysis effort. DOD's Pacific Command (PACOM) has provided logistical support for the US government's humanitarian assistance and civic action programs in Southeast Asia (GAO, September 2005: 11).

State has monitored and reported on incidents and patterns of governmental and societal discrimination, harassment, and abuse in State's annual International Religious Freedom Report and Country Reports on Human Rights Practices. State and USAID have been implementing several efforts such as the Middle East Partnership Initiatives (MEPI), Muslim World Outreach (MWO), and the South Asia Regional programs. In addition, USAID sought to provide an alternative to political expression through extremism. It has also conducted studies on the Muslim world to better understand the root causes of extremism and the dynamics in different regions. The Trans-Sahara Counterterrorism Initiatives have been one example of a US government multiagency effort involving USAID and the Departments of Defense and State (GAO, September 2005: 13).

As to Department of Defense Communications, the department has attempted to shift its focus from traditional military operations and missions to the nonkinetic, information and influence-based activities contemplated within concepts of "strategic communication" (SC), integrating communication and influence guidance throughout operational planning and execution. These activities are primarily supported by three capabilities: (i) Information Operations (IO), and, primarily within IO, Psychological Operations (PSYOP); (ii) Public Affairs (PA); and (iii) Defense Support to Public Diplomacy (DSPD), Military Diplomacy (MD) and Visual Information (VI) also support SC-related activities (GAO, September 2005: 11).

While the term "public diplomacy" as used by the State Department describes communications with foreign populations, strategic communication involves DOD interacting with and influencing foreign publics, military adversaries, partner and nonpartner governments, other US government agencies, and the American people. Beginning in 2007, the DOD created small teams of social scientists and anthropologists, known as Human Terrain Teams, and embedded them in brigades in Iraq and Afghanistan; these teams seek to provide insights into the customs and values of the local population. In response to the 2007 National Strategy for Public Diplomacy and Strategic Communication, which called for relevant agencies to provide information on activities related to the goals and concepts of the Strategy, DOD identified a number of specific SC-related programs and activities among others government-to-government and military-to-military activities, but DOD also described programs for interaction with foreign publics and related activities (GAO, September 2005: 29).

There are other programs in support of the effect to counterterrorist and extremist groups, such as the Trans-regional Web Initiative (TRWI), which directed Combatant Commands (COCOMs) to create regionally focused websites with tailored communications to foreign audiences and measure and assess their effectiveness. DOD also created the Expanded Trans-Regional PSYOP Program (ETRP) to unify and synchronize communications themes and objectives across all theaters, guiding individual COCOM-created communications to selected foreign audiences in each region. ETRP authorized communications through audiovisual, telephonic, and web-based applications. DOD also created Regional Centers for Security Studies for key foreign audiences on regional security issues.

In 2007, DOD created the Office of the Deputy Secretary of Defense for Support to Public Diplomacy to coordinate DOD communications with other government entities. Michele A. Flournoy, appointed in early 2009 as undersecretary of defense for policy, abolished the office and the deputy assistant secretary positions. The Global Strategic Engagement Team (GSET), headed by Rosa Brooks, senior advisor to the undersecretary, became the coordinator for strategic communication strategy and activities within DOD. Brooks has instituted a Global Engagement Coordinating Committee, represented by various organizations that provide crucial support to DOD's strategic communication efforts, including PA, Legislative Affairs, IO, Joint Staff, and the Office of the Under Secretary of Defense for Policy (GAO, September 2005: 31).

The Bush administration took numerous other actions to improve the effectiveness of its public diplomacy. In November 2001, the Bush administration created the Coalition Information Center (CIC), headquartered in the Old Executive Office Building. The CIC executed public diplomacy and coordinated US government agency press conferences and talking points, dispersing them rapidly worldwide. President Bush created the Office of Global Communications (OGC) in July 2002, which replaced the CIC, and had a primary mission of coordinating strategic communications with global audiences; it was disbanded in 2005. National Security Advisor Condoleezza Rice then established the Strategic Communication Policy Coordinating Committee (PCC) in September 2002 to coordinate interagency activities, develop the White House message, and disseminate it abroad (Epstein & Mages, October 31, 2005: 2).

The president also created the Muslim World Outreach Policy Coordinating Committee in July 2004, which replaced the Strategic Communication Policy

Coordinating Committee. The State Department established an Office of Policy, Planning, and Resources in the Office of the Under Secretary for Public Diplomacy and Public Affairs. The Department of Defense created the Office of Strategic Influence (OSI) to oversee military propaganda and other information related to its antiterrorism war, which was then dissolved. The DOD issued the Information Operations Roadmap on the direction that DOD wanted to take in public diplomacy and other information operations. The undersecretary of defense for policy continually defined DOD's role in public diplomacy (Epstein & Mages, October 31, 2005: 2).

USAID has an important role in public diplomacy; the agency works with the State Department through the State–USAID Joint Policy Council and Management Council to coordinate public diplomacy efforts. Other actors, such as a number of congressional mandated NGOs—many founded during the Cold War, such as the National Endowment for Democracy (NED), the Asia Foundation, the East West Center at the University of Hawaii, and the Eisenhower Exchange Fellowship program—perform work in support of US foreign policy objectives. USAID became more involved in public diplomacy after the 9/11 Commission reported to Congress that the population of countries that received the largest US foreign aid voiced very strong anti-American sentiment. Since 2004 USAID has established the State–USAID Policy Council and a Public Diplomacy Working Group; it actively publicized America's humanitarian and development aid initiatives (Epstein & Mages, October 31, 2005: 2).

From 1999 to 2005, the CRS compiled, in chronological order, reports with major recommendations to help the US public diplomacy apparatus win the war on terror. The recommendations suggested that the administration did not sufficiently define the overall strategy for the use of public diplomacy to improve the US image abroad. The GAO reported that the US needed to do a better job at defining its public diplomacy message, that without an interagency strategy there would be complications in conveying consistent messages, and that there was a need to measure public diplomacy success (GAO, July 2003, September 2003, 2004, April 2005, September 2005, 2006, 2007).

Some studies focused on the White House coordination efforts, urging President Bush to issue a directive to strengthen the importance of communication and public diplomacy (Council on Foreign Relations Independent Task Force, 2003). Other reports suggested the creation of an effective instrument of public diplomacy. Six studies referred specifically to the need to redefine the role of the undersecretary of state for public diplomacy. The US Advisory Commission on Public Diplomacy noted that attempts to improve coordination with the creation of the Office of Global Communications and the Strategic Communication Policy Coordination Committee and the Strategic Communications Fusion team were still deemed inadequate (Council on Foreign Relations Independent Task Force, 2004). The commission recommended that the agents and structure needed to be properly coordinated; that public diplomacy messaging must become more strategic and responsive and must reflect the values of target audiences; and that there was also a need to apply some sort of measurement that could ensure the desired effects were being achieved, as public diplomacy had become a national security priority (Epstein & Mages, October 31, 2005: 40).

Some suggested improvements from the commission included the following: that public diplomacy must "Connect," "Modernize," "Focus," "Innovate," Partner," "Train," "Engage," "Adapt," "Invent," and "Measure" (United States Advisory Commission on Public Diplomacy, 2005: 26–7). There was an apparent realization that practitioners of public diplomacy needed to adapt and embrace innovation by constantly engaging appropriate technologies. Instantaneous news and electronic information distribution required proactive communication strategies and lightning-fast response, making a one-size-fits-all approach incapable of working properly.

The commission issued a draft on May 13th, 2005, that highlighted the fact that the US public diplomacy apparatus was facing several chronic and systemic problems: there was no one in charge, there was no national strategy, and public diplomacy was underresourced (United States Advisory Commission on Public Diplomacy, "Reinvigorating US Public Diplomacy: A Review of Recent Studies, Unpublished draft 2 [13 May 2005: 1]). The report viewed that the effectiveness of the Muslim World Outreach Policy Coordinating Committee's (PPC) strategy to work with moderate Muslims and counter extremists—an ambitious plan that aimed to influence trends and ideas within Islam itself—remained to be determined.

When the White House Coordination Committee—Office of Global Communications disbanded in March 2005, the appointments of Condoleezza Rice as the secretary of state and Karen Hughes as the undersecretary of public diplomacy and public affairs signified that the public diplomacy's center of gravity was then located in the State Department. Undersecretary Hughes brought proven communications skills and a direct line to the top. From then on, the State Department also actively participated in the Interagency Strategic Communication Fusion Team and the Interagency Working Group on US Government-Sponsored International Exchanges and Training (United States Advisory Commission on Public Diplomacy, "Reinvigorating US Public Diplomacy: A Review of Recent Studies, Unpublished draft 2 (May 13, 2005: 1). According to the GAO, the fusion team, which supported the Muslim World Outreach PCC, met weekly to discuss proposed and ongoing public diplomacy efforts across the federal government (GAO, April 2005).

The PCC was intended as the principal interagency coordination body for US government communications with foreign publics. The Global Strategic Engagement Center, established by former undersecretary of state for public diplomacy and public affairs James K. Glassman, acts as a subject-matter advisory group for the PCC and also serves as a response unit for counterterrorism communications (Nakamura and Weed, 2009: 32). In 2007, the PCC released the National Strategy for Public Diplomacy and Strategic Communication. The PCC does not convene regularly, while Rosa Brooks, senior advisor to the undersecretary for policy, Michele Flournoy, currently meets regularly with the undersecretary for public diplomacy and public affairs, Judith McHale, to discuss interagency issues and plans for public diplomacy and strategic communication (Nakamura & Weed, 2009: 32).

On January 18th, 2006, Secretary of State Condoleezza Rice announced that to match President Bush's bold mission of supporting democracy around the world, the United States needed an equally bold diplomacy that sought to change the world itself. Secretary Rice, in fact, engaged in an effort to pursue this and attempted to inculcate public diplomacy into both policy design and implementation. The strategic framework for US public diplomacy now consists of three goals: (i) to foster a sense of common interests and values with the people of other countries; (ii) isolate, marginalize, and discredit violent extremists; and (iii) foster a positive vision of hope and opportunity that is rooted in US values (i.e., a belief in freedom, equality, the dignity and worth of every human being) (Nakamura & Weed, 2009: 32).

The way forward

We conclude that the rapid development of communication technologies has contributed to the significant changes in the execution of US psychological warfare. As shown in the case of the US covert operations in Indonesia in the late 1950s, the US government had failed to made linkage of television, or other form of media technology, to security concerns caused by the growth of communism in the archipelago. In order to make a psychological advance against the Soekarno administration—to undermine if not replace the government entirely—Washington relied only on traditional military measures. These include the use of CIA and the US Seventh Fleet to provide arms and personnel to help regional dissidents against the Indonesian government. The US covert operations in Indonesia in the late 1950s were in sharp contrast with US psychological warfare strategies in the aftermath of the 9/11 attacks. Following 9/11, the US utilized various forms of communication media, as well as agencies ranging from the intelligence agencies to DOD, State Department, and USAID, to identify, monitor and counter Islamic extremism propaganda in Iraq as well as other parts of the Middle East and the Horn of Africa. However, despite the change in the practices of US psychological warfare from the Cold War to the post-9/11 era, the US has not taken advantage of advanced communications technologies to reach key audiences and therefore induce change.

We identify three weaknesses of US psychological warfare and some future challenges that may arise due to the persistent difficulties in remedying these weaknesses. First, the US is lacking in understanding of targeted countries. Second, the US government is lagging behind the nonstate actor in using information technologies to influence audiences in foreign countries. Third, the US has not managed to convey decisive and coherent messages to targeted audiences.

First, we argue that the true weakness in the US psychological warfare strategies is, paraphrasing Ben Anderson (Anderson, 2006: 60–9), that operators lack specifics on the countries targeted. And what is more important is the lack of a new message or a projection of the global image of America that respects and recognizes greater global difference (Schwoch, 2009: 171). The US covert operations in Indonesia in the late 1950s, for instance, were made with little knowledge of the internal political development of target countries. Rather than achieving the change that Washington desired, the US found itself supporting the losing side. This problem persists today.

At no time in history has any country enjoyed such dominance, reach, and influence as the US currently enjoys. American power is not simply confined to military dominance, because it includes institutions that govern the global economy where, in reality, American power and influence determines the allocation of international credit and investment flows. In addition, there is also the impact of American culture and soft power (Beeson, 2006). The US in this sense "has the capacity to directly or indirectly influence other parts of the world in a number of levels in a way no other country has" (Beeson, 2006: x). In fact, the US has this power, but the interesting question to pose at this juncture is: has the US used it wisely?

Following 9/11, President Bush refocused US foreign and military policy to the war on terror, also known as the Global War on Terrorism. Consequently, US psychological warfare has been refocused primarily to address threat posed by terrorist/extremist groups. In this context, it is relevant to cite Douglas Lovelace Jr., who asserts that the US at that time felt that it had a national interest in what goes on within Islam—not only in the Islamic world. The ideology of violent extremists gave birth to the idea that the US must be promoting one army in that war—ideological moderates (Lovelace Jr., April 2008).

What was truly required was the use of America's soft power or information campaign to prevent violent factions in the Muslim world, who espouse extremist ideologies, from defining America's image; if this is not achieved, US security will continue to be threatened. In the future, as Peter Krause and Stephen Van Evera argue, the US public diplomacy must attempt to change the terms of debate in the majority Muslim states and emphasize dialogue that is tailored to individual foreign audiences. A tailored message conveys respect (Krause & Van Evera, 2009: 106–34). The failure to understand the domestic condition of targeted countries in the future can not only trigger civil conflicts, which kill many civilians, but can also strengthen anti-American sentiment raised by groups that manage to capitalize the popular patriotic outrage against the perceived US intervention in their countries.

Future US psychological warfare should reach a broader range of audience at targeted countries. The failure of US psychological warfare against communism in Indonesia was due partly to US fixation on dissident colonels. Consequently, Washington ignored the presence of other noncommunist elements, including the army and various political leaders as well as officials serving under Soekarno's administration. This is also the case in the US war against terrorism. By focusing only on Islamic groups, the US has neglected the tolerant and moderate Muslims who do not have any affiliations with such groups.

Future US psychological warfare should include this group of people as part of its targeted audiences in Muslim majority countries.

In 2003, officials began reworking the White House's National Strategy for Combating Terrorism and rewriting the ideas section with bold, new language, calling it *Muslim World Outreach.* This attempt aimed at strengthening the hand of the moderates, as America has not been reaching out to this group. The Muslim World Outreach strategy also stated that the United States and its allies have a national security interest not only in what happens in the Islamic world but also within Islam itself. To succeed, America must rely on partners who share values like democracy, women's rights, and tolerance (Kaplan, 2005).

President Bush approved the Muslim World Outreach strategy, but it stirred up controversy as it went to the core of American belief that, generally, Americans do not mess with freedom of religion (Kaplan, 2005). The dilemma for Americans was that the ideological challenge of today came in the form of religion. Kaplan believed that in crafting their strategy, US officials were taking pages from the Cold War playbook of divide and rule (Kaplan, 2005). Another strategy being pursued was to make peace with radical Muslim figures like the Muslim Brotherhood and the Deobandi sect in Pakistan, whose fundamentalism schooled the Taliban and inspired al Qaeda followers (Kaplan, 2005).

Intelligence operatives were making innovative moves by setting up bogus jihad websites and targeting Arab media. The CIA Office of Transnational Issues created a Global Information and Influence Team to assess key US targets. It also hosted a public diplomacy conference to focus on strategies to influence six nations: China, Egypt, France, Indonesia, Nigeria, and Venezuela (Kaplan, 2005). Under CIA auspices, there were Cyber-Influence Conference Series to explore how to combat terrorist use of the internet (Kaplan, 2005).

The government also operated new Arabic broadcasting services—Radio Sawa and Alhurra TV—that were showing some success. USAID's spending nearly tripled, and more than half of that was then destined for the Muslim world. Along with traditional aid for agriculture and education were the training of political organizers and the funding of independent media, and increasingly those grants were going to Islamic groups (Kaplan, 2005 and USAID Bureau for Policy and Program Coordination, 2006; USAID Strategic Plan for Indonesia 2004–2008, 2004, 2006). Since 9/11, there have been many Islamic projects bankrolled by American taxpayers (Kaplan, 2005 and USAID Bureau for Policy and Program Coordination, 2006; USAID Strategic Plan for Indonesia 2004–2008, 2004, 2006). Often the aid did not have an explicit Islamic theme—for instance, "Muppet Diplomacy." The grants were not secret but were done in a subtle manner (Kaplan, 2005).

In the Horn of Africa, the US military was running a model program that aimed at winning hearts and minds by directly competing with madrasahs. Military officers gathered intelligence on where militants planned to start a religious school, and then they targeted those areas by building up new public schools and the local infrastructure (Kaplan, 2005). Elsewhere, US officials were working quietly, through third parties, to add math, science, civics, and health to the madrasah's curriculum (Kaplan, 2005). Legal experts questioned whether America's growing involvement with Islam was legal, as US taxpayer dollars were going to Islamic radio, Islamic TV, Islamic schools, mosques, and monuments.

There are three broad parallels between the Cold War environment and the present time (Rabasa, Benard, Schwartz & Sickle, 2007: 15). First, the US is confronting a new and confusing geopolitical environment with new security threats. Second, there is the creation of large new US government bureaucracies to combat these threats. Third, there is widespread recognition that the US and its allies are engaged in an ideological conflict. The difference in the Cold War era was that the Soviet Union had state interests to protect, defined geographical bodies, and a clear government structure. At present, the US has to confront nonstate actors that reject the norms of the international system (Rabasa et al., 2007: xvi). The remedy naturally was the promotion of democracy that required the development of civil society. Both democracy promotion and civil-society building faced active resistance by authoritarian regimes and suffered from a lack of tangible performance measurement criteria (Rabasa et al., 2007: xix).

There was some random networking of moderates that was deemed insufficiently considered. Distinguishing between authentic moderates and extremists masquerading as moderates was a problem that needed to be resolved, and the US government has thus far failed to develop clear criteria identifying authentic moderates (Rabasa et al., 2007: 3). Robert M. Gates spoke at the US Global Leadership Campaign in Washington DC on Tuesday July 15, 2008. He asserted that "to truly harness the full strength of America requires having civilian institutions of diplomacy and development that are adequately staffed and properly funded." While the State Department is the recognized lead for US public diplomacy efforts, a number of government agencies engage in communications with foreign publics, such as DOD and USAID. A number of other agencies conduct exchanges under the Mutual Educational and Cultural Exchange Act of 1956 (MECEA), and, with the advent of instant global communication through satellite and Internet media, more agencies must consider the effect of their messaging on foreign publics (Nakamura & Weed, 2009: 28).

The most important problem in nations populated by a Muslim majority is that the non-Islamist agenda is either controlled by a discredited regime and/or has little organized basis in society itself (Brumberg, 2005–06: 48). As a result, the multiplicity of cultural, religious, ethnic, and linguistic cleavages does not have a direct or democratic expression in formal political life. The sharpest divide pits Muslims who want to align politics with religion against those who wish to keep these spheres apart (Brumberg, 2005–06: 48).

Brumberg argues that the solution is found less in the realm of ideas than in support for the quintessentially political process of creating procedures, laws, and institutions that ensure the inclusion of all significant groups in government, because only a constitutional order that defends everyone's right to practice or not practice his or her religion freely can secure a polity that is peaceful (Brumberg, 2005–06: 105–6). The US government, in facilitating the process by effectively leading multiple dialogues and utilizing new global communication technology, can help improve lives and simultaneously craft the American global image.

Second, the US is lagging behind private entities in viewing the importance of technologies to improve its political influence vis-à-vis nonstate and state actors. Since the end of the Cold War, there has been an increasing interest in US efforts to win the hearts and minds of the people during the conflict between Washington and Moscow. Volker R. Berghahn's work, for instance, examines how American and European policymakers, corporate leaders, and intellectuals joined forces to win a battle over culture and ideas. He views it as a two-part struggle: one was the effort to counter Soviet efforts to extend communism, the other the effort to overcome Western European cultural anti-Americanism (Berghahn, 2001). Frances Stonor Saunders' book discusses how, in promoting American writers and poets, the Boston symphony, the Metropolitan Opera, and abstract expressionism, the cultural cold warriors challenged people's view of the US (Saunders, 2000).

By the time of the al Qaeda attacks, the US government had ceded management of America's image abroad to Hollywood producers and rap musicians. In the aftermath, American officials began to ponder how to get their message out. The challenge of presenting the many voices of America, both public and private, could actually be overcome by utilizing advanced technological aids. Many of these tools had either fallen into disuse or required new innovative tools if they were to be adopted. The major public diplomacy breakthrough during the Iraq War involved television. New video technologies, small portable cameras, and satellite uplinks allowed media organization to file live-action reports (Dizard Jr., 2001: 225). The US government seemed to lag behind private entities in recognizing the role of satellite television in its public diplomacy strategy for the Iraq War (Dizard Jr., 2001: 225). Other global electronic networks and the flow of digital information through the Internet and cell phones were also providing similar links to the public. Nongovernmental groups certainly were using these global electronic communications to tilt the balance of political influence away from traditional government activities and were operating beyond the reach of traditional governmental activities (Dizard Jr., 2001: 227).

It is common to assume that there are limited consequences of a wide variety of social actors to access, manipulate, and disseminate information in all its forms (Drake, paper delivered in the National War College/Northwestern, 2001). But as more power will be in nonstate hands, this circumstance will increasingly influence government policy and interactions not mediated by the state (Drake, paper delivered in the National War College/Northwestern, 2001). Information revolution builds upon, democratizes, and magnifies existing social frameworks that challenge state-centered hierarchies. These different actors' capacity to influence US foreign policy through mass media and communication systems is an important factor. Nonstate actors will become more empowered and set standards and rules in certain areas. There will be the globalization of awareness and the rebalancing of relations between states and markets. Investigating how exactly these dynamics complicate hierarchical state power and its diplomatic practices will have theoretical implications for the study of foreign policy. Although it was true that American broadcasting efforts in the Muslim world, especially in Arab countries, would never have the reach of existing outlets such as al-Jazeera and al-Arabiya, the State Department did attempt to expand regional public diplomacy hubs beyond London, Brussels, and Dubai, staffed with spokespersons whose full-time jobs were to appear on regional media outlets with a focus on television (Satloff, 2004: 31).

For future warfare against terrorism, we are inclined to argue that to monitor the effect that transformational diplomacy initiatives have on achieving foreign policy goals, it will be very useful to also investigate how the persuasive discourse of realism can constrain international interaction. This is a rather crucial point, because we will need to assess thoroughly the views of other nations as to whether sovereign governments take issue with the US government agenda. Also, how receptive would people be to American public diplomacy's multifaceted effort to influence the people's views on US policies, culture, society, and values? (Nakamura & Epstein, 2007: 2; Nakamura & Epstein, 2008).

The US has also failed to keep up with private entities in using technologies to improve its influence when dealing with challenges posed by other states. In order to improve its influence in the future, the US could learn from private entities such as the Mandiant Company, which has reported the extent of China and Iran-based cyber espionage and attacks. The Mandiant 2014 report mentioned that "the Chinese government is expanding the scope of its cyber operations" to obtain information that could support its state-owned enterprises and that "China-based advanced threat actors are keen to acquire data about how businesses operate—not just about how they make their products" (Mandiant, 2014: 15–6). In future psychological warfare, the US should not limit its strategies to prepare and address cyber attacks from other countries. Rather, it should also publicize examples of other countries' cyber defense and commercial espionage as well as their efforts to inform foreign public opinion through various means (see Cheng, July 12, 2013). This will boost the US position in the war to win the hearts and minds of the global audience.

Third, we argue that the US has not been successful in conveying decisive and coherent messages to influence its global audience. Given that this problem continues to reoccur, this will be a sticking point for US future psychological warfare in dealing with challenges posed by nonstate and state actors. In the digital information age, the US leader's communicating style must establish an ongoing stream of messages that allow people to sense predictability and also allow each and every organization's resources and natural tendencies to emerge. We can revisit attempts at forging American national strategy that had mixed results as policymakers couldn't agree on the target—worldwide terrorism or Islamic extremism—or on its root causes—poverty, Saudi funding, misunderstood US policies, or something else (Kaplan, 2005). Turf war also came into play as the war of ideas cut across neat lines of responsibility in bureaucratic Washington (Kaplan, 2005).

Perhaps, because the US feels less dependent on the largesse of its partners in the international system, it is therefore less inclined to engage in strategic public diplomacy to systematically address challenges brought by not only non-state but also state actors (Manheim, 1994). In the realm of the war on terror, the intervening states, or more specifically the United States during the Bush administration, do embark on the quest to repair, reinterpret, reform, or fix Islam in a range of official and nongovernmental initiatives. Daniel Brumberg argues, however, that without understanding the intricacies of Islamist politics, US efforts to enlist Islamists and their ideas in the cause of democracy might do considerable harm (Manheim, 1994).

With respect to state actors, the US has not engaged in systematic public diplomacy efforts to address challenges posed by emerging powers such as China and Russia. China's systematic psychological warfare operations in South and East China Sea to challenge the US are already underway (Cheng, July 12, 2013). The General Political Department (GPD) within the Chinese military has been assigned to plan and coordinate the country's psychological warfare operations. Two sub-units within the GDP, liaison and security, are responsible "for influencing the political opinions and attitudes of enemy personnel, collecting and analyzing intelligence on social, political and demographic situations in target countries," as well as for providing training on psychological warfare (Cole and Shan, 26 August 2011). China has used state and Communist Party media to broadcast live military firing exercises and landing and naval drills, denounce the award made by the Permanent Court of Arbitration in the Hague, shape a political discourse regarding the US implementation of "Cold War"—style containment police against China, and inform the growing anti-American mode at its domestic front (Gertz, June 1, 2016; Daiss, May 18, 2016). A leap forward in China's psychological warfare took place in 2013 as China's new PSYOP (Psychological Operations) aircraft, the Gaoxin-7, was unveiled. It "is essentially a flying broadcast station" that can be used to "broadcast messages to demoralize" enemy troops and urge them to surrender (Jensen, August 14, 2013).

The July 2016 Hague tribunal ruling on the South China Sea dispute was overwhelmingly not in favor of China's claim of historic rights to resources in this area (see Guardian, July 12, 2016). Thus, it is expected that in the future, China's psychological warfare against the US will intensify. Gertz noticed that the US has carried out few freedom-of-navigation operations—three from October 2015 to June 2016—and only sporadic surveillance flights in South China Sea (Gertz, June 1, 2016).

Similar to China, Russia has also developed systematic psychological warfare strategies. Russia has developed a new concept, "the information security of society." Russia instituted several changes because, first, the information and psychological struggle has replaced the propaganda agitation of Marxism–Leninism. Second, some theorists consider psychological operations to be an independent form of military operations. Third, the Russians armed forces have also developed military occupational specialists devoted to psychological confrontation. Fourth, some Russians officers have come to view the information psychological struggle as an integral part of information warfare.

In response, Timothy L. Thomas (Thomas, 1997) stated that first, the US at the joint-staff strategic level cannot afford to lose sight of the development in the PSYOP arena. Second, "US planners need to recognize the increasing synchronization of PSYOP and information revolution." In reality, the above proved difficult to realize. In 2014, President Obama gave an interview with the *Economist*. He used it to portray Russia as weak and irrelevant. The *Economist* stated that President Obama's effort was like the entire anti-Ukraine, anti-Russian effort: it seems to have little relationship to the truth (*Economist*, 2014).

During the Obama administration, the concept of strategic communication has grown more expansive; President Obama has authorized harsh legal action against government leakers of US foreign policy and intelligence practices. President Obama has mounted a public campaign against the Russian Network (RT) in the recent attempts at information warfare. NATO has a new temple of information technology called the NATO Strategic Communications Center of Excellence STRATCOM, located in Latvia (North, 2015). US military commanders and national security advisors speak of cyber war, net war, and cyber espionage as the major threats to US national security; however there is no focus on psychological operations (Developing Tomorrow Blog). Strategic Communication is related Information Operations (IO) and Psychological Operations (PSYOPS). It may not be ideology with the Russians anymore, but Russian national interests remain. They have used forged news stories, and there is a new rise in active measures (Developing Tomorrow Blog). In the meantime, Russian has named the US as threat to national security (Fox News, 2016).

One of the more effective Russian weapons is information, as was evident from the recent release of the 20,000 emails from the Democratic National Committee (Gessen, 2016). The Wikileaks release was deemed to have more of a tinge of Russian-style information war with an intent to alter political events

(Hennessey, 2016). Fiona Hill views that there are several reasons Russia might want to stir up trouble in an already contentious campaign: Russian views of the US are at a low point; in the past, the US took advantage of Russian domestic divisions and vulnerabilities; and Russia would like to undermine US international credibility by illuminating the deficiencies in American party politics. President Putin, in particular, knows how to "work with people" and "work with information." Putin also knows how to "play the media" (Hill, 2016). Michael Morell, former acting CIA chief, said that Trump has been unwittingly "played" by Russia (Davis, 2016). Donald Trump "wanted to be complimented and (so) he complimented him and Trump did not understand…that Putin was playing him and Putin thinks Trump is an unwitting agent of the Russian Federation" (Davis, 2016).

As the US yields only piecemeal response to China and Russia's psychological warfare, it sends confusing messages to China, Russia, its allies, and the rest of the world (see Cheng, July 12, 2013). There is a need for the US to convey coherent strategic messages and prepare "to counter such psychological operations" in the future (Cheng, July 12, 2013). Drawing from the historical lesson of US covert operations in Indonesia and the present war against terrorism campaign, in order to present decisive and coherent messages in the future, the US will need to establish a public diplomacy coordinating structure, create a new White House architecture for public diplomacy, revive the USIA, craft a national communications strategy, include public diplomacy in the creation of foreign policy, increase private-sector involvement in public diplomacy, create a corporation for public diplomacy, improve capacity to "listen" to and dialogue with foreign publics, craft messages with greater sensitivity to foreign audiences, revive and enhance successful exchange programs, provide better access to information on American policies and values, and increase readiness and capabilities of public diplomacy personnel (see GAO, July 2003, September 2003, 2004, April 2005, September 2005, 2006, 2007; Nakamura & Weed, 2009; Epstein and Mages, October 31, 2005).

References

Anderson, Benedict R. O'G (1882) "Perspective and Method in American Research on Indonesia." in Benedict R. O'G. Anderson & Audrey Kahin (eds.), *Interpreting Indonesian Politics: Thirteen Contributions to the Debate*—Interim Report Series (Publication No. 62), Cornell Modern Indonesia Project Southeast Asia Program. New York, NY: Cornell University Press.

Barnett, Thomas P. M. (2009). *Great Powers America and the World after Bush*. New York, NY: G. P. Putnam's Sons.

Beeson, Mark. (2006). "Introduction," in Mark Beeson (ed.), *Bush and Asia. America's Evolving Relations with East Asia,* London: Routledge.

Berghahn, Volker R. (2001). *America and the Intellectual Cold Wars in Europe*. Princeton, NJ: Princeton University Press.

Brouwer, Melinda (March 9, 2008). "State Department: Living in the Shadow of the Pentagon." Available at http://foreignpolicyblogs.com/2008/03/09/state-department-living-in-the-shadow-of-the-pentagon/.

Brumberg, Daniel (2005–06). "Islam Is Not the Solution (or the Problem)," *Washington Quarterly*, 29(1) (Winter): 97–116.

Cheng, Dean, "Winning Without Fighting: The Chinese Psychological Warfare Challenge," July 12, 2013. Available at: http://www.heritage.org/research/reports/2013/07/winning-without-fighting-the-chinese-psychological-warfare-challenge.

Cole, Michael & Shan, Shelly, "PRC steps up psychological warfare targeted at Taiwan," August 26, 2011. Available at: http://www.taipeitimes.com/News/front/archives/2011/08/26/2003511686/1.

Council on Foreign Relations Independent Task Force (2003). "Finding America's Voice: A Strategy for Reinvigorating US Public Diplomacy," New York. NY.

Daiss, Tim (May 18, 2016). "China Kicks Off War Games As Taiwan Prepares For New President." Available at: http://www.forbes.com/sites/timdaiss/2016/05/18/psychological-warfare-china-kicks-off-war-drills-as-taiwan-prepares-for-new-president/#606912733a13.

Davis, Maia (August 7, 2016). "Trump Has Been Unwittingly 'Played' By Putin: Former CIA Chief," ABC News.

Der Derian, James. (2001). "The War of Networks," *Theory and Event* 5(4). Available at: http://muse. jhu.edu/journals/theory_and_event/v005/5.4derderian.html.

Developing Tomorrow, A Blog By The Research Analysis Strategy Operation Group "Psychological Operations in the Information Age." Available at: https://developingtomorrow.wordpress.com/2012/03/28/psychological -operations-in-the-information-age/.

Dizard Jr., Wilson. (2001). *Digital Diplomacy: US Foreign Policy in the Information Age*. Washington, DC.: Praeger.

Doeppers, Daniel F. (1972). "An Incident in the PRRI/Permesta Rebellion of 1958," *Indonesia* 14: 182–195.

Drake, William J. (2001). Paper Delivered in the National War College/Northwestern University Symposium "Information Age Diplomacy," April 5-6, 2001.

Economist (August 5, 2014). "Obama Continues Psyop Against Russia, Threatens China. Available at www.antiwar.com.

Epstein, Susan B. & Lisa Mages (2005). "Public Diplomacy: A Review of Past Recommendations," Updated October 31, 2005 CRS Report for Congress, RL33062 Congressional Research Service, Library of Congress, Washington, DC.

Evans III, Bryan. (1989). "The Influence of the United States Army on the Development of the Indonesian Army (1954–1964)," *Indonesia*, 47: 25–48.

Fox News (January 3, 2016). "Russia Reportedly Names US as Threat to National Security for First Times." Available at: http://www.foxnews.com/politics/2016/01/03/russia-reportedly-names-us-as-threat-to-national-security-for-first-time.html.

Gardner, Paul F. (1997). *Shared Hopes, Separate Fears: Fifty Years of US–Indonesian Relations*. Philadelphia, PA: University of Pennsylvania Press.

Gertz, Bill (June 1, 2016). "China's Presses the Fast Forward Button to Dominate the South China Sea." Available at: http://nationalinterest.org/blog/the-buzz/chinas-presses-the-fast-forward-button-dominate-the-south-16416.

Gessen, Masha (July 27, 2016). "The Trump-Putin Fallacy," The New York Review. Available at: http://www.nybooks.com/dai;y/2016/07/26/trump-putin-fallacy-failure-of-imagination/.

The Guardian (July 12, 2016). "Beijing rejects tribunal's ruling in South China Sea case." Available at: https://www.theguardian.com/world/2016/jul/12/philippines-wins-south-china-sea-case-against-china.

Hennessey, Susan (July 25, 2016). "What Does the U.S Government Know about Russia and the DNC Hack," Washington, DC: Brookings Institute.

Hill, Fiona (August 3, 2016). "Order from Chaos: 3 Reasons Russia's Vladimir Putin Might Want to Interfere in the US Presidential Elections,"Washington, DC: Brookings Institute.

Jensen, Aaron (August 14, 2013). "China prepares for psychological warfare." Available at http://thediplomat.com/2013/08/china-prepares-for-psychological-warfare/.

Jones, Matthew. (1999). "Maximum Disavowable Aid: Britain, the United States and the Indonesian Rebellion, 1957–58," *The English Historical Review*, 114. 1179–1216.

Kahin, Audrey R. & Kahin, George McT. (1995). *Subversion as Foreign Policy: The Secret Eisenhower and Dulles Debacle in Indonesia*. New York, NY: New Press.

Kaplan, David E. (April 25, 2005). "Hearts, Minds, and Dollars," in *The US News and World Report*.

Krause, Peter & Van Evera, Stephen (2009). "Public Diplomacy: Ideas for the War of Ideas," *Middle East Policy*, XVI(3): 106.

Kroef, Justus M. Van Der. (1958). "Disunited Indonesia (II)," *Far Eastern Survey*, 27(5): 73–80.

Lafeber, Walter (2000). "Presidential Address: Technology and U.S. Foreign Relations," *Diplomatic History*, 24(1): 1–19.

Laipson, Ellen (2004). "Relating to the Muslim World: Less Is More?" In Adam Garfinkle (ed), *A Practical Guide to Winning the War on Terrorism* (Hoover Press).

Lovelace Jr., Douglas C. (2008). "Forward," in Sherifa Zuhur, "Precision on the Global War on Terror: Inciting Muslims Through the War of Ideas," http://www.StrategicStudiesInstitute.army.mil accessed July 1, 2008.

Mandiant Fire Eye Company (2014). "Threat Report." Available at: https://dl.mandiant.com/EE/library/WP_M-Trends2014_140409.pdf.

Manheim, Jarol. (1994). *Strategic Public Diplomacy and American Foreign Policy: The Evolution of Influence*, New York: Oxford University Press.

McMahon, Robert J. (1981). *Colonialism and Cold War*. New York, NY: Cornell University Press.

Nakamura, Kennon H. & Matthew C. Weed. (Dec. 18, 2009). "US Public Diplomacy: Background and Current Issues," CRS Report for Congress, R40989, The Library of Congress, Washington, DC.

Nakamura, Kennon H. & Susan B. Epstein. (Aug. 23, 2007). "Diplomacy for the 21st Century: Transformational Diplomacy," CRS Report for Congress, RL34141, The Library of Congress, Washington, DC.

Rabasa, Angel, Cheryl Benard, L. H. Schwartz, & Peter Sickle (2007). *Building Moderate Muslim Networks*. Santa Monica, CA: Rand Center for Middle East Public Policy.

Satloff, R. (2004). *The Battle of Ideas in the War on Terror: Essays on US Public Diplomacy in the Middle East*, Washington: The Washington Institute for Near East Policy.

Saunders, Frances Stonor (2000). *The Cultural Cold War: The CIA and the World of Arts and Letters*. New York, NY: New Press.

Scherer, Michael (June 24, 2013). "The Geeks Who Leak." *Time*: 22–29.

Schwoch, James. (2002). "The Cold War, The Space Race, and the Globalization of Public Opinion Polling," paper presented at the International Studies Association Conference, New Orleans, LA, March 24–28, 2002.

Schwoch, James (2009). *Global TV: New Media and the Cold War 1946–69*. Urbana, IL: University of Illinois Press.

Scott, Peter Dale (1985). "The United States and the Overthrow of Sukarno, 1965–1967," *Pacific Affairs*, 58(2): 239–64.

Snow, Nancy (2009). "Rethinking Public Diplomacy," in Nancy Snow & Philip M. Taylor (eds). *Routledge Handbook of Public Diplomacy*. New York, NY: Routledge: 3–11.

Thomas, Timothy L. (1997). "Russia Information Psychological Actions," *Special Warfare*, 10(1): 12–19.

USAID Bureau for Policy and Program Coordination (February 2006). "Summary Assessment of the Islam and Civil Society Program in Indonesia. Promoting Democracy and Pluralism in the Muslim World," PPC Evaluation Brief No. 13, Washington, DC.

USAID Strategic Plan for Indonesia 2004–2008 (July 28, 2004). "Strengthening a Moderate, Stable, and Productive Indonesia," Washington, DC.

United States General Accounting Office Report to the Committee on International Relations, House of Representatives (July 2003). "US International Broadcasting: New Strategic Approach Focuses on Reaching Large Audiences but Lacks Measurable Program Objectives," GAO Report for Congress, GAO-03-772.

United States General Accounting Office Report to the Committee on International Relations, House of Representatives. (September 2003). "US Public Diplomacy: State Department Expands Efforts but Faces Significant Challenges," GAO Report for Congress, GAO-03-951.

United States General Accounting Office Report to the Subcommittee on National Security, Emerging Threats, and International Relations, House Committee on Government Reform (August 2004). "US Public Diplomacy: State Department and Broadcasting Board of Governors Expand Post 9/11 Efforts but Challenges Remain," GAO Report for Congress, GAO-04-1061T.

United States General Accounting Office Report to the Committee on Science, State, Justice, and Commerce, and Related Agencies, Committee on Appropriations, House of Representatives (April 2005). "US Public Diplomacy: Interagency Coordination Efforts Hampered by the Lack of a National Communication Strategy," GAO Report for Congress, GAO-05-323.

United States General Accounting Office Report to Congressional Requesters (September 2005). "International Affairs: Information on US Agencies' Efforts to Address Islamic Extremism," GAO Report for Congress, GAO-03-952.

United States General Accounting Office Report to the Chairman, Subcommittee on Science, the Department of State, Justice, and Commerce, and Related Agencies, Committee on Appropriations, House of Representatives (May 2006). "US Public Diplomacy: State Department Efforts to Engage Muslim Audiences Lack Certain Communication Elements and Face Significant Challenges," GAO Report for Congress, GAO-06-535.

United States General Accounting Office Testimony to the Subcommittee on International Organizations, Human Rights, and Oversight, House Committee on Foreign Affairs, House of Representatives (April 26, 2007). "US Public Diplomacy: Strategic Planning Efforts Have Improved, but Agencies Face Significant Implementation Challenges," GAO Report for Congress, GAO-07-795T.

United States Advisory Commission on Public Diplomacy Report (2004).

United States Advisory Commission on Public Diplomacy Report (2005).

United States Advisory Commission on Public Diplomacy (May 13, 2005). "Reinvigorating US Public Diplomacy: A Review of Recent Studies, Unpublished draft 2.

United States General Accounting Office Report to the Committee on Science, State, Justice, and Commerce, and Related Agencies, Committee on Appropriations, House of Representatives (2005). "US Public Diplomacy. Interagency Coordination Efforts Hampered by the Lack of a National Communication Strategy."

Wilcox LTC USA (Ret.), Greg (2006). "Strategic Transformation: Aligning National Security Policy/Operations," 1–7. Available at: http://dnipogo.org/wilcox/wilcox_strat_trans.pdf.

10 The US and "small footprint" military assistance in colombia

A model for future intervention?

Jorge E. Delgado

Introduction

Since autumn 2014, the challenge posed by the Islamic State (IS) has intensified academic and policy debates in the United States (US), and their European partners in the North Atlantic Treaty Organization (NATO), over the possibility of dispatching "boots on the ground" to assist their allies in the Middle East. It is apparent that US military leaders, since the initial escalation of hostilities, have been aware that to fulfill the stated policy goal of defeating IS, a major commitment of combat troops beyond an advisory role in Iraq and covert assistance to Kurdish and Syrian groups is required. The deployment of combat troops, it has been argued, would be a requisite to consolidate any gains obtained by the aerial bombing campaign waged by NATO since 2014. As the former secretary of defense Robert Gates stated in an interview at the height of the siege of Mosul by IS, "there will be boots on the ground if there's to be any hope of success in the strategy. And I think that by continuing to repeat that [the U.S. won't put boots on the ground], the president, in effect, traps himself" (Levine 2014). Curiously, this statement by Gates marked a 180-degree change in his view, as prior to his retirement in 2011, he was adamant that "Any future defense secretary who advises the president to again send a big American land army into Asia or into the Middle East or Africa should 'have his head examined'" (Levine 2014). Former Secretary Gates's change of opinion is somewhat understandable and demonstrates well how the debate over the need to deploy combat troops, two years after the rise of IS, is still an extremely strategically contentious one. The complexity of the debate is perceptibly due to the intractability of the conflict in Syria and Iraq, the limits of the aerial bombing campaign, and the growing threat of homegrown Islamic terrorism as evidenced after the attacks in Paris, San Bernardino, and Brussels. But a central narrative underpinning the ongoing debate over the need for "boots on the ground" to confront IS and the proposal in the past two years of alternative modes of military intervention is evidently a disillusionment with the "big footprint" counterinsurgency (COIN) waged by the US and its partners in Iraq and Afghanistan.

"Small footprint" military intervention—the dispatch of advisors and cash to bolster the efforts of indigenous forces—is seen as a promising alternative in that it remedies the major shortcomings of recent "big footprint" COIN expeditions waged by the US: the opposition expressed in domestic public opinion with a small appetite for large expenditures and significant casualties in remote battle spaces. In this context, Colombia has been frequently invoked by pundits, US officials, and military practitioners as a success story that has offered a consoling narrative. Take, for instance, the response offered by retired US Army Colonel Peter Mansoor, who served as General David Petraeus' executive officer during the Iraq War, to the question of what victory in Afghanistan would look like after the termination of the ISAF mission mandate:

> At this point the best the United States can hope for is to support an Afghan government that can keep the country together after 2014 and convince the Taliban that it cannot win the war in any conceivable time frame. [...] For a view of what winning might look like, look at Colombia. A decade ago the country seemed on the verge of disaster with the FARC on the ascendancy, [...]. Good Colombian presidential leadership and U.S. support were the keys to victory."
>
> (Szoldra 2014)

While Mansoor rightfully points out the importance of good local political leadership and good strategic judgment by the US in accounting for the alleged success in Colombia, it is crucial to underline that these two elements combined were peculiar to the "host" country's political context. In other words, key contingent factors make Colombia a unique case. Not only it is a weak analogy to assess the role played the US during the same time frame in Afghanistan or Iraq; it is fundamentally a poor case model of "small footprint" intervention to inform policy options today in the Middle East, and for that matter, elsewhere in the world in the foreseeable future. The purpose of this chapter, then, is to make a case against the unsuitability of making the US intervention in Colombia a case model.

Colombia as a success case and the limits of replication

Since 2002, the South American country has seen the longest period of sustained military activity against one of the many insurgencies that have contested the established political order since the early 1950s. The campaign against the insurgents of the *Fuerzas Armadas Revolucionarias de Colombia* (Revolutionary Armed Forces of Colombia [FARC]), as well as the implementation of a robust counternarcotics policy and the dynamics of US security assistance backing both efforts, has attracted the attention of scholars and practitioners around the world. Since the re-emergence of COIN as a topic of academic interest after the Iraq surge and the quest for diverse experiences to inform military intervention models by the US and their allies in the Middle East and Africa, the Colombian case has been brought in the spotlight as a success story. "Colombia's dramatic gains serve as a beacon of hope for other countries that today are besieged by the violence and blight of insurgency and drug trafficking," stated a recent study by the Joint Special Operations University of the US military. "They demonstrate that nations can, with internal political will and persistent American support, overcome seemingly intractable obstacles." (Moyar, Pagan & Griego, 2014: 49).

According to supporters of the "small footprint" advising model, *Plan Colombia*—a US$6 billion program of US assistance to swell and restructure the Colombian military and police launched in 1999—translated into an interagency effort that imposed exacting human rights standards, reinvigorated Bogotá's relationship with its citizens and received mutual endorsement from both the Colombian military and the government.[1] The US Department of State also assisted Colombia in the reform of its weak criminal justice system, to improve the link between the justice system and the security services, and to promote and protect human rights and prosecute violators. US assistance also focused on the creation of government capacity to implement effective programs to design and implement stabilization programs, assist internally displaced persons, support the millions of internally displaced persons, and disarm and reintegrate surrendered insurgents (US Department of State 2010).

Counterinsurgency expert David Spencer from the National Defense University maintains that US military assistance to Colombia offers 20 key lessons for dealing with internal irregular conflict at the strategic and operational level (Spencer 2012: 6–7). But in general, such lessons simply amount to preconditions like "leadership," "political consensus," "institutional strength," "well-regulated local forces," or "inability of opponents to adapt," which are almost impossible to find in battlefields such as Mali, Syria, or Libya, not to mention Iraq and Afghanistan. Moreover, the attempt to extract replicable "lessons learned" from the Colombian case to implement in other theaters risks decontextualizing that conflict. Critics of counterinsurgency, like strategic theorist M. L. R. Smith and David Martin Jones, sustain that "lessons learned" literature directed to the formulation of doctrine amounts to "recipe books that exhort eager neophytes merely to follow a specific set of tactics, permeated with sage operational advice, to achieve the perfect counterinsurgency" (Smith & Jones 2010: 436). In other words, template military solutions tend to minimize the uniqueness of each strategic context, and the politics and underlying values that shape the societal and cultural complexity from which conflicts arise are considered to be contingent matters.

One could argue further that the overemphasis on the study of Colombia's latest operational and technical progress—developed in great part via US assistance—has minimized, among other factors, the analysis of the politics that shape local strategic thought; that is, paraphrasing John Stone, how an actor's judgments and views on how the employment of varied techniques—the means of war—are to be translated into political effects (Stone 2011: 4). In other words, tactical proficiency and innovative military techniques count for naught if there is no correct understanding of the uniqueness of the Colombian context, and if the politics and underlying values that influence the diverse actors' definitions of ends and selections of means are relegated to matters of secondary importance. In this respect, beyond the quality and utility of the material support provided by the US to enhance Colombian capabilities, the key to the Colombian security services' success against the FARC rebels in the last decade has been contingent factors, in particular the existence of a legitimate and strong local leadership and institutional capacity that was able to define the political and strategic parameters for success.

Colombian exceptionalism

Scholar Douglas Porch, a military historian and counterinsurgency expert, has made the case for Colombian exceptionalism since 2008, that is, some years before the country became the COIN and military assistance success model it has turned into recently. His research on the Colombian conflict highlighted that in the context of the US-led "Global War on Terror," the South American country was exceptional in the sense that it was in itself a country that conformed to Western liberal values. "Colombian democracy, while restricted and imperfect," wrote Porch, "formed part of that nation's birth-right. Indeed, Colombia has one of the most impressive records of democratic continuity in Latin America." (Porch 2008: 127). This has meant that the country's political development has, despite serious security challenges, led to the consolidation of robust and legitimate institutions and sufficient government capacity to design and sustain counterinsurgency and counterterrorism efforts over a long period of time. More importantly, it has been the essential condition facilitating the consolidation of the US–Colombia security partnership. In short, Colombia's politics have fostered a culture of civilian control of the military and facilitated processes of professionalization of the armed forces that are on a par with those of the majority of Western democracies.

However, the factors that account for Colombia's exceptionalism are not only associated with its political context but also with the nature of the threat faced by the Colombian state. The character of the insurgent groups that have challenged the authority of the state since the second half of the twentieth century has been distinctive, and ironically, as Porch has argued, has accounted for most of the disagreements over security and defense policy formulation in the country's recent history. Precisely, since the advent of the "Global War on Terror," there has been disagreement over the nature and causes of Colombia's conflict as they are not anchored in ethnic or religious divisions or discontent, and ideological lines in the country are even less distinguishable (Porch 2008: 129). In other words, and paraphrasing Samuel Huntington, it is clear that the South American country does not suffer from a "clash of civilizations" interfering in the US–Colombia security partnership. This is a key point to take into consideration when considering the type and level of engagement required by Washington in other areas of the world, as strategic scenarios vary enormously. FARC and IS are not comparable simply by the fact that they have used similar irregular warfare methods to attempt to achieve their goals.

Any meaningful evaluation of US military assistance to Colombia during the last two decades cannot disregard the unique political context of that country and how it favored the intervention of Washington. At the policy and operational level, a legitimate government with a history of civilian control of the military and the existence of government capacity to conduct interagency planning bolstered the effects of US security assistance aimed at augmenting Colombian proficiencies. The rest of this chapter will emphasize how these contingent factors developed in Colombia, discuss how they accounted for the success of US assistance, and further explain why they are not invariably transferrable.

Local leadership: Organizing and guiding the "host" country's "big footprint"

Álvaro Uribe was elected to the presidency of an established functioning democracy—unlike most Middle Eastern and African countries—after a landslide victory in May 2002. The Colombian electorate had become increasingly frustrated following fruitless 4-year-long peace negotiations with the insurgency, which had witnessed a serious deterioration in security. Uribe had sufficient legitimacy and political will to pursue an aggressive policy toward the insurgency. As a US Military Group officer recalled in 2010: "Uribe was the only president who eventually realized that the FARC isn't that competent. He kicked the military in the ass. It's as if he said: 'You've got popular backing, political support, and equipment—what are you waiting for?'" (Porch & Delgado 2010: 277). Another US official who witnessed Uribe's strong leadership style recounts in the following his relationship with the Colombian military's senior officers in the early days of his presidency:

> President Uribe summoned his military commanders for what is known in military parlance as a 'come to Jesus' meeting, and told them in no uncertain terms that it was time to get out from behind their desks and into the field with their troops—where they belonged. In addition, the distractions of costly education and training junkets to the United States and elsewhere would henceforth be curtailed. He also admonished military commanders to do their jobs or resign. Commander of the Armed Forces General Jorge Enrique Moral Rangel underscored Uribe's goals and stated that the new mission of the armed forces would be to 'prevent, rather than react to the attacks by the guerrillas against the Colombian population and the regional and local government infrastructure'.
>
> (Taw 2014: 152)

While Uribe's success at the polls represented a public endorsement for a more aggressive posture in dealing with the FARC, it took time for the incoming government to translate the will to fight into a workable action plan. Given the political and financial magnitude of Uribe's ambitions, he was forced to rely on economy-conscious civilian technocrats to reshape the Colombian defense establishment, which traditionally had been dominated by the army (Porch 2008: 45). Furthermore, the new president sought to raise at least US$ 1bn of new funding within weeks of the inauguration, which not only represented an enormous amount by Colombian standards but also was seen as a requirement for the country to assume responsibility for financing its own security in the long run (US State Department 2002). The Colombian military expanded its numbers from 145,000 in 2000 to 236,000 by 2008, and the security and defense budget increased from $5.7 in 2003 to $10.4 billion in 2012 (Schipani 2013).

A key actor in Uribe's strategic planning team was the Ministry of Defence advisor Sergio Jaramillo, a polished, cultivated senior civil servant who had studied philology and classical languages at Oxford, Cambridge, and Heidelberg universities and who was familiar with the works of Robert Thompson. Tasked with drafting the government's COIN policy, known as Democratic Security (DSP), Jaramillo opted for the classic "oil spot" approach to COIN. Jaramillo help institute the Malayan emergency "traffic-light system," which defined the status of the areas to be pacified: insurgent-dominated red areas, yellow areas in transition, and green or stabilized areas. He also established a division of labour between the military and the police (Delgado 2015: 415) A main feature of Jaramillo's territorial control approach was to assuage grievances in the recovered areas by providing previously nonexistent basic services in the less-populated periphery, where the insurgency thrived. Indeed, Jaramillo believed that Colombia's security problems stem directly from the historical absence of effective government in peripheral areas, which the DSP aimed to supply. "More than facing an enemy," he has explained, "what you are actually doing is trying to re-establish the rule of law. You want to win the battle of governance; you want to show that you are the legitimate authority" (Pfanner *et al.*, 2018: 823).

The early development of President Uribe's security policy demonstrates how, 5 years before the rise to orthodoxy of COIN doctrine in the US with the experience in post-invasion Iraq and the leadership of General Petraeus, Colombian officials were already framing the campaign against the FARC insurgency in according to the doctrine's basic tenets. US Embassy officials in Bogotá declared that the DSP was "a promising step forward that clearly "proposed a 'clear-hold strategy' with the Colombian FUDRA as the leading element." This, they argued, demonstrated that "Uribe's team and the GOC [Government of Colombia] had experience with strategic thought and organization"[2] (US Department of State 2002).

Indeed, in the wake of 9/11, increased support from the US government, meant to showcase Colombia as its western hemisphere's poster child for the "war on terror," put muscle behind Uribe's government effort against the FARC, given that the main focus from the onset was on a military "surge" to confront the guerrilla. The George W. Bush administration allowed Uribe to use key military assets, originally allocated for counternarcotics operations under former President William Clinton's counternarcotics based *Plan*

Colombia, in "counter-terrorism" operations (Powell 2002). Most notably, the Colombian military was upgraded with over 200 helicopters, which increased mobility; a reinforced intelligence apparatus; and the development of special operations forces (SOF) and joint operations capabilities (Delgado 2015: 413).

Likewise, Washington was eager to assist the Colombian government in the refinement of the main operational concepts of the DSP. The US Embassy in Bogotá insisted that it was necessary to ensure that "the Colombian strategy dovetails with the USG effort." Given that there was a common understanding that "Colombia's security problems boiled down to the absence of Government authority, especially in the outlying coca producing areas where the terrorist are active" (US Department of State 2002), the main goal of the effort was the strengthening of the rule of law and the establishment of police presence. The US assisted through the framework of *Plan Colombia* to help the country incorporate interagency coordination techniques to intervene in the areas with insurgent presence and promote economic development.

In this context, thanks to a determined local political control over the military with Uribe, which had been historically absent in the country, the government was able to strategically focus these operational and tactical enhancements offered by US assistance. The Colombian case, in this regard, shows that the "small footprint" intervention is likely to be successful only when occurring along a well guided and organized host country's "big footprint." Whereas the US military entered into a wide-ranging engagement with the Afghan national army, the US did not have to build up a new indigenous armed force in Colombia. Militarily, the objective of US assistance in Colombia over the past 15 years was focused on improving the clear-cut operational deficiencies of an established and strong armed force capable of bearing the brunt of the responsibility. As the US Southern Command (SOUTHCOM) had set out in at the onset of *Plan Colombia*, the assistance simply aimed to allow the Colombians to better shape "their operating environment to conduct decisive joint operations by virtue of improved training and modernization" (USSOUTHCOM 2000).

However, when trying to understand the reasons why the Colombian military has been so receptive to US assistance more generally, it is also necessary to take into account that the association between the countries goes far beyond *Plan Colombia*. A longstanding historical relationship has effectively influenced Colombia's strategic outlook and its responsiveness to the latest package of military assistance.

A long engagement with limited objectives: The Colombian military relationship with the US in historical perspective

In contrast to its immediate neighbors in recent decades, like Venezuela, Ecuador, Brazil, and Nicaragua, Colombia has been robustly pro-US. The Colombian military officer corps has for their part, on the whole, proven receptive to US security assistance; eager to modernize, at least on a technical and doctrinal level; and able to adapt tactically and operationally to the evolving security environment. Receptivity to US military assistance is a result of a long-standing relationship that dates from World War II. "The Colombian army has been the United States' staunchest ally, I mean the staunchest. There's practically been a direct line of communication between the Colombian and the U.S. Army," claimed General Fernando Landázabal. a former chief of the armed forces in the early 1990s. (Dudley 2004: 34). Such a view may well encapsulate what Douglas Porch and Christopher Muller defined as the "Colombian military's emotional dependence on US support" during their five-decade-long COIN effort (Porch & Muller 2008: 181). The Colombian's characterization of the military relationship in terms of a vital alliance has placed the US as an essential actor in their structure of incentives for its definition of ends and selection of means. Washington, in short, has been viewed as a basis of political will and resources for pursuing specific courses of action by the military leadership.

Since 1939, when the Colombians welcomed for the first time US military advisors and instructors to the war department and the *Escuela Militar* (the Army cadet School), replacing the German military mission that had been active since the beginning of the twentieth century, their presence in the country became a permanent feature. Initially with the objective of securing the Panama Canal and preventing further German encroachment in the looming world war, the US effectively concluded that Colombia could not be neglected and assumed an active policy of military assistance that promptly lured the Colombians to its side (Coleman 2008: 13–15). After the war and up to the late 1950s, US defense policy towards Colombia was based on the precepts of Hemispheric Defense, which emphasized the strengthening of the "conventional" capabilities of the military and its modernization under American military parameters of organization, training, and doctrine. Again, the US was focused on protecting the Panama Canal, this time against possible diversionary attacks from the Soviet Union, a mission in which they trusted the Colombians could assist.

In the early Cold War, the Colombian army's participation in the Korean War under US sponsorship, helped consolidate the objectives of Hemispheric Defense,[3] paving the way for the consolidation of a "Military Assistance Program" that secured a constant flow of material and the establishment of a "Military Assistance and Advisory Group" that secured further tutoring throughout the 1950s. Moreover, Colombian participation in Korea activated the direct line of communication between both militaries cherished by General Landazábal. Assistance by the US military in training had, by the end of the decade, assisted the Colombians fairly well in professionalizing its cadres. According to a report of the British embassy in Bogotá, relating a visit of the itinerant military attaché to the *Escuela Militar* and other service schools in the Spring of 1959, "the organization of studies both military and academic, and of practical military training, drawn up by the American mission, seemed excellent" (Foreign Office 1959).

With the initiation of rural insurgency at the turn of the decade, the returning Korean war veterans began to utilize their contacts in the US military not only to expedite assistance in terms of materiel, training and planning but also as political support for their strategies to confront the growing threat *vis-à-vis* the Colombian government during the 1960s. The priorities of the US toward Latin America soon changed with the advent of the Cuban revolution, as they concluded that the region would not be a target in a "conventional" war but rather would fall victim to subversion. This meant a change in the nature of Washington's cooperation, transitioning from Hemispheric Defense to counterinsurgency. This process translated into a significant increase in assistance to develop the Colombian army's capabilities for this sort of warfare and to engage in nation-building tasks.

Nonetheless, the extent of US military assistance to Colombia since the 1960s responded to Washington's own strategic goals, which were in essence limited. The prevention of communist takeover in Colombia would not require the ultimate defeat of the insurgencies; rather, the goal would be to stabilize the country, which set the reasonable limit of the effort. Since the height of the Cold War, with the growing fears of communist subversion, the "drug war" in the mid-1990s, and even in the wake of 9/11 counterterrorism mantra, the FARC has not been recognized by Washington as an existential national security threat. In short, different US administrations through the decades have made it clear that Colombians are the ones who have to assume primary responsibility for the financing and maintenance of their own security. This was the basic assumption behind the formulation of *Plan Colombia* in 1999. So, given the political context and the unique circumstances of the conflict, the only policy option for Washington in terms of Colombia has been "small footprint" advising. It is an illusion to consider it a model that can be used elsewhere.

Conclusion

This chapter has offered a brief outlook of key contingencies that shaped US military intervention in Colombia. It began by highlighting a series of structural factors that illustrate the unsuitability of the "small footprint" approach in Colombia to inform what is happening in the Middle East. A fundamental point is that in Colombia, there is a clear absence of the ethnic or religious divisions that affect most Middle Eastern and African countries. Like most of Latin America, Colombia is in essence a Catholic/Christian country where the majority of the population subscribes to Western values; it also prides itself in having the longest record of unchallenged democratic rule in the region. In this respect, the Colombian people have not been patently resistant to Washington's political goals, and in particular when it extended *Plan Colombia* from a counternarcotics program to one of counterterrorism aimed at "the overall goal of helping Colombia become a prosperous democracy that respects human rights and the rule of law, free from narcotrafficking and terrorism" (US Department of State 2002).

Parallel to this, it is important to take into consideration that security assistance always happens in a unique historical and national context. In an evolving political and strategic landscape in which local priorities and influences predominate, security assistance also assumes the burden of "host" government weaknesses as well as strengths. Even the most accommodating of governments like that of Bogotá are shaped by their own underlying values and act in their own political interests, not those of the US. It is quite evident that when looking carefully at the balance sheet of the "Global War on Terrorism" after a decade of intervention in Iraq and Afghanistan, Colombia appears to show how, with little of taxpayers' money and no need for "boots on the ground," Washington was able to advance its policy goals by assisting a foreign government in dealing with its own security problems with the right tools and techniques. However, attention given to the Colombian experience as a model of "small footprint" assistance appears to leave out the evolution of its military strategic thinking throughout more than half a century of involvement in a protracted internal conflict. If anything, this chapter is a call for prudence before any attempts to use the Colombian case as a model for future intervention by the US by avoiding generalizations and exploring the case study within its unique historical and political context.

Notes

1 Examples of this literature include Stephen Watts, et.al, *The Uses and Limits of Small Scale Intervention*, RAND Co: Santa Monica, 2012; Linda Robinson, *The Future of US Special Operation Forces*, Council on Foreign Relations, Special Report No. 66, April 2013; Dana Priest, 'Covert Action in Colombia. U.S. Intelligence, GPS Bomb Kits Help Latin American Nation Cripple Rebel Forces,' *The Washington Post*, 21 December 2013.

2 The *Fuerza de Despliegue Rápido*—FUDRA is a 4,000-man-strong elite army unit with its own air support assets.

3 It is important to note that Colombia was the only Latin American country to send combat troops to fight under the United Nations banner in Korea.

References

Andres Schipani, 'Security for Export', 'The New Colombia' supplement, *Financial*

Bradley Coleman, *Colombia and the United States: The Making of an Inter-American Alliance, 1939–1960* (Kent OH: The Kent State University Press, 2008).

Colin Powell, 'Targeting Terrorist Leaders in Colombia', [Heavily Excised], Secret, Cable 008083, September 6 2002. Accessed via Digital National Security Archive on 7 July 2013.

David Spencer, *Lessons from Colombia's Road to Recovery, 1982–2010*, Strategic Issues in US/Latin American Relations, CHDS Occasional Paper May 2012 Volume 2 Number 1.

Douglas Porch and Christopher Muller, 'Imperial Grunts Revisited: The US Advisory Effort in Colombia', in Donald Stoker, *Military Advising and Assistance: From Mercenaries to Privatization, 1815–2007* (New York, NY: Routledge, 2008).

Douglas Porch and Jorge Delgado, '"Masters of Today": Military Intelligence and Counterinsurgency in Colombia, 1990–2009', *Small Wars & Insurgencies*, 21:2, (2010), 277–302.

Douglas Porch, 'Preserving Autonomy in Conflict: Civil–Military Relations in Colombia', in Thomas Bruneau and Harold Trinkunas, *Global Politics of Defense Reform* (Basingstoke: Palgrave Macmillan, 2008).

Foreign Office United Kingdom, 'Army of Colombia'. Code AL file 1201, Minutes, Confidential, FO 371, 1959, National Archives, Kew Gardens.

Jennifer Taw, 'Problems and Potential: The Role of the Police in the Colombian Insurgency', in Christine Fair and Ganguly, S (eds.), *Policing Insurgencies. Cops as Counterinsurgents* (New Delhi: Oxford University Press, 2014).

John Stone, *Military Strategy. The Politics and Technique Of War* (London: Continuum, 2011).

Jorge E. Delgado, 'Counterinsurgency and the Limits of State-Building: An Analysis of Colombia's Policy of Territorial Consolidation, 2006–2012', *Small Wars & Insurgencies* 26:3 (2015).

M. L. R. Smith and David Martin Jones, 'Grammar but No Logic Technique is Not Enough—A Response to Nagl and Burton', *Journal of Strategic Studies* 33:3 (2010), 436

Mark Moyar, Hector Pagan, and Will R. Griego, *Persistent Engagement in Colombia*, JSOU Report 14-3, The JSOU Press, MacDill Air Force Base, Florida 2014.

Paul Szoldra, 'The Most-Fascinating Insights from the Man Closest to General Petraeus During the Iraq "Surge"', *Business Insider*, 3 January 2014.

Sam Levine, 'Robert Gates Says Obama Will Need To Use Boots On The Ground To Defeat ISIS', *The Huffington Post*, 17 September 2014.

Stephen Dudley, *Walking Ghosts: Murder and Guerrilla Politics in Colombia*, (New York: Routledge 2004). *Times*, 4 June 2013.

Tony Pfanner, Nils Melzer, and Kathleen Gibson, 'Interview with Sergio Jaramillo Caro', *International Review of the Red Cross* 90:872 (December 2008).

US Department of State, 'Developing a Colombian National Security Strategy—The Strategic Roadmap for USG Cooperation', unclassified cable, 4 July 2002. Accessed via Digital National Security Archive on 7 July 2013.

US Department of State, *FY 2010 Mission Strategic Plan, Mission to Colombia*, 2. US Mission Strategic Plan.

USSOUTHCOM 'Operations Order for Plan Colombia', secret cable, excised copy, 5 December 2000, accessed via Digital National Security Archive on 7 July 2013.

11 The utility of force in cities

Future challenges of US Army urban stabilization efforts

Antônio Sampaio

Introduction

The history of armed conflict is tightly linked to humanity's gradual but inexorable movement towards urban centres. Some of the first large-scale clashes between organised armies recorded in written history took the form of sieges or battles for control of cities in Mesopotamia or Egypt around 5,000 years ago. Some of the earliest archaeological evidence of organised warfare comes from collapsed walls and sling bullets in Hamoukar, near the Syrian border with Iraq—evidence of heavy bombardment (University of Chicago, 2005). Its destruction is reminiscent of the battles that devastated some of Syria's biggest cities starting in 2011—an eerie reminder of urban warfare's enduring vocation for widespread destruction. Between Hamoukar and the twenty-first-century Syrian civil war, urban armed conflict has been nearly constant in human history. Perhaps because of the almost always devastating impact of urban armed conflict, the city has been almost unanimously understood in contemporary military documents as an unforgiving, risky, and dystopian geography to be avoided and feared. Fear has led to a reluctant approach to the complexity of the urban environment and ultimately to a shortage of depth in considering the utility of armed force in cities.

Recent attempts to rethink Western militaries' approaches emphasise the inevitability of operating in densely populated areas: "It is inevitable that at some point the United States Army will be asked to operate in a megacity and currently the Army is ill-prepared to do so," warns an Army Strategic Studies Group document (Harris et al., 2014, p. 3). It will come as bad news for the world's metropolitan dwellers (now over half of humanity) that most military thinkers believe that high-intensity urban battles are unavoidable, citing examples such as Stalingrad, Grozny (both Russian cities destroyed by war), and the block-to-block clashes in Fallujah, Iraq (Turse, 2007). The city as a "battlefield" has been a frequent feature in studies of this issue (for instance, Ashworth, 1991, p. 116). The UK Ministry of Defence's Future Operating Environment 2035 (UK Ministry of Defence, 2015, p. 24), after issuing several warnings about the "challenging" urban environment, displays an illustration of an infantryman holding a rifle overlooking a sprawling city with high-rising buildings burning and a thick column of smoke rising to a red sky.

Faced by such a dystopian future, military thinkers have recommended to forces in the US and other Western powers approaches that would shield their soldiers during raids into developing-world cities and allow them to "control villages or cities" (Desch, 2001), "shaping urban perceptions" and become a "social-engineering tool" (Glenn et al., 2007, p. XV). There has been a pattern of short-term focus to planning for metropolitan centres, arguing for technological and tactical adjustments needed to subdue and destroy the enemy's forces (Edwards, 2000, p. XIV).

However, a predominantly kinetic focus to urban operations overlooks the strategic importance of cities (a fact that many official military documents recognise in their introductions) and the increasingly endemic and protracted character of urban armed conflict. Whereas US forces were braced for a fierce urban resistance and street fighting in key Iraqi cities during the 2003 invasion, that fear did not materialise (Hills, 2004, p. 8). Instead, the state enemy dissolved and different, nonstate, threats arose months later to pose a protracted challenge. The Iraq context illustrates the emerging shape of urban armed conflict in the twenty-first century: protracted, labour-intensive, and multidimensional security efforts such as peacekeeping, stabilization efforts in support of fragile states, counterterrorism, and anti-crime pacification. The enemies, in turn, have not only been native to the cities where the conflict is being conducted but also they often do not need to leave their own neighbourhood to conduct disruptive attacks on state security forces. This is the case with urban militias, improvised rebel forces and highly violent gangs—increasingly common actors in violent settings such as Mogadishu, Bangui, Port-au-Prince, Karachi, and Rio de Janeiro, among other developing-world cities.

This chapter looks at the strategic level of armed action in urban environments. The main argument is that the utility of armed force in densely inhabited metropolitan areas (large and mega-cities) has been reduced as global urban complexity and population has increased. A short-term focus that emphasises continuation of historical uses of force no longer achieves meaningful results. Army operations in cities are now immersed, sometimes unwillingly, into an urban strategic environment that interacts as part of a system—around which a variety of tools for statecraft have to be orchestrated in order to steer an undesirable situation (violence, instability, nonstate armed threat) into a more acceptable balance. The chapter will explain how recent trends point to the need for an expanded temporal and spatial focus—the before and after of operations, the challenges and spill-over effects of unstable peripheries. First it is important to take a brief look at the strategic context underlying urban operations and then at the evolution of military thinking on urban areas.[1]

Expanding the strategic horizon

The rising tempo of armed conflict in urban centres has had profound implication for the understanding of modern conflict and responses to it. Cities are front and center at the "hybrid war" concept—the trend towards asymmetry and nonstate armed action spanning different modes of warfare. According to its main proponent, Frank Hoffman, hybridity's most likely setting is "the burgeoning cities of the developing world"—which he calls "contested zones"—featuring complex terrain chosen by adversaries facing technological and firepower disadvantage in order to "draw out conflicts, protract their duration and costs and sap American will" (2007, p. 15). Humanitarian organizations tasked with responding to the social consequences of war have picked up these trends: the International Committee of the Red Cross estimated in 2015 that 50 million urban inhabitants worldwide are affected by armed conflict and laid out guidelines to understand humanitarian impacts of progressive deterioration of essential urban services (ICRC, 2015a, p. 5).

Despite stressing the increasing frequency and complexity of the problem, Western militaries have not fully embraced the specificities of operating in living, breathing cities. There has been a disconnect between the assessment of urban armed conflict as an increasingly protracted, endemic, and socially complex endeavour and the limitations self-imposed by strategic thinkers on how to act there—forming a recurrent emphasis on battle tactics, zero-sum calculations and technology as a means to increase lethality. Sustained military usefulness in such places requires an embrace of the concept of the urban strategic dimension as a separate realm of strategic action in the twenty-first century, requiring very different approaches, which may sometimes depart from those derived from Clausewitzian and mostly rural understandings of warfare.

General Sir Rupert Smith, for whom this relates directly to state strategy, has most prominently addressed the problematic of "utility of force" at the strategic studies literature. "It is necessary," he wrote, "to understand that in many of the circumstances into which we now deploy, our forces as a military force will not be effective" (2006, p. 10). Governments and militaries, he added, are still thinking on old paradigms that lead them to apply "massive and impressive" force, but the "utility of force" is minimal (Ibid., p. 5). What is beginning to be articulated by military and nonmilitary writers such as David Kilcullen, Alice Hills, and Michael Evans is that the same miscalculation has consistently plagued recent military deployments in urban areas. Hence the need to consider the broader social, political, and economic dynamics that affect the urban environment as a site of strategic action—strategy being understood as "an expression of the aim and its links to the overall purpose and the context of the conflict, together with the limitations on action that flow from that political purpose in the circumstances" (Ibid., p. 13). Rupert Smith also notes that a good strategy "achieves its object without the necessity for battle" (Ibid., p. 13). The application of the principles and implications of a revised understanding of military strategy in the twenty-first century should be articulated alongside the understanding of the general characteristics of development, stability, and the overall functioning of large and mega-cities.

Urban forms of armed conflict are defined by dynamics linked to the nature of cities as systems, whose functioning is determined by delicate relationships between several actors, areas, and services. The utility of force in cities, therefore, should be reconsidered in accordance with the understanding of the urban space as a complex strategic environment. This argument implies two changes from mainstream thinking on urban military operations: on one hand, it implies a restricted utility of military power (understood as kinetic, usually high-intensity, operations and methods) to achieve objectives in densely inhabited areas. It reinforces the need for cooperation with a wide range of nonmilitary actors. On the other hand, it expands the toolbox available for national and local policy makers in stabilizing violent areas through the calibration of security and development means.

If the assessment that armed force has a limited role to play in most urban contexts sounds disconnected from the real-world challenges facing the US or any other Western military power, it is worth quickly considering another field of military activity. Few concepts have encouraged more sophisticated approaches to strategy than that of information operations. In the early 2000s, following a broad consensus in the US government that information and cyber space were an "emerging domain," the Army's Information Operations Doctrine, FM 3-13, removed the term *information warfare* and replaced it with a much broader concept, the *information domain*: "The information environment is the aggregate of individuals, organizations, or systems that collect, process, or disseminate information; also included is the information itself" (US Army, 2003, p. 1–2).

The broadening of the conceptual lens for how to deal with the social construct of information as a complex environment has meant a redefinition of "the entire spectrum of information operations" (Rid & Hecker, 2009, p. 59]. It has meant a constant awareness of "the entire information environment" and how missions affect and are affected by it (US Army, 2003, p. 1–2). This analytical perspective has expanded the usefulness of the literature on information operations for state strategists by providing a context in which information operations "activities fit in as part of a whole-of-government approach to effectively shape the information environment" (Joint Chiefs of Staff, 2014, p. IV-11).

The embrace of complexity and whole-of-government approaches should be on our minds as we consider the specificities of another manmade space, the urban. The point here is not to compare approaches in very different domains but to contrast the widely accepted embrace of complexity and coordination in the mainstream security and defense literature on information operations and the relative lack of nuance in understandings of what the complexities of the urban environment entail (which is just now beginning to be remedied, as will be demonstrated below).

The new urban environment

The recent impetus for military thinking on cities is linked to the unprecedented speed in urban population growth in the developing world. Developing countries will concentrate almost all of the 2.5 billion new urban dwellers expected by 2050 (United Nations, 2014, p. 1). Africa, beset by various types of armed conflict, will see a threefold increase in its urban population by then (Ibid., p. 11). Some fragile and conflict-afflicted countries, already struggling to implement governance and manage the demands of metropolitan citizens, will be at the forefront of this trend—the cases of Nigeria, the Democratic Republic of Congo, and Pakistan (Ibid., p. 12). The global population living in slums will grow by 500 million people between 2013 and 2020 (International Bank for Reconstruction and Development /The World Bank, 2013, p. 12). The urban population skyrocketed by 500 percent between 1950 and 2014—and it is expected to almost double until 2045, reaching 6 billion inhabitants (United Nations, 2014, p. 1).

Another, less studied, urban trend is the centrality of large and mega-cities for the management and coordination of the global economy—what Saskia Sassen calls "global cities" (2005, p. 34). Admittedly, these crucial hubs are predominantly in developed countries and are therefore not usually affected by high population growth rates common in developing-world cities. However, the quickly expanding and frequently violent cities of the developing world have also carved their places as hubs for trade, information, and human flows through the so-called South–South networks. Urban areas not only account for 80 percent of the global production of goods and services (The World Bank, 2013) but have also been identified as crucial sites for the economic development of low-income countries—a classification that includes many of the fragile settings facing internal armed conflict (Spence, Annez, & Buckley, 2009).

This unprecedented growth in size and strategic importance of metropolitan areas has had several spatial, cultural, and social impacts that are relevant for the purpose of applying armed force–for instance, mega-cities (those with 10 million or more inhabitants) are now sprawling agglomerations where specific trouble areas cannot be easily isolated from their surroundings or controlled through a "loose cordon," as Hahn II and Jezior have suggested (1999). Nor can they suffer their services, economies, and human flows to be disrupted, either by state military action or nonstate attacks (such as terrorism) without severe risk of political ripple effects, economic impacts, and social disturbance.

None of these trends are inherently threatening to national or international security. But they do correlate with worrying evidence on the weakening of state presence in dense slums or peripheries of large or mega-cities that have faced fast and unmanaged growth. David Kilcullen, who has vast experience in advising the US Armed Forces in Iraq, says "no-go areas" that tend to emerge in the gaps of state authority in peri-urban zones "allow safe havens for criminal networks or non-state armed groups," sometimes acting on grievances linked to their current situation, other times importing older sources of tension from rural areas (2013, p. 40). The diagnosis of an environment in which different sources of tension pile up and erupt into violence seems to be backed up by a variety of sources, which frequently point to this trend as a challenge to development and humanitarian work. UN Habitat has warned of the alarming frequency of "autonomous and no-go areas" in slums and other informal settlements, lying "outside local and central government control" and constituting both a security and developmental problem (UN Habitat, 2014, p. 35). The term *fragile city* has been used by the World Bank and the United Nations to encapsulate dysfunctional cities facing what Robert Muggah calls "endemic urban violence," brought about by the aggregation and accumulation of risk factors that "overwhelm local coping systems and reshapes identities, agency and social relations" (2012, p. 39).

The preoccupation, stated in strategic documents from several Western military powers, with the rising likelihood that they will participate in urban operations seems to be one firmly backed up by recent data. The growing social and economic importance of cities as hubs for mega-populations and economic activity, accompanied by dysfunctional governance, have preoccupied communities of practice in the development, humanitarian, and security fields. This preoccupation, in turn, has been enshrined in the United Nation's agenda for Sustainable Development Goals (chosen in 2015 to replace the Millennium Development Goals): Goal 11 directly links security with the concepts of inclusiveness, resilience, and sustainability as key global challenges for the next 15 years. What is not clear at all, however, is how militaries will calibrate responses to these trends.

The dark dimension of war

The difficulties of the urban environment for state combatants are legendary. For Sun Tzu, they were places that the good strategist would do best to avoid.[2] Despite the military search for increased speed and lethality, dense, chaotic, and sprawling developing-world cities have often defeated modern armies. In a comprehensive study of urban warfare by the US Strategic Studies Institute from 2011, the urban is described as "not a happy domain": army companies fighting to take control of city blocks take an average of 30–45 percent of casualties during a 12-hour fight. Such casualties and stress essentially render the company useless in the aftermath of the battle (Posen, 2001, p. 153); hence the high financial, human, and logistical costs of classic urban warfare. This historical consensus casts a shadow on urban military thinking. Titles of past studies by US think tanks provide ample evidence, which can be depressing or amusing depending on how the reader interprets the likelihood of their predictions to come true: "Dark, Dense, and Dangerous" (Hoffman, 2013), "Mars Unmasked" (Edwards, 2000), "Combat in Hell" (Glenn, 1996).

In the 1990s, the efforts to understand future urban operations relied on the promise of technology. In 1999, the US Army put forward the project for an Urban Warfighter System, with the aim of achieving "the transformation of the current infantry soldier into a truly lethal urban warrior, equal parts man and machine" (Hahn II & Jezior, 1999). An article in the US Army War College, ambitiously setting out the "urban warfighter of 2025," carried the thesis that "high-technology weapon systems will fundamentally alter the course of urban warfare" (Ibid.). Understandably, this technologically minded planning for future operation trickled down to independent analysts, who tended to emphasise the role of technology in isolating and shielding infantrymen from the urban chaos, as reflected in this 1999 Strategic Comment:

> The US military's future urban operational concept thus recommends the creation and use of relatively small units equipped with high-technology systems. When necessary, high-speed armoured vehicles will move forces into or through urban areas to defeat pockets of strong opposition. To locate and destroy enemy positions, while minimising American casualties, will require heavy reliance on unmanned aerial vehicles, unmanned ground vehicles (UGVs), a wide variety of sensors and an improved "urban version" of the Land Warrior System (LWS).
>
> (IISS, 1999)

The results were disappointing. The first large-scale exercise to test high-tech gear designed to finally provide soldiers the upper hand in urban warfare, the Urban Warrior Advanced Warfighting Experiment, conducted by the US Marine Corps in March 1999 in California, indicated casualty rates of up to 70 percent in a few hours (Ricks, 1999). The reality then began to sink in: "We didn't come up with any silver bullets" then-Marine Corps' chief of staff Col. Gary Anderson told the *Wall Street Journal* (ibid.).

Since then, the emphasis on fabulous technologies coming "to the rescue" of militaries in cities has greatly diminished in strategic documents. The term *Military Operations on Urban (or Urbanized) Terrain* (MOUT) (US Marine Corps, 1998) has been used to describe tactical practices emphasizing deadly battles as the predominant shape of future urban operations. One of the studies pointing out the urgency of adapting to the new urban challenges, entitled "The Changing Face of Urban Operations," carries a grim conclusion: "Killing the enemy's troops will probably remain the most efficacious way to defeat" enemies' will to fight in urban terrain (Edwards, 2000). Alice Hills reminds us that the US military itself, in its Joint Urban Operations (JUO) doctrine from 2002, considered that "MOUT has strong connotations of urban ground combat at the tactical level" (Hills: 2004, p. 6).

Conceptual changes later took place that emphasised the multidimensional character of urban security operations, intended to equip the local-level commanders with the versatility, autonomy, and rules of engagement to be able to defeat adversaries in the hostile city of the future. Since the city is the site of complex and protracted forms of armed conflict with nonstate groups, the concept of the "Three-Block War" gained traction during the 2000s to explain the variety of military challenges in such areas. Its proponent, Gen. Charles C. Krulak, described it as "contingencies in which Marines may be confronted by the entire spectrum of tactical challenges in the span of a few hours and within the space of three contiguous city blocks" (1999). The missions associated with this view of the multitasking urban warrior are high-intensity combat, humanitarian relief, and peacekeeping (including mediation between warring factions), all within three blocks (Boot & Kirkpatrick, 2006). A fourth block, information operations, was introduced years later (Mattis & Hoffman, 2005). But something wasn't quite right. The concept gradually disappeared during the second half of the 2000s. The Canadian Army had made the concept one of its fundamental strategic concepts following the apparent traction it had gained with its US counterpart, but dropped it after a few years (Dorn & Varey, 2009, p. 43).

The attempt to tilt the military balance towards state forces by adopting an "everything-at-once" approach misunderstood tactical versatility with strategic understanding. Above all, it brought together types of mission—humanitarian relief and peacekeeping—which had originated and been more comfortable outside of purely urban contexts, neither of which fit well in the same block with high-intensity combat. In summary, and as a critical review of the issue later stated, "by emphasizing the means (warfighting), the Three-Block War concept risks losing focus on the goal (peace)" (ibid., p. 40). Indeed, fitting three different types of mission into one doctrinal concept does not solve the challenge of understanding the role that armed force can play or what it can eventually achieve in urban environments.

Flexible space?

One of the world's most experienced urban military forces has developed further the understanding of how social aspects such as architecture and population density affect strategy in the metropolitan space. Israel has had to adapt to the reality, dreaded by colleagues elsewhere, of predominantly urban-based forms of armed conflict. In the beginning of the twenty-first century, its operations have been based in densely and utterly chaotic urban zones such as Gaza City, which registers one of the highest population densities in the world; and the refugee camp city of Nahr al-Bared, which illustrated the new urban forms arising from refugees and wars in the Middle East (Shteiwi, 2013).

Urban combat in Israel's recent wars has again displayed the destructive vocation of this form of conflict, as well as the perennial capacity of nonstate armed groups to adapt to densely built environments. In the words of the commander of an Israeli Armoured Corps school training cadets for urban warfare, they are up against a "vanishing enemy," operating entirely on foot, which makes it harder to identify and even harder to strike; "if you miss, the enemy vanishes" (Lappin, 2014). The Israeli Defence Force (IDF) has also understood that the nonstate armed forces that they have been up against (especially Palestinian Islamist group Hamas and the Lebanese Hezbollah) have almost perfectly fused with densely built areas, making the entire urban sprawl its base and exploring its three-dimensional aspects. Hamas, for instance, reportedly employs 1,000 operatives to excavate tunnels in the Gaza Strip in their perpetual preparation for Israeli assaults (The Jerusalem Post, 2016).

In response, IDF military thinkers came to the conclusion that adaptation to the urban environment also means, on the state military's side, adapting to a "flexible space." As an officer in charge of operations against terrorist groups in the Palestinian city of Nablus explained, the IDF adopted an "inverse geometry" approach, which Israeli architect Eyal Weizman called "walking through walls" (Weizman, 2005): space, the officer said, "is nothing but your interpretation of it":

> We interpreted the alley as a place forbidden to walk through, and the door as a place forbidden to pass through, and the window as a place forbidden to look through, because a weapon awaits us in the alley, and a booby trap awaits us behind the doors. This is because the enemy interprets space in a traditional, classical manner, and I do not wish to obey this interpretation and fall into his traps. Not only do I not want to fall into his traps, I want to surprise him! This is the essence of war. I need to win.
> (Weizman, 2005)

Walls, therefore, can be destroyed and turned into pathways. Explosives and large hammers were then used to excavate these routes through shops and homes, with families being forcefully locked into one room and left there for hours, sometimes days. This has been seen as a relatively gentle form of warfare in comparison to the destruction inflicted on the Palestinian refugee camp of Jenin in 2002, during which 10 percent of the buildings were destroyed and thousands of people died in what the IDF classified as "the suicide bombers' capital" (Beaumont, 2002).

Israel's emerging interpretation of the urban built environment is flexible in a highly kinetic way. It understands that achieving power in the urban environment means adapting your methods to more effectively address its social and spatial context. But it fails to understand that the military-insurgent/terrorist relationship is not the only one occurring in the operational area (which very often encompasses entire cities with millions of inhabitants). Operation Defensive Shield, under which both the Nablus and Jenin offensives were conducted, was considered a success in seriously damaging terrorist infrastructure, apprehending weapons and explosives and killing hundreds of militants (Alon, 2002).

From the operational level, therefore, the actions were fairly successful. However, the warning made above about a confusing demarcation between the means (war) and the goal (peace) seems applicable to the "flexible space" experiment, just as it was for the Three-Block War. This operation, and similar ones conducted in the Gaza Strip in 2008–09, has been followed by recurrent Israeli urban incursions, reinforced by checkpoints (especially in the West Bank) and fortified settlements. On the other hand, terrorism seemed to diminish in the years following the 2007 and 2008 operations in the West Bank and Gaza, but then it rebounded. The Global Terrorism Database reports 330 terrorist incidents in the Palestinian Territories between 2008 and 2014, with a sharp rise starting in 2012. The Palestinian Territories and Israel have both been hit by rockets and terrorist attacks. Operational success, therefore, has not fully translated into longer-term stability nor human security.

Conflict loop in cities

The resort to recurrent incursions of the same urban areas over and over again has been a long-term pattern in some cities. Western militaries, and the US in particular, have been caught in recurrent loops of conflict and stabilization duties in urban terrain. In Iraq, the Army and the Marine Corps have been locked in unexpectedly long-lasting urban missions—for instance patrolling, repelling nonstate enemies, and sometimes even reconquering cities. These operations were placed within the framework of counterinsurgency but required forces to secure densely inhabited areas for years in a highly urbanised country with 70 percent of its population in cities (Central Intelligence Agency). Fallujah, the site of not one but two ferocious battles in 2003 and 2004, was patrolled for years afterwards by the US Marine Corps—only to be taken by the so-called Islamic State in 2014. More than 10 years after the US-led invasion of Iraq, urban warfare has returned to the agenda because the Islamic State targets cities as both hubs for trade and fiscal income as well as symbolic sites of "state" authority and power (Giovanni, Goodman & Sharkov, 2014).

Commentators and US military authorities have also observed the growing militarization of anticrime efforts next door, in Latin America. A city that came to symbolize the "conflict loop" in urban areas is Rio de Janeiro—where urban warfare became the status quo approach for decades with the military police's frequent raids into crime-controlled slums. The ideological "pull factor" is not present when it comes to gangs' "criminal enclaves" (Sullivan, 2011) as it is for the Islamic State. Instead, criminal gangs fight turf wars to wrest key slums in order to control profitable illegal revenues, coordinate criminal operations, and connect transnational criminal operations (IISS, 2015). Nevertheless, both in Rio and in Iraq, military forces were repeatedly called upon to break armed resistance and secure urban areas from nonstate armed groups. Current demographic trends only reinforce the likelihood of the "contested" character that have characterised cities throughout history to continue, but this time with "police, military, criminal, and paramilitary forces" being thrown in the mix (Sullivan & Elkus, 2011).

Despite the pattern of "conflict loop" and long-term deployments in urban spaces, there has been little depth and ambition in military attempts of preparing for these missions. The US Joint Urban Operations doctrine published in 2013 makes great strides in understanding that the "entire urban environment must be addressed simultaneously and systematically" (Joint Chiefs of Staff, 2013, p. VIII). However, while emphasizing "joint operations," the document's remit is still mostly on ground-level commanders and what it calls "battlespace." For instance, disrupted public utilities are analyzed according to the obstacles they create to firepower or the way that an enemy can disrupt or contaminate the water flow in order to harm the state actor's cause (Ibid., p. II-8). This is understandable because, as a military doctrine document, it should not be expected to broaden its focus towards a holistic, whole-of-government approach. Besides, as Alice Hills reminds us, revision was not a priority for

US forces given the fact that recent wars and foreign operations have either avoided high-intensity urban battles or resulted in operational (although not necessarily strategic) success, such as in Baghdad (Hills, 2004, p. 24). This reinforces the view, cited by Hills, that "a 'premium' is placed by today's military on flexible force structures capable of dealing with the unexpected" rather than understanding the nuances of urban environments [Ibid., p. 8].

Therefore, we are stuck in an ironic situation when it comes to the role of armed forces in cities: whereas numerous military and academic documents point to an increased likelihood of military missions in urban environments of heightened difficulty, a strategy-level revision has not occurred. Instead, concepts such as the Three-Block War, flexible spaces, and programs like Urban Warrior have marked military thinking on cities. They have in common a focus on articulating the need for and conduct of highly kinetic operations to defeat enemies in cities. This is an important focus for armed forces in specific circumstances, for instance, the capture of a major city in a friendly country by a powerful threat such as an extremist/terrorist group (i.e., the Islamic State in Iraq). There is no evidence that this type of threat and the immediate short-term need for military action of that type will cease to exist anytime soon. But this focus no longer feels sufficient given the tendency towards protracted low-intensity urban conflict in rapidly urbanizing regions and the resulting calls for adaptation of traditional policy tools to this changed geographic and demographic context.

However, whereas some patterns of urban warfare remain remarkably constant throughout time, there is an emerging consensus that the role of cities for military strategies has changed. As Alice Hills has highlighted, Western militaries frequently "rediscover that there is very little that is fundamentally new about urban war," but cities are now linked to broader military goals that are not advanced by destroying densely inhabited areas (Ibid., pp. 24-25]. Hills cites the case of the Israeli invasion of Lebanon, during which rural operations went well but political defeat started to take shape when television coverage showed the bombardment of Beirut (Ibid., p. 14). A related problem was raised by the US Strategic Studies Institute in a review of modern urban battles. The history of MOUT has been marked by failures or, sometimes, by "successes" with disastrous consequences (Desch, 2001, p. 6); the French managed to win the 1958 Battle of Algiers in Algeria after widespread violations of human rights, and the Russians conquered the Chechen city of Grozny in 2000 after essentially depopulating it and destroying a great deal of it (Ibid., p. 6).

The urban strategic environment

A growing number of voices have been calling for a further shift in paradigms. The International Committee of the Red Cross has highlighted the incapacity of armed forces in most urban conflict situations to avoid the use of explosive weapons with a wide-impact area (2015)—the same systems that are widely used in open battlefields. Ironically, it agrees with most military and nonmilitary urban thinkers in diagnosing a rapidly rising curve of urban warfare "over the last century [...] thereby exposing civilians to greater risk of death, injury, and displacement" (Ibid., p.2). But it detects that these "conflicts continue to be waged with weapon systems originally designed for use in open battlefields" (Ibid., p. 2). The use of wide-area impact explosives in open battlefields brings significantly different questions than when they are used in densely populated areas. For instance, in addition to the immediate humanitarian impacts of dropping a bomb, there are larger and longer cumulative impacts of direct and indirect effects on urban services (ICRC, 2015b, p. 9). This also carries significant consequences for a military commander: if employees of critical municipal or humanitarian agencies fail to show up for work because of conflict-related insecurity (an indirect impact of military action), services and aid relief are likely to diminish, public health deteriorates, and popular support for a government or external force is likely to suffer.

There are also indications that Western militaries will not be the sole actors, or even the dominant ones, in foreign deployments to cities. There have been calls for a doctrinal rethink of concepts such as stabilization and peacekeeping in order to address their growing overlap with urban missions. These missions emphasise whole-of-government approaches, a deep understanding of the holistic character of missions and are sometimes led by civilians linked to international organizations such as the United Nations or the African Union. The framework of stability operations has been frequently cited in connection to urban areas. Pickett and Best (2014) argued in an article titled "Stability is the Best Offense" that the US Army has the tools and resources to stabilize "ungoverned but globally-connected cities" but remains "blind" to economic and social root causes of armed threats. Others have warned against a rigid interpretation of stabilization and its proposed application at complex and context-sensitive cities as if they were "*terra nullius*" (Muggah & Jütersonke, 2012, p. 322). More recently, voices within the United Nations system began to highlight the almost shocking fact that the 104-page-long "comprehensive review" of peacekeeping published in 2015 did not mention the words "urban" or "cities" at all (Bosetti et al., 2016); despite the occurrence of deadly urban violence, gunfights, high-profile kidnaps and situations akin to urban warfare in capital cities of major peacekeeping host countries such as the Central African Republic and Haiti.

At the same time, there seems to be growing criticism towards the narrow scope of the prevailing military thinking on cities—coming sometimes from mainstream strategic studies voices. Michael Evans, a former army officer and leading thinker on urban armed conflict, has criticised the aforementioned report by the US Army's Megacities Concept Team for focusing its urban studies too narrowly on megacities as essentially dangerous and dystopian sites that will inevitably become key battlefields of the future. Instead, he says "the quickest way to degrade American combat power will be to deploy large numbers of troops into a megacity without a thorough examination" of complex trends (Evans, 2015, p. 42). Similarly, Alice Hills emphasises in her comprehensive book on Future War in Cities the ample offer of tactical analyses available involving the urban space, but a "reluctance of governments (as opposed to insurgents or terrorists) to acknowledge, let alone explore, the potential strategic utility of cities" (Hills, 2004, p. 223). Both authors stress the importance of a first step consisting of a widening of the analytical lens in order to draw from disciplines beyond defence and security.

Equally problematic seems to be the tendency of interpreting the recurrent pattern of urban armed conflict—the conflict loop—as a circumstantial detail of broader military strategic endeavours such as counterinsurgency (as mentioned above in the case of Iraq). David Kilcullen has highlighted this unidimensional (or undisciplinary) tendency in Western militaries, noting that it is linked to institutional reasons: it is more useful for government, military, and law enforcement agencies (as well as for some academics) to refer to "theories of conflict framed around a single threat—insurgency, terrorism, piracy, narcotics, gangs, organized crime" (Kilcullen, 2013, p. 16). That is where the current narrowness of military interpretation becomes most visible because, as Kilcullen puts it, in the real world of complex, adaptive, and social systems that characterise cities, "there has never been, and never will be, a single-threat environment" (Ibid., p. 17).

The approach of Western militaries to urban warfare so far, therefore, has led to increasing tensions between the short-term, tactical, and compartmentalised military thinking and the complexity of cities. "Complexity" is used here to refer to the linkage between several demographic and security trends pointed out above—reflected in recent academic criticism towards the perceived narrow focus of recent military explorations of the urban environment.

The status quo of the study and conduct of urban warfare by Western militaries, therefore, predominantly considers the city a battlespace—an area for the application of previously designed tactics and ready models of operations, often with an aspiration to increase the lethality of state militaries against a non-state enemy. In other instances, such as when US-led coalitions in Iraq and Afghanistan were taking place, counterinsurgency teachings were applied across the board. This approach carries in its heart the principle of isolating and defending areas from insurgents (Ucko, 2013)—something that is impractical in dense and sprawling areas, not to say counterproductive (would the population of a neighbourhood or slum support the actor that is cutting them off from transportation networks, relatives, entertainment, and job opportunities?). As was pointed out above, combat tactics and counterinsurgency analyses are important and should be in dialogue with studies of the urban environment. But their application in recent theatres seems to have been motivated less by careful strategic thinking and more by a scarcity of—or reluctance towards—broadened perspectives.

The role of developing-world cities in foreign military interventions has fundamentally changed: from a battlespace to a strategic environment. The term *strategic environment* is not a well-defined one. It has been used mostly to define the global-level geopolitical context. It has also been defined as the "realm in which the leadership interacts with other states or actors to advance the well-being of the state." The relationships, opportunities, threats, and interactions being exchanged between actors within a strategic environment "influence the success of the state in relation to the physical world, other states and actors" (Yarger, 2006, p. 17). As systems, cities are socially constructed spaces, with their densely built areas being a material envelope for a deeper set of human and institutional connections that are the true drivers of the success or failure of any given security or stabilization mission.[3] These connections are now of such importance that ensuring a smooth running of them is a strategic imperative to secure the well-being of a state or the longer-term sustainability of any stabilization, peace-building, counterterrorism, counternarcotics or warfare effort—all of which have traditionally employed military force to some extent.

Cities, as we have shown above, have been a frequent stage for operations intended to defeat nonstate armed groups. But through the intervention into what is essentially a delicate social system, the military force also becomes intertwined with this system and affected by its changes—its ability to control, pacify, or stabilise the environment is influenced by how stable or otherwise the metropolitan area has become since its intervention. In other words, choosing to operate in cities also engages armed forces in a struggle to secure and stabilise disputed areas—most often disputed peripheries—a task that requires a variety of security, developmental, humanitarian, and political tools.

This growing tension between prevailing military approaches and the social complexity of urban areas implies a disconnect between the tools and the objectives of states in the urban strategic environment. The tactical approach so far has focused on the short-term mission needs—enter, clear, patrol, build. The more frequent reality—and likely scenario for the future—is that of protracted situations coming closer to (but rarely fitting precisely into) stabilization, peace-keeping, enforcement, and other forms of low-intensity conflict with nonstate armed groups. These circumstances imply a close contact with, and sometimes even direct responsibility over, the nitty-gritties of the functioning of the metropolitan system. The main point is that the urban environment requires an expansion of the prism of analysis.

The flexible utility of force in cities

The utility of force in such environment is changing and becoming more fragmented rather than clear-cut. It is now linked to a more complex system, which implies longer deployments, more attention to geographical nuances, and flexibility in interpreting the role of military power. In the absence of an effective method of achieving classic battlespace victories or counterinsurgency-style isolation and control of populations without disrupting the urban system, governments and militaries have to be prepared to expand the time-frames and the toolbox for operating in cities.

In order to interpret the utility of military power in today's urban environment, the social dynamics of the city are the key element of analysis. This is different from approaching the urban space as a predominantly physical or topographical concept that underlines "everything-at-once" approaches such as the Three-Block War or the search for increased lethality in the US Army's defunct Urban Warrior project. These approaches have often recognised the complex social dynamics in cities but have sought to bypass or overcome them by prescribing techniques to adapt existing concepts to a densely inhabited area. For instance, the US Joint Doctrine on Urban Operations acknowledges that "Operations are time-consuming" (Joint Chiefs of Staff, 2013, pp. I-5-I-6). But a closer look at this bullet point reveals that the unit of analysis in that case was the battle (Stalingrad, Khorramshahr, and Hue were cited) and the time-frame involved was the beginning and end of a particular clash. Again, there is nothing wrong with this approach, which may well prove useful for Western militaries. But the trends and observations noted above point to other, more socially embedded, dynamics as becoming increasingly relevant to the point of justifying a change in the calculations of armed forces.

The logical next step would be to draw a more in-depth understanding of cities as strategic sites for state interests in the twenty-first century. Even though full-scale urban warfare remains a relevant and useful concept, the urban space will likely be the stage for many tiers of policy objectives that have involved the use of military power to several degrees. Military power, in other words, can be useful as part of a broader state strategy in order to guide targeted troubled communities into stability, peace and even prosperity. The focus on the urban space as a strategic environment, where several options are available to states and their militaries, would allow for important shifts in current mainstream thinking.

First, it would allow for the expansion of the time horizon used in conceiving security and defense issues in urban areas. The urban security landscape (or "threat environment" to use a military jargon) should not be conceived of in terms of peace and conflict, good guys or bad guys. The availability of different degrees of military power can be useful in helping civilian authorities to prevent conflict and, at other times, to stabilise areas where violence has been reduced. Many cities, especially in fragile or post-conflict countries, will see gradually escalating armed violence, brought about by a variety of spoilers, militias, vigilantes, and even criminal groups—some of which might not be immediately identified as a threat but could contribute to destabilization. This has been the case, for instance, of ethnic militias in Bangui, Central African Republic; of gangs in Port-au-Prince in Haiti; and of ethnic and criminal violence in Karachi in Pakistan—unconventional urban conflict situations that eventually led to military interventions.

In some ways, such tasks are considered within stabilization or peacekeeping missions. But as was pointed out above, they lack the explicit articulation with urban-specific fields of policy making, such as resilience, design, planning, and service provisions—although linkage is growing with other fields such as humanitarian action and policing. The most important strategic consideration is to recognise the prolonged aspect of current urban armed conflict and input a time-sensitive and gradual approach that evolves through different stages of intervention, from the early signs of significant armed escalation through the arrival of the military force to gradual political stabilization and economic recovery.

Second, in situations of ongoing armed violence, it is possible to think of targeted interventions focusing on problem areas—disputed or unstable peripheries. These areas are not only more frequently affected by serious armed violence but also tend to accelerate violent spill over into the broader urban system—which partially explains the rise of "gated communities" resorting to walls to separate themselves from surrounding ganglands or other forms of armed violence. Peripheries here are meant not solely as areas geographically distant from city centres but also as areas economically—and socially—marginalised and underserviced in comparison to more affluent areas—the case of slums and other informal agglomerations.

Focusing on unstable peripheries can allow for a manageable intervention before escalation and a spillover of violence over larger areas of the city. Medellín in Colombia and Rio de Janeiro in Brazil have applied such approaches to specific slum areas, using armed force as an initial robust state entrance into guerrilla- or gang-controlled areas, tightly linked with peacebuilding and urban improvements such as libraries, parks, and cable cars linking the cities to the broader metropolitan infrastructure and economic opportunities. Both cities have had problems mustering resources for what is a labour-intensive and prolonged deployment close to peace enforcement and stabilization tasks. But both cities have seen important reductions in their homicide rates (Sampaio, 2015). As John Sullivan and Adam Elkus (2011, p. 7) have argued, there is a temptation among military planners to focus on and prepare for cordoning off slums and "control them through periodic raids"—something that, as they argue, would amount to temporary attempts to alleviate violent situations. Instead, evidence from these and other developing-world cities point to unstable peripheries responding better to interventions with a light military footprint and a comparatively heightened developmental component (Muggah, 2015).

The application of Western military power is of limited utility and has different effects depending on the context of the security challenge being considered. In order to allow for a broadened time horizon and more flexible policy options, civilian and military intelligence should jointly attempt to conceive early warning systems in highly fragile or threatened environments. The focus on unstable peripheries could allow a more targeted and doable approach. An early and targeted approach allows for more flexibility in aiding local and national governments, as well as international forces, to apply tools such as training, security-sector reform, building urban resilience, planning, design, and technology as tools to bring a whole-of-government approach into the urban security and defense effort.

Underlying these changed approaches is the imperative of minimizing—or reducing the likelihood of—full-scale warfare in cities. This is the emerging utility of force in the twenty-first century's global urban environment. Changed demographic and conflict dynamics observed in cities have led to a gradual (but slow) acceptance of the goals of stability and peacebuilding as being key strategic goals in urban areas—which military power can sometimes contribute to and expand by supporting transitions to policing and other civilian bodies.

This is a longer-term approach. But it is not something detached from the current role of militaries. The occurrence of unstable peripheries with their nonstate armed groups is, as we saw above, a challenging feature of developing-world metropolitan areas. Armed forces have been and will continue to be involved in securing such areas. But accepting the key goal of stability, conflict prevention, and long-term peace rather than the classic—and improbable—military victory imperative (understood as the complete defeat of a well-defined enemy) looks like a more suitable approach to today's complex and dense metropolitan environments.

Notes

1 Emphasis is placed on the US and other Western military forces—the latter is understood as those of western Europe and Israel.
2 "Laying siege to a city is only done when other options are not available" (Sun Tzu (translation), 1999).
3 Urban system comprises "a city represented as a set if interacting subsystems or their elements," with those subsystems being held together by feedback dynamics (Batty, 2008, pp. 3, 5).

References

Alon, N. (2002, April 10). Operation defensive shield: The Israeli actions in the West Bank. The Washington Institute. Retrieved from http://www.washingtoninstitute.org/policy-analysis/view/operation-defensive-shield-the-israel-actions-in-the-west-bank

Ashworth, G. J. (1991). *War and the city*. London: Routledge.

Batty, M. (2008). "Cities as complex systems: scaling, interactions, networks, dynamics and urban morphologies," Working Paper Series. UCL Centre for Advanced Spatial Analysis. [Online]. Available at: <http://discovery.ucl.ac.uk/15183/1/15183.pdf>.

Beaumont, P. (2002, April 19). Army denies frenzy of destruction in Jenin. The Guardian. Retrieved from http://www.theguardian.com/world/2002/apr/19/israel

Boot, M. & Kirkpatrick, J. J. (2006). Beyond the 3-block war. Council on Foreign Relations. Retrieved from http://www.cfr.org/united-states/beyond-3-block-war/p10204

Bosetti, L., Cooper, H., de Boer, J. & Munshey, M. (2016, April 12). Peacekeeping in cities: Is the UN prepared? United Nations University, Centre for Policy Research. Retrieved from http://cpr.unu.edu/peacekeeping-in-cities-is-the-un-prepared.html

Central Intelligence Agency(CIA). (2015). *The world factbook*. Retrieved from https://www.cia.gov/library/publications/the-world-factbook/fields/2212.html

Desch, M. C. (2001). *Soldiers in cities: Military operations on urban terrain*. Carlisle Barracks, PA: Strategic Studies Institute.

Dorn, A. W. & Varey, M. (2009). The rise and demise of the "three block war". *Canadian Military Journal*, 10(1), 38–45.

Edwards, S. J. (2000). *Mars unmasked: The changing face of urban operations*. Santa Monica, CA: Rand.

Giovanni, J., Goodman, L. M. & Sharkov, D. (2014, June 11). How does ISIS fund its reign of terror? *Newsweek*. Retrieved from http://europe.newsweek.com/how-does-isis-fund-its-reign-terror-282607?rm=eu

Glenn, R. W. (1996). *Combat in hell: A consideration of constrained urban warfare*. Santa Monica, CA: RAND.

Glenn, R. W., Paul, C., Helmus, T., Steinberg, P. (2007). *"People make the city": Joint urban operations, observations, and insights from Afghanistan and Iraq*. Santa Monica, CA: Rand.

Global Terrorism Database. Consulted on 06 May 2016. Retrieved from https://www.start.umd.edu/gtd/search/Results.aspx?start_yearonly=2008&end_yearonly=2014&start_year=&start_month=&start_day=&end_year=&end_month=&end_day=&region=10&asmSelect0=&country=155&asmSelect1=&dtp2=all&success=yes&casualties_type=b&casualties_max=

Hahn II, R. & Jezior, B. (1999). Urban Warfare and the Urban Warfighter of 2025. Parameters, Summer 1999, pp. 74–86, Retrieved from http://strategicstudiesinstitute.army.mil/pubs/parameters/Articles/99summer/hahn.htm

Harris, M., Dixon, R., Melin, N., Hendrex, D., Russo, R., and Bailey, M. (2014). *Megacities and the United States Army: Preparing for a complex and uncertain future*. Arlington, Strategic Studies Group.

Hills, A. (2004). Future war in cities: Rethinking a liberal dilemma. London: Frank Cass.

Hoffman, F. G. (2007). *Conflict in the 21st century: The rise of hybrid wars*. Arlington, VA: Potomac Institute for Policy Studies.

Hoffman, F. G. (2013, October 10). Into the cities: Dark, dense, and dangerous. *Small Wars Journal*. Retrieved from http://smallwarsjournal.com/jrnl/art/into-the-cities-dark-dense-and-dangerous

International Bank for Reconstruction and Development /The World Bank (2013). *Global monitoring report 2013: Rural-urban dynamics and the millennium development goals*. Washington DC: International Bank for Reconstruction and Development /The World Bank.

ICRC (2015a). *Urban services during protracted armed conflict: A call for a better approach to assisting affected people*, International Committee of the Red Cross: Geneva.

ICRC (2015b). *Explosive weapons in populated areas: Humanitarian, legal, technical and military aspects*. Expert Meeting, Chavannes-de-Bogis, 24–25 February 2015.

IISS (1999). The future of urban warfare, Strategic Comments, 5:2, 1–2. Retrieved from http://iiss.tandfonline.com/doi/pdf/10.1080/1356788990524

IISS (2015). Fresh gang violence in Rio despite pacification efforts. 2015(25). Retrieved from https://www.iiss.org/en/publications/strategic%20comments/sections/2015-1f4d/fresh-gang-violence-in-rio-despite-pacification-efforts-45e2

Joint Chiefs of Staff (2013). Joint publication 3-06: Joint Urban Operations. Washington, DC.

Joint Chiefs of Staff (2014). Joint publication 3-13: Information Operations. Washington, DC.

Kilcullen, D. (2013). *Out of the mountains: The coming age of the urban guerrilla*. London: Hurst.

Krulak, C. C. (1999, January). The strategic corporal: Leadership in the three block war. *Marines Magazine*. Retrieved from http://www.au.af.mil/au/awc/awcgate/usmc/strategic_corporal.htm

Lappin, Y. (2014, December 27). How the IDF is training the next generation for urban warfare. *The Jerusalem Post*. Retrieved from http://www.jpost.com/International/Security-and-defense-Training-the-next-generation-for-urban-warfare-385794

Mattis, J. N., Hoffman, F. G. (2005). Future warfare: The rise of hybrid wars. *Proceedings*, 132(11). November 2005.

Muggah, R. (2012). Researching the urban dilemma: Urbanization, poverty and violence. Ottawa: IDRC.

Muggah, R. (2015, January 15). Fixing fragile cities: Solutions for urban violence and poverty. *Foreign Affairs*. Retrieved from https://www.foreignaffairs.com/articles/africa/2015-01-15/fixing-fragile-cities

Muggah, R. & Jütersonke, O. (2012). Rethinking Stabilization and Humanitarian Action in "Fragile Cities". In Perrin, B. (Ed.), *Modern warfare: Armed groups, private militaries, humanitarian organizations, and the law* (311–327). Vancouver: UBC Press.

Posen, B. R. (2001). Urban operations: Tactical realities and strategic ambiguities. Desch, M. C. (ed.). Carlisle Barracks, PA: Strategic Studies Institute.

Pickett, M. S. & Best, A. (2014, September 1). The army's next mission: Stability is the best offense. *Small Wars Journal*. Retrieved from http://smallwarsjournal.com/jrnl/art/the-army%E2%80%99s-next-mission-stability-is-the-best-offense

Ricks, T. E. (1999, October 12). Urban Warfare: Where Innovation Hasn't Helped. *The Wall Street Journal*. Retrieved from http://www.wsj.com/articles/SB939692529523950426

Rid, T., & Hecker, M. (2009). *War 2.0: Irregular warfare in the information age*. Westport, CT: Praeger Security International.

Sampaio, A. (Oct. 12, 2015). "Violent cities need whole government solutions," BeyondBrics. [Online]. Available at: <https://www.ft.com/content/2d3b21e9-a974-3622-a4a9-036f9d349ae0>.

Sassen, S. (2005), The global city: Introducing a concept. *Brown Journal of World Affairs*, XI(2), 27–43.

Spence, M., Annez, P. C., & Buckley, R. M. (2009). *Urbanization and growth*. Washington, DC: Commission on Growth and Development.

Shteiwi, Z. (2013, March 5). Lebanon: Can Nahr al-Bared's market make a comeback? Al-Akhbar. Retrieved from http://english.al-akhbar.com/node/15151

Smith, R. (2006). *The utility of force: The art of war in the modern world*. London, UK: Penguin.

Sullivan, J. P. (2011). *From drug wars to criminal insurgency: Mexican cartels, criminal enclaves and criminal insurgency in Mexico and Central America. Implications for Global Security*. Paris: Fondation Maison des Sciences de l'Homme.

Sullivan, J. P. & Elkus, A. (2011, September 26). Command of the cities: Towards a theory of urban strategy. *Small Wars Journal*. Retrieved from https://www.google.co.uk/url?sa=t&rct=j&q=&esrc=s&source=web&cd=1&cad=rja&uact=8&ved=0ahUKEwj8gOrJ1sXMAhXrJcAKHadBBS4QFgghMAA&url=http%3A%2F%2Fsmallwarsjournal.com%2Fprintpdf%2F11533&usg=AFQjCNEd92dOR3sqPSwYSGvDJR-Kg9c-ZpQ&sig2=XNRpmRGoqW1-KaS1GTl_7Q&bvm=bv.121421273,d.ZGg

Sun Tzu (translation posted in 1999). *Sun Tzu's The Art of War: Original, accurate, and complete translation of all 13 chapters*. Retrieved from https://www.sonshi.com/original-the-art-of-war-translation-not-giles.html

The Jerusalem Post (2016, April 07). *Report: Hamas taps over 1,000 terror operatives to dig Gaza tunnels. The Jerusalem Post. Retrieved from* http://www.jpost.com/Arab-Israeli-Conflict/Report-Hamas-taps-over-1000-terror-operatives-to-dig-Gaza-tunnels-450556

The University of Chicago News Office (2005). University of Chicago-Syrian team finds first evidence of warfare in ancient Mesopotamia. Retrieved from http://www-news.uchicago.edu/releases/05/051216.hamoukar.shtml

The World Bank (2013). Developing countries need to harness urbanization to achieve the MDGs: IMF-World Bank report. Retrieved from http://www.worldbank.org/en/news/press-release/2013/04/17/developing-countries-need-to-harness-urbanization-to-achieve-mdgs-imf-world-bank-report

Turse, Nick (2007). Slum fights: The Pentagon plans for a new hundred years' war. Retrieved from http://www.tomdispatch.com/post/174847

Ucko, D. H. (2013, November 7). Clear-hold-build-fail? Rethinking local-level counter-insurgency. War on the Rocks. Retrieved from http://warontherocks.com/2013/11/clear-hold-build-fail-rethinking-local-level-counterinsurgency/

UK Ministry of Defence (2015). Future Operating Environment 2035. London: Ministry of Defence.

United Nations, Department of Economic and Social Affairs, Population Division (2014). *World Urbanization Prospects: The 2014 Revision—Highlights*. New York: United Nations.

UN Habitat (2014). *The state of African cities 2014: Re-imagining sustainable urban transitions*. Nairobi: UN Habitat.

US Army (2003). *Information operations: Doctrine, tactics, techniques, and procedures*. Washington,DC.

US Marine Corps (1998). *Military Operations on Urbanized Terrain (MOUT)*. Washington DC: Department of the Navy.

Weizman, E. (2005). Walking through walls: Soldiers as architects in the Israeli/Palestinian conflict. Conference lectured at the symposium "Archipelago of Exception. Sovereignties of extraterritoriality" CCCB 10–11. November 2005. Retrieved from http://www.publicspace.org/en/text-library/eng/b018-walking-through-walls-soldiers-as-architects-in-the-israeli-palestinian-conflict

Yarger, H. R. (2006). Strategic theory for the 21st century: The little book of big strategy. Carlisle Barracks, PA: Strategic Studies Institute.

Part 3

New and reinvigorated technology

12 Toward robotic warfare

Łukasz Kamieński

In April 2016, a search for "drone warfare" on Google returned almost 550,000 results, while Amazon.com listed 205 book titles. The proliferation of scholarly and popular writings on the topic directly reflects the explosive growth of unmanned aerial vehicles (UAVs) over the last two decades. And there is much more to come, as we are still very much in the midst of the military robotics revolution. Some defense analysts go so far as to argue that Predator, the first and most famous high-tech, remotely operated UAV in the United States (US) armada, "may someday be to drone warfare what the V-2 was to long-range ballistic missiles: a crude, but important, first step in a new era of warfare" (Weinberger, 2014).

But neither "killer drones" nor aerial vehicles constitute a large percentage of the present-day US stock of armaments; nor, though dominant, are they the only facet of the ongoing robotization of war. A wide range of increasingly autonomous and remotely controlled air, land, and sea systems of various size, shape, motion mode, speed, and capability is being operated by all branches of the services. With the US relying more heavily on unmanned systems, the demand for a new paradigm of warfare grows greater, and a new mindset is already slowly emerging from policy papers, the services' roadmaps, and expert debates. This chapter aims to reflect on the role of the unmanned revolution in the US way of war in the twenty-first century.

Welcome to the age of autonomous military technology

Ever since Pioneer successfully went through a baptism of fire in the first Gulf War (1990–1991), marking the first time when remotely piloted aircraft (RPA) were used on a large scale in armed conflict, the design, development, and application of unmanned systems have been expanding at a breathless rate. The War on Terror, in particular the operations in Afghanistan in Iraq, became the real springboard for the dynamic robotization of US warfare.

In 2001–2011, the Department of Defense's (DoD) expenditure on drones increased six-fold compared to the previous decade, with the government's budget for research and development (R&D) in this area growing significantly at about 23 percent annually. In 2001, the US military had fewer than 50 UAVs at its disposal. As the demand increased by 300 percent every year since the beginning of Operation Iraqi Freedom in 2003, in 2008 the armada was already composed of 5331 drones, while in 2011 it exceeded 7000, rising to over 11,000 in 2015 (Singer, 2009, 37; Thurnher, 2012, 79). Land and sea-based systems have undergone similarly remarkable growth, though in lesser numbers.

What are the main explanations for such a dynamic robotization of the US military? First, technological development creates new capabilities. Rapid advances in electronics, software, and power supply have allowed for increasingly smaller, faster, multifunctional, and more durable systems. Particularly crucial has been the progress in real-time satellite navigation and communication, which enables the remote control of vehicles and the flow of the information they gather. Second, the US way of war has been driven by the need to collect all available data for achieving a dominant "situational awareness." RPAs are the most cost-effective response to this continuously growing demand for intelligence, surveillance, and reconnaissance (ISR). Unmanned systems are, after all, the offshoot of the transformation occurring in the US military in the aftermath of the Gulf War in the 1990s. Robotization is thus equally capabilities and demand-driven. Third, the asymmetrical character of contemporary armed conflicts blurs the distinction between combatants and civilians and calls for more accurate methods for discriminating between enemies and noncombatants. Remote systems offer the benefit of persistent and accurate ISR, allowing the identification of enemy targets and monitoring of populations over lengthy periods of time. Fourth, one of the greatest advantages of unmanned autonomous systems (UASs) is the massive lowering of risk to service members. Sending remotely operated technologies reduces the loss of personnel during combat, thus meeting the key demand of the "force protection" paradigm. This, in turn, makes it possible to keep domestic public aversion to US casualties in check and bring down the political costs of military operations overseas. Fifth, because the US has been the leader in military innovation, it is axiomatic that this process furthers robotization. Holding a hegemonic status in military R&D lures a dominant power into a "technological trap"—if it does not want to lose its edge, it must keep mastering new technologies. Inevitably, then, US warfare is becoming progressively robotized, with other actors following suit. And finally, the robotics industry is highly interested in the growing and lucrative government demand for unmanned systems, yet the role of the military–industrial–robotics complex should not be overplayed. All these factors will continue to feed the unmanned (re)evolution of US warfare in the years to come.

Current applications

Over the last two decades, the roles played by unmanned systems have proliferated greatly. UAVs are most commonly used for ISR missions to collect image and video data, which is transmitted in real time back to ground operation centers. Drones conduct terrain reconnaissance, observe enemy forces, monitor infrastructure, identify targets, and carry out post-attack battle damage assessments. By laser-pointing targets, they add to the precision of air and ground strikes. Long-endurance systems such as Global Hawk can provide an overall picture of how things develop over time. Small drones (such as the hand-launched RQ-11 Raven) and land-based vehicles (such as the XM1216), which are perfect for "around the corner" short-time reconnaissance missions, provide timely and often lifesaving information to tactical ground units. Hardly noticeable microdrones are perfect for spying as they can, for example, monitor buildings from the inside. Equipped with sophisticated ultrasensitive sensors, unmanned platforms can find hidden enemies by tracking down an electromagnetic field or body heat. Furthermore, they are used in search and rescue missions to find survivors on the battlefield. When loaded with special sensors, they can detect biological, chemical, or radiological weapons and contamination. For these very reasons, UASs will be of special importance in the urban warfare of the future.

Killer drones have gone from the world of science fiction to the battlefield and back to the realm of popular culture, for example in Daniel Suarez's techno thriller *Kill Decision* (2012) or Andrew Nicoll's *Good Kill* (2014). Arming UAVs has brought about one of their most significant but also most controversial new roles: lethal attack. In late 2001, Predator was equipped with laser-guided Hellfire missiles and began regular combat missions eliminating al-Qaeda operatives and affiliates in Afghanistan, Pakistan, Somalia, Yemen, and Syria. In the future, in some theaters of war (such as Pakistan, Yemen, and Somalia today) combat drones will constitute the dominant weapon system, while in others they will play a supportive role for land operations (such as they have done in Afghanistan and Iraq).

Ground robot-sappers are not only designed for tracking and detonating explosives but are also useful for terrestrial reconnaissance. In Afghanistan and Iraq, they proved so effective in clearing improvised explosive devices (IEDs) and roadside bombs, thereby making war zones safer for servicemen, that the demand for counter-IED robots (mainly PackBots and Talons) grew rapidly, with an overall number of about 5000 vehicles deployed in these two theatres of operation by 2007 (Lin et al., 2008, 5).

Beneath the waves, unmanned underwater vehicles (UUVs) also play many roles of course, such as surveillance and reconnaissance, communication, and mine detection and disposal.

To sum up, because unmanned platforms can perform better than humans, they have found plentiful applications. The advantages of using UASs are abundant: they do not lose effectiveness despite long periods of activity; they do not have to rest, except for refueling/recharging; they do not forget orders; they never suffer from combat trauma; and drones that are not affected by gravitational pressures can fly at speeds that would incapacitate men (high-g maneuvers). In short, as the DoD puts it, "unmanned systems provide persistence, versatility, survivability, and reduced risk to human life" (*Unmanned systems*, 2013, 20). Emerging technological breakthroughs will only further multiply the advantages of uninhabited over manned platforms.

Development trends and new capabilities

Given the US lead in unmanned technologies (despite budget constraints, it accounts for more than 50 percent of total worldwide UAV procurement and R&D, though the competition is growing fast), considerable advances in military robotics should be expected in the near future.

Miniaturization, leading to the production of more mobile and less conspicuous machines, has probably been the most crucial trend. Microvehicles (such as Extender, Dragon Eye, or Swallow) capable of flying at very low altitudes will provide extremely precise reconnaissance. The DoD seeks further miniaturization, not only to make remote systems smaller but also less energy-consuming and lighter. This will, in turn, enable them to carry multiple and more advanced payloads for multimission capabilities, thereby opening windows for new applications and deployment tactics (*United States*, 2014, 36). In the more distant future, the emerging advancements in nanotechnology may take remote systems into the nanoscale (e.g., nanodrones or even "dust robots").

Modularity is the new leading principle governing the developments in equipment for unmanned platforms. A wide assortment of payloads will be brick-like parts, that is, easy to plug in to various systems. Modularity, defined as "the ability to mix and match weapons and sensors to meet given mission requirements on a platform," can be achieved mainly through standardization of armament and sensor interfaces (*United States*, 2014, 36). This solution has already been developed by the Air Force for piloted aircraft (the Universal Armament Interface and Universal Sensor Interface). "Plug and play" properties will allow new equipment to be added and novel system configurations to be achieved with little need to modify the platform itself, thus making them rapidly customizable for different applications at lower costs. Tactical and operational planning and procurement apart, modularity will enhance multimission capabilities and the integrability of different platforms. Modularity is, in short, key for both the future interoperability of autonomous systems and for conducting joint manned–unmanned missions.

Small and micro UAVs, such as the hand-launched Raven, will most probably be produced on demand depending on specific mission requirements. In February 2016, the US Army announced its program "On-Demand Small Unmanned Aircraft Systems," aimed at producing drones composed of off-the-shelf and custom 3D printed parts in less than 24 hours (McNally, 2016). Thus we will see a much greater flexibility in the employment of UASs in combat.

The new unmanned capabilities are also becoming available as a consequence of three parallel technological advancements: (1) the production of low power-consuming devices, (2) the extension of battery life, and (3) improved propulsion. Research is underway into new sources of powering, including hydrogen-fueled propulsion, nuclear isomers, and laser power-beaming. Microdrones have been loaded with electromagnets for recharging their batteries off sources of electricity such as outlets or light bulbs. As early as 2007, the Defense Advanced Research Projects Agency (DARPA) announced its project for an ultraendurance reconnaissance drone that could stay aloft for up to 5 years. Also, as projected by the Air Force UAV vision for 2013–2038, the Highly Energy-Efficient Turbine Engine may "enable fuel-efficient, subsonic propulsion to support future extreme endurance and range requirements, including embedded engines incorporating complex inlets and exhausts" (*United States*, 2014, 87).

Improving the survivability of drones by bestowing them with credible defensive capabilities is also a high priority. Research goes in two directions. The first is into making UAVs less visible by applying the new breed of antiradar stealth technology based on emerging hyperstealthy metamaterials, that is, composites of remarkable capabilities that bend electromagnetic waves and make them flow round an object (Marsh, 2010). The second is into equipping drones, and later probably also ground systems, with powerful laser antimissile weapons. Such is the goal of DARPA's "Endurance" project, carried out by Northrop Grumman and Lockheed Martin. In the not-too-distant future, UASs may not only be able to shoot down missiles in self-defense but also take on a new type of mission as a battlefield antimissile platform.

As scientists increasingly look to nature for inspiration, we will see more biologically inspired systems in terms of shape (i.e., bugs and dogs), locomotion (i.e., wings that mimic hawks and insects), maneuver techniques, and bio-sensors. DARPA's "Fast Lightweight Autonomy" program, for example, intends to develop small and agile drones that mimic birds and insects and are able to "fly autonomously through complex, cluttered environments" (Fast Lightweight Autonomy, 2014). But there is also a more awkward dimension of unmanned technology—the rise of "insect cyborgs" suitable for surveillance in especially challenging environments such as the interiors of buildings (Shadbolt, 2015; Anthes, 2016). The aim of DARPA's programs such as "Hybrid Insect Micro-Electro-Mechanical Systems" is to take control of bugs through implanted electronics. Military-sponsored experiments have succeeded in creating remotely controlled beetles and cockroaches, but also a robo-rat (Shadbolt, 2015). DARPA also seeks to seize control of sharks, as their unique senses could "be exploited to provide data feedback in relation to enemy ship movement or underwater explosions" (Brown, 2006). In the future, will unmanned systems be assisted by remotely controlled animal-cyborgs? This is as plausible as it is unsettling.

Emerging and future applications

Innovations will enable the next generation of air, land, and underwater UASs, loaded with greater arrays of more sophisticated, smaller, and lower power-consuming equipment, to perform new missions.

Drones carrying extremely lightweight and long-endurance jamming devices may be used to disrupt an enemy's use of the electromagnetic spectrum and break its communications. UASs will play an important role in enhancing US electronic warfare capabilities, not only by overwhelming an adversary's radar and antiaircraft systems but also by adding to the protection of friendly facilities and equipment. They might be crucial, particularly when working as a dynamic team, in providing long-range wireless communication relays in battlefield contexts and "acting as a gateway manager of multiple communications to enhance C2 connectivity and span of control" (*United States*, 2014, 52; Hauert et al.). Such a communication backup, based on an unmanned collective cooperating system, could prove vital for sustaining operational effectiveness.

Uncrewed underwater torpedo-like vehicles with wings can be a perfect weapon for hunting hostile submarines. They have long endurance, are nearly silent, and consume extremely low levels of power, for instead of having a propeller they glide using small changes in buoyancy to covert vertical motion to horizontal, or they draw energy from temperature differentials in the ocean. The vision of such vehicles patrolling large areas is highly plausible, as the US Navy has been successfully working on a number of glider projects, such as the "Littoral Battlespace Sensing-Glider" (Hambling, 2016). Undersea and sea vehicles will likely have very long endurance, enabling them to operate for up to 5 years or until mechanical failure. This will bring an overwhelming potential for persistent surveillance of oceans, undersea infrastructure, enemy ships, and submarines. Advanced experiments have also been conducted with the DARPA-sponsored completely robotic ship Sea Hunter, designed to patrol the seas, track submarines, and carry out countermine tasks. If successful, the program could transition to the US Navy by 2018.

The first armed unmanned ground vehicle, called *Sword*, was introduced on the battlefield in 2007 to patrol streets of Baghdad. As the army will in future probably deploy plentiful robots on the first line of fire with an enemy force, the riskiest encounters will be taken over by "robotic bullet catchers." Bob Sadowski, the army roboticist, believes that such autonomous vanguards will become standard equipment for ground forces, maneuver formations in particular (Vergun, 2016). So imagine the following scenario in 2030: to take control of a city, ground forces send "robotic screens" for the dangerous initial attack. As one author speculates, "patrols of reconnaissance/observation robots and teams of robots capable of remote fire—all of which are all semiautonomous devices—will arrive ahead of the troops. Their task is to " 'thin out' the situation before 'conventional' forces intervene" (Malis, 2014, 41). Unmanned vehicles will be best suited also to serve as decoys to confuse the enemy by undertaking mock maneuvers. And after the successful capture of an enemy site, robots will be used to provide continuous and effective control of spaces and spots. Moreover, ground forces will be further robotized at the lower level of echelon, as the army already seeks to make pocket-sized drones commonly available to soldiers from the squad level up and plans to begin their larger procurement as soon as in 2018 (Finley, 2016).

Autonomous systems will also increasingly be used to perform logistics missions, particularly those for which the use of a manned platform is impossible or too dangerous. While aerial UASs have a weight-limited capacity, autonomous land mules will accompany troops, significantly lightening soldiers' combat loads. The most advanced contemporary examples are designed by Boston Dynamics under contracts with DARPA: BigDog, Legged Squad Support System (LS3), and Spot (BigDog; LS3). Mechanical quadruped mules can overcome obstacles; climb over steep slopes; and walk on rocks, in snow, ice, and water perfectly, keeping their balance all the while. Although due to its disadvantageous lawnmower-like noise the LS3 has not been deployed by the Marines, a smaller, lighter, and more silent version called *Spot* has been tested for battlefield use. Smaller and quieter versions are already under development. Lockheed Martin's Squad Mission Support System (SMSS) is a more traditional-looking six-wheeled unmanned ground vehicle designed for transport and logistical support for light, early entry, and special operations forces. It operates autonomously or works via remote control, teleoperation, or manual control modes (Lockheed Martin, 2014). Driverless tactical convoy vehicles have also been tested, and it is very likely that in the not-too-distant future, self-driving platforms will conduct routine supply missions. Overall, autonomous vehicles will reduce or even "eliminate the need for convoys and large numbers of ground crew" (Galliott, 2015, 203).

The vision of robots who save the lives of their wounded human fellows is no longer merely science fiction. What Larry Nieven imagined in his novel *World of Ptavvs* (1966) and called "autodoc" has been designed as a prototype DARPA-founded "Trauma Pod." An automated medic will be able to evacuate an injured serviceman from a war zone and carry out patient-care procedures. The robot could then send medical records to a human surgeon, who could perform life-saving telesurgery (Garcia et al. 2009; 137). Although experiments with a prototype proved successful, the challenge remains to miniaturize the platform to a size deployable on the battlefield. Overall, as former "Trauma Pod" project manager Richard M. Satava believes, "it is highly likely that the majority of the medical evacuation will eventually occur on a UAV" (Satava, 2006, 5). Autonomous machines could vastly improve the chances of survival for seriously injured combatants, especially in inaccessible locations.

To conclude, robots have already moved beyond their early "dull, dangerous, and dirty" missions and will find their place in a broad array of novel battlefield tasks. Robotics will change the landscape of ground battlefield as autonomous machines will take on more active combat support and combat functions. With the rise of "attack robots," UASs will go beyond the present-day tasks performed by "sensor robots" (i.e., out of the corner reconnaissance, target identification and designation, explosives detection and clearing, and protection of sites and bases) and "servant robots" (i.e., transportation, logistics, and medical evacuation). As one author predicts, by 2020–2025 it is most likely that 70–80 percent of all robots on the land battlefield will be sensors, 10–20 percent servants, and 5–10 percent attack (Yakovleff, 2014, 249). In 1992, a study of the US Army *STAR 21: Strategic technologies for the Army of the 21st century* projected that while the core twentieth-century weapons system of land war had been the tank, in the twenty-first century it was apparently to be the unmanned system (*STAR 21*, 1992, 119). This prediction ultimately proved correct. Yet, the transformation is still going on, and on the march now are not mere unmanned machines but networks of robots capable of collaborating with one another to form (semi)autonomous tactical units.

The rise of unmanned tactical formations?

"Swarming" describes a process whereby individual entities, usually insects, cooperate by following simple rules that make them behave as a large but decentralized system. Interactions between elements lead to "intelligent" group behavior. Interestingly, from the interplay between autonomous machines and the environment, some patterns of collective collaborative actions can emerge, similar to those in colonies of bees, ants, wasps, or crabs. Swarm robotics is based on the assumption that complexity can arise from simple behaviors displayed in a group, and that with little initial human input, autonomous units might form a networked self-organizing system. Hundreds of devices can perform as a single force. Thus uninhabited vehicles, especially small ones, could cooperate to work as a team, forming a networked tactical formation. Autonomous swarm capabilities would also enhance the safety of a group, as teamwork tends to increase mutually protective behavior. A swarm of military UASs could perform more effectively and be used for missions unachievable for single vehicles monitored by a single operator, such as in electronic warfare. Special machine-to-machine interfaces will, for example, enable robot colonies "to function in a self-forming rule-/role-based airborne network to create virtual large array antennas or cyber and kinetic effects that over-saturate an adversary's defensive systems at a relatively lower cost" (*United States*, 2014, 40).

In August 2014, the US Navy conducted a test with five small uninhabited boats that steered and took action by themselves while engaging with an enemy ship. The experiment proved that swarm-boat capabilities will enable multiple unmanned vessels to provide active ship protection. When needed, they can become offensive, too (Lendon, 2014). The Air Force also takes swarming seriously. The 2016 strategic vision for small UASs identified two key aspects of next-generation drones: one is their greater autonomy; another is their ability to work together. Under the "Gremlins" program, DARPA seeks to develop volleys of low-cost, reusable small drones of great operational flexibility and with "coordinated and distributed capabilities." Launched from large aircraft, Gremlins will carry out their missions over great distances and then be retrieved in mid-air by a transport aircraft (Friendly "gremlins," 2015; Lamothe, 2016). A swarm of hundreds of micro/nano sensors collecting and streaming multifarious data and swarms of robomunitions are, however, still science fiction visions of tomorrow.

Not only will UASs operating in teams become a norm, but also a new type of jointness will develop between a variety of uninhabited ground, surface, subsurface, and air systems. So, as Alexander Kott and his colleagues rightly argue, "A critical challenge of the mid-21st century will involve successfully managing and integrating the collections, teams, and swarms of robots that would act independently or collaboratively as they undertook a variety of missions" (Kott et al., 2015, 23). With the deployment of autonomous tactical formations, robotization of land warfare will bring changes in combat methods and tactics. The emergence of a robot-based ecosystem operating in swarms will also call for adjustments in battle command as well as in rules of engagement.

Swarming, in short, is one of the key concepts around which future tactics for robotic systems will develop, since it can allow the US military to disperse combat power and "overwhelm the enemy through mass" (quoted in Lamothe, 2016).

Greater autonomy

Progress in swarming capabilities is closely related to advancements in machine autonomy, and there are three main reasons why the military increasingly seeks more autonomous technologies.

First, a gradual increase in autonomy has been inherent to technological development. Military autonomization is not a revolutionary turnover but progresses rather evolutionarily (the examples of common combat automation include, after all, autopilots and satellite navigation systems). Science and technology allow for greater machine autonomy, while new innovations, such as ultra-long endurance or hypersonic flight, call for it.

Second, remotely operated vehicles are susceptible to hacking and signal jamming. Losing control of a platform in the midst of its mission threatens the success of the operation. The best way to reduce this risk is to delegate more authority to machines, thus decreasing the amount of information flow between a system and its human operator. Therefore, bestowing machines with the capabilities to autonomously process the data they collect reduces vulnerabilities and increases operational effectiveness.

Third, in terms of computational speed and information processing, machines vastly outrun humans. While unmanned systems perform much better in the first two nodes of John Boyd's OODA cycle (*orient* and *observe*), autonomization will allow them to independently perform the remaining two phases of the loop (*decide* and *act*). The basic idea behind Boyd's concept is that the fastest tempo of operating all steps is essential to winning a conflict. And it is the human-in-the-loop that retards the entire performance. Biologically limited abilities apart, human skills regress without enhancement, while machines operate at a constant level of high efficiency. Keeping humans in the decision-making loop, therefore, degrades combat efficiency for, as Michael W. Byrnes argues, "Humans average 200–300 milliseconds to react to simple stimuli, but machines can select or synthesize and execute maneuvers, making millions of corrections in that same quarter of a second. Every step in OODA that we can do, they will do better" (Byrnes, 2014, 57).

Because it is not possible to program machines for all eventualities, their ability to learn independently of human input is critical for their greater autonomy. Scientists agree that the best way to obtain artificial intelligence (AI) is through the "bottom-up approach," or machine learning, that is, making robots discover the world and act in the way children's cognition develops. Autonomous behavior emerges over time along with machines interacting with the environment, with humans, and with other machines; with their observing the world, mimicking, and receiving feedback on their actions. Swarming will significantly contribute to the development of greater autonomy, since UASs operating in a network will be able to share the lessons learned, "making partnered aircraft smarter by every engagement" (Byrnes, 2014, 54). Such "group learning" capabilities may be closer to appearing on the horizon than one might suspect. Scientists at Brown University, for example, have been working on a project to allow robots that learn tasks to upload their data to the cloud for sharing with other machines. If this ability is important for civilian robots, it is critical for military systems. Around 2020–2025, robots may be able to teach each other, and this might make programming each one separately unnecessary (Schaffer, 2016).

Autonomy, being a matter of degree, will only increase—from human-assisted or low (when an operator controls some actions and movements, such as taking off and landing, and gives directions that the system follows), through human-delegation or task autonomy (when an operator instructs the robot by giving it general tasks) to human-supervised or full autonomy (when a machine sets and performs tasks on its own without any human input). When operators become supervisors, when men merely monitor machines, then "human-in-the-loop" is replaced by "human-on-the-loop." What raises the most fundamental dilemmas here is the delegation of lethal authority to a machine, no matter how "intelligent" it might be. In 2012, responding to these ethical concerns, the DoD adopted "Directive 3000.09: Autonomy in Weapon Systems," which allows for the development of only nonlethal autonomous systems and introduces a 10-year moratorium on lethal autonomy. Although the directive sets limits on robotic warfare, the policy can be waived by senior officials at the Pentagon, and many experts believe that full autonomous lethal systems will be available in the 21st century. As Deputy Defense Secretary Robert Work observed "We believe strongly that humans should be the only ones to decide when to use lethal force. But when you're under attack, especially at machine speeds, we want to have a machine that can protect us" (quoted in Lohn, Parasiliti, Welsser IV, 2016). Whatever the issues, there seems to be a consensus that in the case of "killer robots," humans must maintain the ability to change the level of a machine's authority and rewrite its decisions. As Christopher Coker optimistically concludes, "Given the complexity of war is ever increasing there will always be a need for human supervision, for people to take the decisions that computers cannot" (Coker, 2013, 141). Fortunately, war will remain, as Thucydides put it, a "human thing."

A challenge: Force structure and service culture

In terms of military service cultures, UASs pose the greatest challenge to the Air Force. Unmanned systems hardly influence the essence of what it means to be an infantryman, marine, or seaman because they are simply recognized as cost-effective and risk-reducing force multipliers. This explains why the US Navy has already become the leader in unmanned systems, with such highly sophisticated drones as the X-47B (Salty Dog) capable of both operating from an aircraft carrier and refueling while airborne, which will immensely improve maritime situational awareness. The Air Force, however, is built around the air-mindedness for which the very act of flying is connected to the nature of being a pilot. The problem is important as it touches the service's culture, identity, and structure. It is indicative that one-third of pilots would rather leave the service than operate RPAs (Byrnes, 2014, 66). By 2020, the Air Force expects to have the smallest force structure in its history (*United States*, 2014, 17). Already in 2010, for the first time it trained more drone than fixed-wing pilots. This growing unmanned armada will bring about a demand for more UAV operators. These changes will further transform the force structure.

The phenomenon of combat drones generates a range of important, though often contested, issues that need to be understood and dealt with to ease the assimilation of RPA crews into traditional Air Force organization, modus operandi, and strategic culture. Drone-based warfare introduces aviation to a new, virtual domain in which the experience of piloting is much more abstract than flying an aircraft. Interestingly, there are vivid parallels between the evolution of manned aviation, from surveillance to lethal strikes, during the First World War and the present-day development of an unmanned air force, from the early artillery observation of the trenches to planes carrying rudimentary bombs, and from the first Pioneer reconnaissance missions in the Gulf War to the weaponizing Predators in 2001. Given these evolutionary patterns, one can expect that "the capabilities of UAVs will continue to develop and increasingly play a part in all air power roles" (Lamonte, 2009). Therefore, a specific framework of "drone-mindedness" will need to emerge. This new dimension of Air Force military culture will have to involve some characteristics that are specific to remote warfare.

Drone operators ("cubicle warriors") do a highly routinized job that is devoid not only of physical risk but also of such a central feature of warfare as unpredictability. Their work at the virtualized cockpit in a cargo container has more in common with a fixed-hour office job than with combat. Waging war from afar transforms combat operations into telewarfare—with "wireless pilots" teleoperating UAVs and being telepresent in remote locations due to real-time visual feedback. Paradoxically, however, while allowing for complete physical separation, drones decrease "cognitive distance in sensor mediated combat" (Blair and Helms, 2013, 28).

Risk-free remote warfare brings novel psychological burdens. Although the main reported source of stress for drone operators is overwork due to long hours and frequent shift changes, fighting without leaving home generates a previously unknown kind of stress. Personnel routinely return home after operating drones on killing missions, and this "war plus a family" model may generate psychosocial costs. While operators are physically distanced from the war zone, they are physically present at home and have to encounter daily family problems. Constant and radical shifts in contexts between the virtual life of combat and private family life often creates a severe cognitive dissonance of living in two worlds simultaneously and can generate "residual stress" (Royakkers and van Est, 2010, 293). Also, despite the popular but unproven and debatable thesis that drone killing is like video-gaming (the "Playstation mentality"), operators are not entirely emotionally disengaged. They tend to get attached to the people they observe, sometimes for days, before targeting. As one author notes, "One of the paradoxes of drones is that, even as they increase the distance to the target, they also create proximity" (Abé, 2012). If combat is "blood, sweat, and tears," remote warfare removes blood and sweat, but it cannot entirely remove the tears. It may dehumanize the enemy and depersonalize combat, yet it does not necessarily detach "cubicle warriors" emotionally. A DoD report concluded that drone operators experience mental health problems like depression and anxiety at similar rates to pilots of manned aircraft deployed in Iraq or Afghanistan. In a survey of nearly 840 RPA operators, as many as 46 percent of Reaper and Predator pilots and 48 percent of Global Hawk sensor operators reported "high operational stress" (Bumiller, 2011). Thus, the Air Force needs not only to introduce more effective training programs for drone pilots to meet the force structure demands (in November 2015 it was announced that civilians would also be used to pilot drones), but should also pay very special attention to their psychological preparation in order to help them deal with the mental costs of remote warfare.

David J. Blair and Nick Helms, aircraft pilots turned drone pilots, argue that while airmen have traditionally seen technology in terms of capabilities (as amplifiers of human skills for greater effectiveness), "cubicle warriors" are generally inaccurately seen through a cybernetic approach. This puts the system in the center and considers humans as mere "subsystems within larger sociomechanical constructs" (Blair and Helms, 2013, 23). Such a predominant cybernetic view of drone pilots can be explained by the US "intoxication" with remote unmanned systems and their peculiar fetishization, the increasing autonomy of UASs, the slow and reluctant integration of their crews into the existing force structure, and the lack of mature tactics and strategy for drone warfare. Blair and Helms articulate the way forward for adjusting UAV pilots to the Air Force's cultural fabric and structure: "the RPA must come into its own as a culture of Airmen by means of a capabilities-based view of technology" (Blair and Helms, 2013, 25). A new unmanned-air-mindedness can only be built within a human-centric, not a system-centric, approach.

Conclusion: Super warriors plus robots

Maybe one day, war will look like a duel between autonomous machines that do combat better than humans. The emergence and deployment of lethal autonomous weapons systems will mark a profound transformation in warfare. Yet, in a nearer future, contrary to many dystopian science-fiction visions, robots will not entirely replace humans, but they will certainly significantly reduce their presence on the battlefield. With fewer humans and more ubiquitous UASs, military force will be based on close man–machine cooperation. Human–machine teaming—with robotic systems vastly expanding and augmenting human abilities—will become one of the characteristics of US warfare. Flesh-and-blood soldiers will have to work in a symbiotic relationship with mechanical "soldiers." The authors of the report on the battlefield of the future correctly anticipate that "The principal Army unit operating in 2050 will be mixed human-robot teams" (Kott et al., 2015, 19). A good, yet still rather crude, contemporary example is the Air Force's "Loyal Wingmen," in which an entirely unmanned F-16 accompanies manned F-35 aircraft to form a man–machine battle network. Most importantly, however, the ultimate effect of the robotization of warfare is its growing pressure on soldiers. Biological limitations of mind and body make servicemen "the weakest element" of the war machine. Because operations are conducted at breathtaking speed, nonstop, and usually at night, soldiers, pilots in particular, experience what Paul Virilio famously referred to as the "violence of speed" (Virilio, 2006, 153, 167). A murderous "operational tempo" requires military personnel to perform at the edge of human capability. That is why the US military continually searches for biotechnological means to finely adjust humans to superefficient machines. As high-tech unmanned systems can perform faster, longer, and more efficiently, soldiers must maintain persistent combat readiness and the ability to regenerate quickly. They need greater stamina and improved alertness, or, in a word, "super-soldier" capabilities. Psychopharmacology (e.g., a new generation of stimulants called eugeroics); neurobiology (e.g., transcranial brain stimulation); cyborgization, which fuses the technological and the human (e.g., implantable biochips, man-machine interfaces, and mechatronic exoskeletons, which are in fact wearable cobots meaning robots that physically interact with humans); as well as genetics and genomics have been extensively researched for their potential to enhance the warrior (in terms of physical endurance, vision, hearing, cognitive skills, mood, etc.). The biotech revolution has already found a place in the military as it offers the possibility of reconfiguring man to partner effectively with robots. Biotechnology can, in fact, help achieve the goal set out in February 2014 by the Air Force's vision of the robotic aircraft of the future: "Manned and remotely piloted systems will be seamlessly integrated and synchronized in airspace, command and control architectures, information sharing, and purpose to optimize airpower and increase capability across the full range of military operations for the combined force" (*United States*, 2014, 13). These two aspects of the current US military transformation—robotics and biotech—should no longer be seen as parallel but rather as complementary. Still, I hope that the confidence expressed by the US *Field Service Regulations* in 1944 that, despite all the advances in technology, the value of a single man remains decisive, will be retained.

References

Abé, N. (2012, December 14). Dreams in infrared: The woes of an American drone operator. Spiegel, Retrieved from http://www.spiegel.de/international/world/pain-continues-after-war-for-american-drone-pilot-a-872726.html.

Anthes, E. (2016, February 17). The race to create "insect cyborgs." *The Observer*, Retrieved from https://www.theguardian.com/science/2013/feb/17/race-to-create-insect-cyborgs.

BigDog—the most advanced rough-terrain robot on earth, Retrieved from http://www.bostondynamics.com/robot_bigdog.html.

Blair, D. J. and Helms, N. (2013, July–August). The swarm, the cloud, and the importance of getting there first. What's at stake in the remote aviation culture debate. *Air & Space Power Journal*, 27 (4), 14–38.

Brown, S., (2006, March 1). Stealth sharks to patrol the high seas. *New Scientist*, Retrieved from https://www.newscientist.com/article/mg18925416-300-stealth-sharks-to-patrol-the-high-seas.

Bumiller, E. (2011, December 18). Air Force drone operators report high levels of stress. *New York Times*, Retrieved from http://www.nytimes.com/2011/12/19/world/asia/air-force-drone-operators-show-high-levels-of-stress.html?_r=0.

Byrnes, M. W. (2014, May–June). Nightfall. Machine autonomy in air-to-air combat. *Air & Space Power Journal*, 28 (3), 48–75.

Coker, C. (2013). *Warrior Geeks. How 21st-Century Technology is Changing the Way We Fight and Think About War*. New York: Columbia University Press.

Fast Lightweight Autonomy (FLA). (2014, December 22). Retrieved from https://www.fbo.gov/index?s=opportunity&mode=form&id=38c56cd0dec12a5d3ae9b95849c102e8&tab=core&_cview=0.

Finley, K. (2016, April 4). The Army wants you to make its soldiers pocket-sized drones. *Wired*, Retrieved from http://www.wired.com/2016/04/us-army-wants-make-tiny-drones-soldiers/?mbid=social_fb.

Friendly "gremlins" could enable cheaper, more effective, distributed air operations. (2015, August 28). Retrieved from http://www.darpa.mil/news-events/2015-08-28#.

Galliott, J. (2015). *Military Robots. Mapping the Moral Landscape*. Burlington VT: Ashgate.

Garcia, P., Rosen, J., Kapoor, C., Noakes, M., Elbert, G., Treat, M., Ganous, T., Hanson, M., Manak, J., Hasser, C., Rohler, D., and Satava, R. (2009). Trauma pod: A semi-automated telerobotic surgical system. *The International Journal of Medical Robotics and Computer Assisted Surgery*, 5(2), 136–146.

Hambling, D. (2016, February 27). Underwater drones may make hiding a nuclear missile sub harder. *New Scientist*, Retrieved from https://www.newscientist.com/article/2078466-underwater-drones-may-make-hiding-a-nuclear-missile-sub-harder.

Hauert, S., Leven, S., Zufferey, J.-C., and Floreano, D. The swarming micro air vehicle network (Smavnet) Project, Retreived from http://lis2.epfl.ch/CompletedResearchProjects/SwarmingMAVs.

Kott, A., Alberts, D., Zalman, A., Shakarian, P., Maymi, F., Wang, C., and Qu, G. (2015). *Visualizing the Tactical Ground Battlefield in the Year 2050*. Adelphi, MD: US Army Research Laboratory.

Lamonte, J. (2009, October). The future of UAVs: Concepts and considerations, Retrieved from http://www.raf.mod.uk/role/thefutureofuav.cfm.

Lamothe, D. (2016, February 8). Why drone swarms will buzz to the forefront in the new Pentagon budget. *The Washington Post*, Retrieved from https://www.washingtonpost.com/news/checkpoint/wp/2016/02/08/why-drone-swarms-will-buzz-to-the-forefront-in-the-new-pentagon-budget.

Lendon, B. (2014, October 13). U.S. Navy could "swarm" foes with boats. CNN, Retrieved from http://edition.cnn.com/2014/10/06/tech/innovation/navy-swarm-boats.

Lin, P., Bekey, G., and Abney, K. (2008). *Autonomous Military Robotics: Risk, Ethics, and Design*. San Louis Obispo: California Polytechnic State University.

Lockheed Martin receives contract for SMSS-KMAX cooperative teaming demo (2014, February). Retrieved from http://www.lockheedmartin.com/us/news/press-releases/2014/february/mfc-021914-lockheed-martin-receives-contract-smss-kmax-cooperative-teaming-demo.html.

Lohn, A., Parasiliti, A., and Welser IV, W. (2016, February 8). Should we fear an AI arms race? *Defense One*, Retrieved from http://www.defenseone.com/ideas/2016/02/should-we-fear-ai-arms-race/125670.

LS3—Legged Squad Support Systems, Retrieved from http://www.bostondynamics.com/robot_ls3.html.

Malis, C. (2014). New extrapolations: Robotics and revolution in military affairs. In Ronan Doaré, Didier Danet, Jean-Paul Hanon, Gérard de Boisboissel (Eds.). *Robots on the Battlefield. Contemporary Perspectives and Implications for the Future* (pp. 33–47). Fort Leavenworth, Kansas: Combat Studies Institute Press.

Marsh, G. (2010, November 1). Going stealthy with composites, Retrieved from http://www.materialstoday.com/composite-applications/features/going-stealthy-with-composites.

McNally, D. (2016, February 16). Army selects 3-D printed unmanned aircraft systems concept for future experiment, Retrieved from http://www.army.mil/article/162445/Army_selects_3_D_printed_unmanned_aircraft_systems_concept_for_future_experiment.

Royakkers, L. and van Est, R. (2010). The cubicle warrior: the marionette of digitalized warfare. *Ethics of Information Technology*, 12 (3), 289–296.

Satava, R. M. (2006, May). How the future of surgery is changing: Robotics, telesurgery, surgical simulators and other advanced technologies, Retrieved from http://depts.washington.edu/surg/biointel/Future-of-Surgery-0606.pdf.

Schaffer, A. (2016). Robots that teach each other. *MIT Technology Review*, Retrieved from https://www.technologyreview.com/s/600768/robots-that-teach-each-other.

Shadbolt, P. (2015, January 14). Robo-wings: Military drones that mimic hawks and insects. CNN, Retrieved from http://edition.cnn.com/2015/01/14/tech/mci-drone-robohawk-robofly.

Singer, P. (2009). *Wired for War. The Robotics Revolution and Conflict in the Twenty-First Century*. New York: The Penguin Press.

STAR 21: Strategic Technologies for the Army of the 21st Century (1992). Washington D.C.: National Academy Press.

Thurnher, J. S. (2012). Legal implications of fully autonomous targeting. *Joint Forces Quarterly*, 67 (4), 79.

United States Air Force RPA Vector. Vision and Enabling Concepts 2013–2038. (2014, February). Washington DC: Headquarters US Air Force.

Unmanned Systems Integrated Roadmap. (2013, January 1). Washington DC: Department of Defense, *Retrieved from archive.defense.gov/pubs/DOD-USRM-2013.pdf*.

Vergun, D. (2016, March 7). Robots: eliminating the first contact with an enemy force. Retrieved from http://www.army.mil/article/163654/Robots__Eliminating_the_first_contact_with_an_enemy_force.

Virilio, P. (2006). *Speed and Politics*. Los Angeles: Semiotext(e).

Weinberger, S. (2014, May 17). The ultra-lethal drones of the future. *New York Post*, Retrieved from http://nypost.com/2014/05/17/evolution-of-the-drone.

Yakovleff, M. (2014). Battlefield robotization: Toward a new combat ecosystem. In Ronan Doaré, Didier Danet, Jean-Paul Hanon, Gérard de Boisboissel (Eds.). *Robots on the Battlefield. Contemporary Perspectives and Implications for the Future* (pp. 243–260). Fort Leavenworth, Kansas: Combat Studies Institute Press.

13 The continuing evolution of US nuclear weapons policy

B. D. Mowell

Introduction

A range of considerations are forcing policymakers to re-evaluate the future of the United States (US) nuclear weapons arsenal. The Cold War thaw and subsequent collapse of the Soviet Union were perceived by many to represent an opportunity for nuclear disarmament or at least nuclear arms reduction and nonproliferation within the framework of a more stable international security regime. Thus far, such optimism has largely proven to be unfounded as reductions in weapons inventories have been somewhat modest relative to the expectations of many, and potential nuclear threats posed by competitor states and rogue regimes have arguably increased since the 1990s. The shortcomings of nonproliferation efforts have permitted the emergence or progress toward development of nuclear weapons capabilities among several countries in recent decades, including anti-Western and potentially unpredictable regimes in North Korea and the nuclear threshold state of Iran. The threat of nuclear technology or materials being transferred via black market channels to rogue regimes or terrorist organizations perhaps looms greater today than at any point during the Cold War. Within this evolving climate of uncertainty, the US nuclear weapons inventory is aging with the physical infrastructure and the technology and tactics behind it reaching or exceeding intended life expectancy. Technological developments such as hypersonic missiles and increasingly compact warhead design present policy makers with the option of modernizing the US nuclear weapons inventory; alternatively, the US may potentially yield advantages to competitor states. This chapter provides an overview of the evolution, complex contemporary dynamic, and possible future directions of US nuclear weapons policy.

Historical background

Becoming the first nation to develop and deploy nuclear weapons helped to establish US political and military hegemony at the end of World War II (WWII). While research and development continued after Japan's surrender, actual warhead production in the years immediately following the war was modest, having reached a total of only 170 warheads by 1949 (Norris & Kristensen 2010). The souring of relations with the Soviet Union and Moscow's first successful atomic bomb test in 1949, followed by the subsequent rapid expansion of the Soviet nuclear weapons program, including their development of the hydrogen bomb by 1955, triggered an accelerated buildup in the US inventory throughout the 1950s and 1960s. Schwartz (2011, 45) notes that peak annual production of nuclear weapons in the US exceeded 7000 in 1959–60 with nearly 19,500 new warheads constructed between 1958 and 1961, a rate of approximately 25 per workday. The US nuclear stockpile reached its historical highpoint in 1966 when around 32,000 total warheads—mostly older and/or lower-yield tactical designs soon to be made inactive—were in inventory (Ibid). The nuclear arms race (see Table 13.1) continued for many years as the Soviets sought to achieve parity and as of 1978 actually overtook the US in total number of nuclear weapons (Norris & Kristensen 2010). Moscow has since maintained overall numerical superiority, though as noted in Figure 13.1, the US *strategic* nuclear weapons inventory continued to increase for another decade, peaking at around 14,000 active warheads by 1988.

US nuclear weapons strategy evolved during the Cold War both in response to expanding Soviet capabilities and to newly emerging technology. Both US and Soviet strategic nuclear weapons platforms were initially one-dimensional in that they were dependent upon bomber aircraft to penetrate enemy airspace and deliver payloads in the event of war. By the 1950s, missile technology had improved sufficiently to allow single warhead nuclear payloads. However, the range of early ballistic missiles was limited to a few thousand miles at best, sufficient for US missiles to strike portions of the USSR from bases in proximity to Soviet territory (i.e., Turkey), but initially not sufficient for Soviet missiles to reach the US. Moscow sought to partially offset the early numerical advantages held by the US in total number of warheads and also the geographical advantages held by the US/North Atlantic Treaty Organization (NATO) in placing forward military bases with conventional or nuclear striking capabilities in proximity to Soviet territory. In 1955, the Soviets became the first nation

Table 13.1 Estimates of US and Soviet/Russian strategic and tactical nuclear weapons inventories, 1950–2010

Year	*US*	*USSR/Russia*
1950	299	5
1960	18,638	1,605
1970	26,008	11,643
1980	24,104	30,062
1990	21,392	37,000
2000	10,577	21,500
2010	5,000*	12,000

*US figure excludes approximately 4000 retired warheads awaiting disassembly as of 2010 (Adapted from Norris & Kristensen 2010).

to develop the capability to launch nuclear missiles from specially designed surfaced submarines. By 1963, a new generation of Soviet subs could launch nuclear payloads while still submerged, allowing for maximum stealth while on patrol. Similar technology had already been under development in the US, which had fielded its own nuclear-capable submarine fleet by the early 1960s. The advent of intercontinental ballistic missiles (ICBMs) in the same decade meant that missiles no longer needed to be launched from forward positions, which were potentially more vulnerable to enemy attack. The US nuclear triad strategy entailed distributing potent strategic capabilities among bombers, missiles, and submarines, ensuring the survivability of sufficient numbers to facilitate an annihilating nuclear counterattack in the event of an enemy first strike. The latter threat of mutually assured destruction underpinning a strategy of nuclear deterrence served as the cornerstone of US nuclear weapons doctrine during most of the Cold War and drove the nuclear arms race with the USSR.

The decline and ultimate collapse of the Soviet Union combined with the budgetary priorities of the Clinton administration saw US military spending slashed by approximately one-third in the 1990s (Walker 2014). Many areas of the national defense budget were impacted during the period, including the aging elements of the nuclear triad, which were approaching the end of their serviceable lifespan and were increasingly in need of modernization. The lowered prioritization of nuclear defense was also in part attributable to the Strategic Arms Reduction Treaty (START I), which took effect in 1994 after having been signed three years earlier. The treaty restricted both the US and Russia from having more than 6000 warheads and a combined total of 1600 ICBMs and bombers, eliminating more than 75 percent of the world's strategic nuclear weapons at the time. As the likelihood of nuclear confrontation initially declined in the post-Cold War era, so did some support for maintaining a large

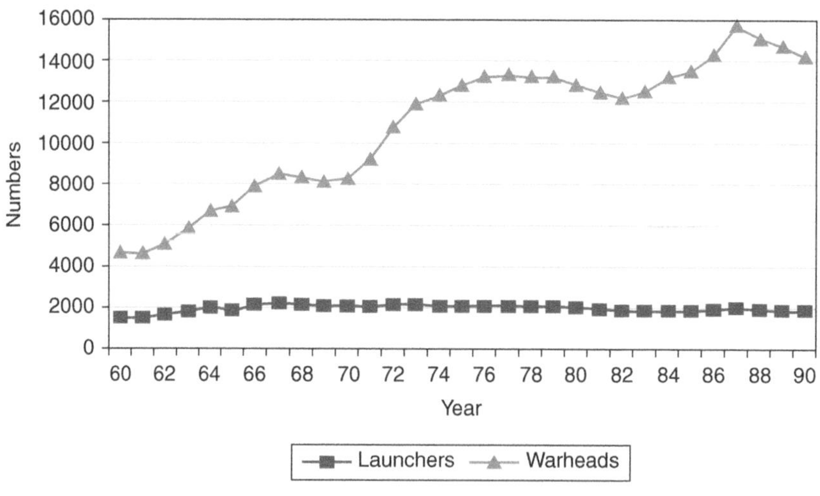

Figure 13.1 US strategic nuclear weapons, 1960–1990

Source: Woolf 2015.

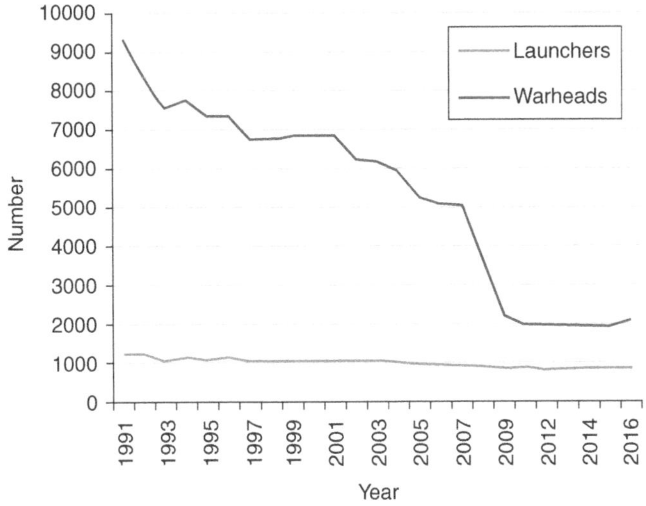

Figure 13.2 US strategic nuclear weapons, 1991–2015

Source: Woolf 2015.

and diverse arsenal of nuclear deterrents. Within the post-Cold War US military, some internal support for nuclear weapons waned as an ever-shrinking defense budget placed conventional weapons/defense programs in budgetary competition with the goals of maintaining and upgrading the nuclear triad.

However, contemporary geopolitical developments are prompting concern regarding the aging stockpile and platforms that comprise US nuclear weapons capabilities. The resurgence of aggressive Russian foreign and military policy, the rapid expansion and modernization of Chinese military capabilities and ambitions, and the concern over burgeoning nuclear weapons and other weapons of mass destruction (WMD) programs among other smaller nations, including Iran and North Korea, highlight the need to prioritize US nuclear readiness. Admiral Cecil Haney, head of US Strategic Command, has publicly stressed that modernization of each element of the nuclear triad is essential in order to prevent competitor nations such as China or Russia from feeling that they are poised to gain an advantage over US military capabilities (Mehta 2016a). Many Pentagon leaders and their supporters in the federal government have echoed such need. Speaking to the House Armed Services Committee in March 2016, Army Chief of Staff General Mark Milley stressed his conviction that the US nuclear triad kept the peace during the Cold War and would play a vital role in doing so in the future (Cox 2016). In examining the role of nuclear weapons in the twenty-first century and the need to maintain nuclear deterrents, Roberts (2015) argues that at present no other nuclear-armed states appear willing to join the US in reducing stockpiles and that unilateral reduction in nuclear inventory by the US would be potentially harmful to its security interests as well as that of its allies.

Budgetary considerations

Presently, the debate over the status of US nuclear weapons policy in Congress is centered largely upon financial considerations. Since developing the world's first atomic bomb during WWII, the US has produced and deployed over 70,000 nuclear weapons at an inflation-adjusted cost of over $5 trillion dollars (Schwartz 2011). In a 2015 study, the Center for Strategic and Budgetary Assessments (CSBA) estimated that the total cost of modernizing and maintaining the nuclear triad at more than $700 billion over the next 25 years (Mehta 2016a). Another analysis conducted by the James Martin Center for Nonproliferation Studies estimated the cost at replacing or upgrading systems and maintaining readiness to be more than $30 billion on average annually for the next 30 years, a combined total of approximately 1 trillion dollars (Wolfsthal et al. 2014). The CSBA analysis breaks down to approximately $279 billion for warhead development, $228 billion for naval delivery systems, $95 billion for land-based delivery systems, $74 billion for air-based systems, and $30 billion for upgrading command and control systems (Mehta 2016a).

The enormity of the latter numbers is diminished when placed within the broader context of overall defense/government spending. In 2015, the declared US defense budget totaled $601 billion out of $3.43 trillion in overall annual federal expenditures. Even if the highest cost estimates are used, the modernization of the US nuclear weapons program constitutes a comparatively small amount of the total defense budget at around 5 percent of DoD spending. Proponents of the program note that for the level of deterrence provided, US nuclear weapons capabilities could be perceived as a relative bargain within the overall defense budget and also that such a deterrent is still needed in an increasingly anarchical and unpredictable post-Cold War world in which competitor states such as China and Russia as well as rogue regimes seek to further develop their nuclear capabilities either numerically or technologically (i.e., Spring 2013; Mehta 2016b).

Nuclear weapons concerns

The possibility of use in war is not the only risk posed by nuclear weapons. The more than 70-year history of nuclear weapons in the US is replete with examples of accidents or mistakes that could have proven disastrous. One of the more famous incidents occurred in 1958, when a B-47 bomber damaged by an accidental mid-air collision on a training flight ejected an inert H-bomb near Tybee Island, Georgia. The bomb, which the Air Force asserted did not have nuclear capability for the training exercise, was never found and remains buried in deep sediment somewhere in the coastal area. More recently, in 2007, a B-52 flight intended to deliver 12 aging, conventional cruise missiles to a military repository for retirement and disassembly mistakenly transported missiles with active nuclear warheads without any proper nuclear weapons safety protocols across country from North Dakota to Louisiana before the error was discovered the following day.

Theft of materials that could be utilized by a terrorist group or rogue state to create nuclear weapons or radiological "dirty" bombs remains a significant concern and was a prominent topic at the 2016 Nuclear Security Summit. Paradoxically, reduction of nuclear weapons inventories does not contribute to a solution of the theft issue in the short term as it actually facilitates the transfer of disassembled nuclear materials to storage or processing centers, which, in many cases, may be less secure than nuclear weapons installations. The larger and more widely dispersed the stockpiles of nuclear materials, the greater the risk of accident or theft. Of the world's highly enriched, weapons-grade uranium, 96 percent is used by military or other government agencies rather than civilian programs, but even the nuclear material housed at presumably high-security military and government facilities in many nations is not completely invulnerable to theft (Sanger & Broad 2016). Between 1993 and 2013, there were 419 known cases of smuggled or stolen nuclear material globally, a large percentage of which were linked to Russia or other former Soviet territories (Koren 2013). Russian economic turmoil and decline in internal security characteristic of the period following the collapse of the Soviet Union afforded ample incentive and opportunity for corrupt engineers, military officers, and others to attempt the theft and sale of nuclear material. A 2014 study by the Nuclear Threat Initiative (NTI) found that significant progress had been made in the US and globally in enhancing security of nuclear stockpiles but stressed that additional effort was necessary. The organization recently ranked the US stockpile as among the world's better secured but called for more frequent and thorough vetting of personnel with access to nuclear materials (NTI 2016). Increased efforts have also been undertaken to prevent the smuggling of nuclear material into the US including installation of radiation detectors at ports and other customs checkpoints.

Many advocates of reform in US nuclear weapons policy note that the evolving security climate in the post-Cold War era no longer necessitates a traditional nuclear posture or as large an active inventory of nuclear deterrents. The global ideological conflict of the middle and late twentieth century has been replaced by smaller-scale regional security issues, for which nuclear weapons are not ideally suited as deterrents. While Russia maintains a larger nuclear arsenal and has engaged in aggressive foreign policy machinations and military interventions in recent years, the realistic possibility of a nuclear confrontation remains minute. Likewise, while some tensions exist with China concerning the South China Sea and other contentious issues, the possibility of nuclear conflict is negligible, and China is believed to possess fewer than 300 nuclear weapons in its active inventory (Norris & Kristensen 2010). Maintaining large inventories of nuclear weapons has not prevented nuclear proliferation in emerging nations such as India and Pakistan or efforts to acquire nuclear weapons by threshold states such as Iran, nor would a large strategic inventory necessarily be suitable to deter or respond to the actions of a nuclear or other WMD-armed terrorist group or rouge regime led by an irrational actor.

Evolution and future prospects for arms reduction, nonproliferation, and disarmament

Calls for complete nuclear disarmament and a ban on nuclear weapons are not new. The first such opposition originated in the final days of WWII and developed into a small but growing movement as the public became aware of the devastation wrought at Hiroshima and Nagasaki (Wittner 2009). The atomic bombs dropped on Japan hastened the end of the war and very likely reduced the overall military and civilian casualties that would have resulted from a conventional invasion of the Japanese mainland. The latter perception, combined with the new Cold War fear of armed conflict with the communist block and the concomitant recognized need for potent deterrents, largely kept the US anti-nuclear-weapons movement confined to a small percentage of the population for decades. The end of the Cold War and collapse of the Soviet Union helped revitalize the nuclear disarmament movement.

While the realistic likelihood of nuclear disarmament or even significant inventory reduction in the foreseeable future is not great, some progress toward tempering the nuclear arms race between the largest powers has been made. A series of bilateral treaties between the US and Moscow beginning with SALT (Strategic Arms Limitation Treaty) I in 1972 and SALT II in 1979 first placed caps on the numbers of warheads and weapons platforms both countries deployed. The START (Strategic Arms Reduction Treaty) I agreement, signed in 1991, reduced the total number of strategic nuclear weapons in the US and the former Soviet Union to a maximum of 6000 warheads distributed over 1600 long-range missiles and bombers. START II, which was signed in 1993 but never implemented, sought further reductions in both nations' inventories, capping the total number of warheads at 3000–3500. An overview of US strategic forces implemented under START I and planned but never realized under START II is provided in Table 13.2. The New START Treaty, signed in 2010 and scheduled to remain in effect through 2021, further reduces the number of actively deployed strategic warheads by around half (see Table 13.3) and also creates a new inspection regime, but it does not reduce the large number of retired, inactive weapons in the stockpiles of either country.

Significant difference of opinion exists as to the necessary size of the US nuclear weapons arsenal and the size and nature of its delivery systems as evidenced in the following examples. The Council on Foreign Relations, in a 2010 study, advocated inventory reductions that significantly transcend the New START Treaty signed by the US and Russia in 2010. The position paper supported bilateral reductions of stockpiles to 1000 warheads or less, a total that includes both strategic and tactical weapons. Several key long-standing obstacles to achieving such a goal were outlined in the study, including the realities that (1) Russia maintains a much larger arsenal of tactical weapons than the US and would accordingly have to bear a disproportionate share of cuts to tactical inventories and (2) that both countries remain at odds over numbers and types of land-based missiles (Zenko 2010).

Table 13.2 US Strategic nuclear weapons and platforms under START I and START II

	Implemented under START I		*Planned (not implemented) for START II*	
System	*Accountable Launchers*	*Accountable Warheads*	*Launchers*	*Warheads*
Minuteman II ICBMs	500	1,200	500	500
Peacekeeper ICBMs	50	500	0	0
Trident I Missiles	168	1,008	0	0
Trident II Missiles	264	2,112	336	1,680
B-52H (ALCM)	97	970	76	940
B-52H (non-ALCM)	47	47	0	0
B-1 Bombers	90	90	0	0
B-2 Bombers	20	20	21	336
Total	**1,237**	**5,948**	**933**	**3,456**

*START I specified that each bomber not equipped to carry the more sophisticated ALCMs (Air Launched Cruise Missiles) count as a single warhead, even though maximum payload was up to 16 bombs; each bomber equipped to carry ALCMs counts as 10 warheads despite maximum capacity of 20 ALCMs.

**B-1 bombers were still included/counted under START I but were no longer equipped for nuclear missions (Woolf 2008).

A 2013 Cato Institute report, positing that the US has a much larger nuclear weapons inventory than is needed in the post-Cold War era, called for the elimination of the nuclear triad in favor of a much smaller arsenal utilizing submarines as the exclusive delivery platform. The study argued that such an approach would potentially save more than $20 billion annually and would suffice as a deterrent, as it is highly unlikely that the entire US nuclear submarine fleet or even a majority thereof could be neutralized by a first strike (Friedman et al. 2013). The authors note that eliminating two of the nuclear triad's three elements may not be politically feasible and, as an alternative, a "dyad" approach comprised of both submarines and a reduced inventory of land-based missiles could be implemented, though at reduced savings.

Other minimalist approaches recommend even more austere stockpiles than the previous examples, arguing that a few hundred warheads in active service would be sufficient. A joint study by the Federation of American Scientists and Natural Resource Defense Council argued that the triad should be maintained as a diverse delivery system but that a combined total of 500 warheads would serve as an adequate deterrent at a fraction of the cost of the current system (Gard & Terryn 2015). A 2010 Brookings Institution study argued for deep cuts, contending that a vastly smaller arsenal of only a few hundred nuclear weapons would serve all realistic US defense needs (Feiveson 2010). One of the lowest estimates to date concerning the number of weapons that could continue to fulfill defensive needs as a deterrent came from a study conducted by several members of the Air University faculty, in which a proposed 311 warheads would be distributed across a vastly compacted delivery

Table 13.3 US strategic nuclear weapons and platforms under new START treaty estimates of forces—2010 caps on forces under new START by 2018*

	Launchers	*Warheads*	*TotalLaunchers*	*Deployed Launchers*	*Warheads*
Minuteman III	450	500	454	400	400
Trident	336	1,152	280	240	1,090
B-52	76	300	46	42	42
B-2	18	200	20	18	18
Total:	**880**	**2,152**	**800**	**700**	**1,550**

*Terms of the New START Treaty specify that the US will retain 14 Trident submarines, 2 of which will be in overhaul. Each sub will have 4 of its current 24 launchers eliminated, meaning each sub will count as 20 launchers. All 450 Minuteman III launchers will be retained, though only 400 will hold missiles (Woolf 2015).

triad: 100 ICBMs, 12 nuclear submarines, and 19 bombers (Gard & Terryn 2015). The architects of the latter studies concede that for such drastic inventory reductions to be possible, the Russian government (and ideally all of the world's other existing nuclear powers should also agree to proportional reductions) would have to take reciprocal action, a commitment that does not appear to be forthcoming from Moscow, making such suggestions more theoretical than practical at present.

It is worth noting that the argument has been postulated that successful arms reductions treaties do not automatically lead to improved relations (i.e., US–Russian relations post-New START) and, significantly, that maintaining larger nuclear arsenals does not necessarily correlate to greater risk or incidence of nuclear alerts (Costlow 2015). In fact, as reflected in the data compiled by a nuclear policy analyst at the National Institute for Public Policy and provided in Table 13.4 below, in general, the larger the US nuclear weapons inventory, the less frequently US or Soviet nuclear forces were placed on alert. Such data supports the contention of many military planners that a large nuclear deterrent is effective at maintaining peace or at least crisis stability and preventing runaway escalation, as neither side would choose to risk engaging in military conflict that could expand to nuclear exchange.

The Nuclear Nonproliferation Treaty (NPT) implemented in 1970 sought to prohibit signatory nations that did not possess weapons at that time from obtaining them and also ostensibly encouraged the five states that did possess nuclear weapons at the time of the treaty's implementation—the United States, Britain, China, France, and Russia—to pursue inventory reduction and ultimately nuclear disarmament. The United Nations describes the accord as a "landmark international treaty whose objective is to prevent the spread of nuclear weapons and weapons technology, to promote cooperation in the peaceful uses of nuclear energy and to further the goal of achieving nuclear disarmament" and notes that it is the only binding multilateral international commitment toward the ultimate goal of eliminating nuclear weapons (UNODA 2016). However, many scholars (e.g., Garvey 2013) have noted that the NPT has fallen short of its original aims and may not be a sufficient vehicle for future progress regarding nuclear weapons. While some inventory reduction has occurred, none of the five original nuclear states have disarmed and no realistic likelihood exists for that goal to be realized in the foreseeable future. To date, 190 states have become signatories of the treaty, including South Africa, which abolished its nascent nuclear weapons capabilities to do so. However, the NPT has not prevented several states from developing nuclear weapons since 1970; India, Israel, and Pakistan, none of which are signatories to the treaty; and North Korea, which was originally a signatory but later renounced and withdrew from the NPT, have all developed nuclear weapons since the agreement was implemented. Rogue regimes including Iran, Iraq, and Libya have also previously sought to covertly develop nuclear weapons despite having signed the NPT. Regarding Iran, the NPT and international pressure has thus far merely delayed rather than eliminated their nuclear weapons program. The fact that vast majority of the world's approximately 200 countries have not attempted to develop nuclear weapons to date is likely not as much a reflection of the effectiveness of the NPT but rather an indication that all but a few countries have thus far either lacked the financial/technological ability or the interest to do so.

Readiness/Alert level

As of 2016, the US and Russia each maintain nearly 1000 nuclear missiles on high alert, also commonly known as *hair-trigger* alert (Union of Concerned Scientists 2016). This readiness policy wherein missiles could be fired within minutes was implemented during the Cold War, ostensibly in order to ensure the readiness of sufficient numbers of nuclear weapons to counterattack in the event of a first strike by an enemy. Since the end of the Cold War, both the US and Russia have disavowed the first strike option, pledging only to use nuclear weapons in response if attacked. However, Russian leaders have suggested recently that their nuclear doctrine should be modified to permit the

Table 13.4 Correlation between US nuclear arsenal size and frequency of US or Soviet nuclear forces being placed on alert status

Decade	*Approximate Average Size of US Nuclear Arsenal*	*Number of Times US or Soviet Nuclear Forces Placed on Heightened Alert*
1945-49	49 weapons	2 US; 0 USSR (A-bomb developed in 1949)
1950s	3,500 weapons	4 US; 0 USSR
1960s	27,000 weapons	4 US; 4 USSR
1970s	26,000 weapons	1 US; 1 USSR
1980s	23,000 weapons	0 US; 1 USSR

Source: Costlow 2015.

option of pre-emptive/first nuclear strikes against enemy states, which may pose threats to Russia.

Maintaining such a heightened state of readiness potentially increases the likelihood of human error and launch via misinterpretation of actions. Risk of both could be diminished by increasing the time and steps required to launch. Although calls for decreasing the readiness/launch status of US and Russian nuclear missile ordinances have been issued from various sources, including the Union of Concerned Scientists, such an agreement would need to be bilateral in nature, and no indication of such willingness exists on the part of either country at present. In fact, in apparent response to international outcry and pressure related to its aggressive posturing in the South China Sea, Chinese military leadership has advocated placing at least a large portion of its roughly 250 nuclear weapons on hair-trigger alert, which would be an unprecedented change in that nation's nuclear weapons readiness policy (Kulacki 2016).

Tactical nukes

Whereas strategic nuclear weapons have longer range and larger payloads and are designed to destroy larger targets such as entire cities, tactical (also known as *nonstrategic*) nuclear weapons have shorter range and yields small enough to be used tactically on the battlefield. Platforms for the delivery of tactical nuclear weapons have included artillery, gravity bombs, mines, torpedoes, and missiles of a wide variety of design and range. While tactical weapons were a significant element of the US nuclear arsenal during the early Cold War, in recent decades, US tactical inventories have been significantly reduced, reflecting a strategy shift favoring strategic nuclear weapons. Presently, the US has around 760 tactical weapons, approximately 150–200 of which are deployed in Europe with NATO forces (Belgium, Germany, Italy, the Netherlands, Turkey) with the balance on US soil (Foradori 2013; Woolf 2016). Recent Russian actions in Eastern Europe have facilitated calls for a larger number of tactical deployments in Europe, though others advocate the withdrawal of existing tactical nukes as a gesture to diffuse the deteriorating European security climate. While estimates of the number of tactical weapons in Russia's current inventory vary widely between 1000 and 6000, what is clear is that they have not reduced their tactical arsenals in levels commensurate with US reductions (Woolf 2016). This is a reflection of both (1) difference in weapons doctrine between the two countries, wherein Russia more heavily emphasizes tactical nuclear capabilities at present; and (2) a lack of attention provided to tactical weapons in previous arms control agreements, which have focused almost exclusively on reductions in strategic nuclear weapons—perhaps reflecting historical US prioritization of a reduction in the numbers of Russian strategic nuclear weapons that pose the greatest direct threat to US territory.

The inventory disparity between the US and Russian tactical arsenals, the need to counter technological advances, and the apparent doctrinal commitment to the use of such weapons by competitor states are cause for concern, and many analysts have advocated renewed prioritization of tactical nukes as a future direction in US military doctrine in order to allow for a measured response to an attack by low-yield weapons (e.g., Lieber and Press 2009; Murdock et al. 2015; Weisgerber 2015). Although Russia may be poised to take the technological lead, both the US and Russia have made advancements in tactical warheads and delivery systems, including downscaled and controlled yield, improved accuracy, and possibly controlled/variable blast intensity, all of which would permit more precise application with significantly less collateral damage. Critics argue that low-yield, highly accurate tactical nukes may make use more tempting, increasing the likelihood of escalated nuclear conflict. Also, the compact size and (at least with regard to Russia) sheer numbers of tactical nuclear weapons may pose more opportunities for theft of such devices, which could find their way into the hands of terrorist groups or rogue regimes. The future direction of US policy regarding tactical nuclear weapons remains uncertain and is likely to be the subject of debate among lawmakers for the foreseeable future.

Conclusion

New designs in nuclear weapons are allowing for increasingly compact warheads and improved accuracy of targeting and delivery systems. Emerging technology may permit controlled blasts of nuclear payloads in which the intensity of the strike can be increased or decreased at will. China, Russia, and the US are actively developing hypersonic missile technology in which the speed, low flight altitudes and maneuverability (rapid course changes) of the next generation of weapons would make them nearly impervious to most existing antimissile countermeasures. Indications exist that the US is entering a new technological arms race, not only with regard to nuclear weapons development but also concerning the development of next-generation missile shield technology and other antinuclear countermeasures. While adding a new chapter to the history of the international nuclear arms race is not desirable, the US and its allies can ill afford to yield nuclear advantages out of complacency or misplaced optimism in a vision of a future peace that may not be shared by competing powers.

Early in his first term as president, Barack Obama expressed support for a world free of nuclear weapons, or at minimum, significant reductions in active inventories. At least initially, basis for optimism regarding such a goal existed. Reductions in US and Russian strategic nuclear forces were accomplished by the New Start Treaty in 2010, followed by an additional US gesture reducing the number of warheads on its land-based missiles from as many as four per missile to a maximum of one. Rather than reciprocating the latter gesture, the Russian government proceeded with plans for new nuclear weapons technologies, including advanced multiple-warhead ICBMs, space weapons that could destroy enemy satellites in the event of nuclear war, and submersible drone dirty bombs designed to detonate off the coast of major cities and send massive clouds of radiation ashore, rendering cities uninhabitable (Broad & Sanger 2016). Geopolitical circumstances in recent years have dashed realistic hopes of denuclearization or substantive arms reductions in the foreseeable future. Heightened military tensions and other international developments have strained relations and sowed an atmosphere of increasing distrust between the US and its principle competitor states, Russia and China, and both nations appear committed to developing new technologies, strategies, and counter-measures with regard to conventional as well as nuclear capabilities. While events of recent years appear to have likely guaranteed the continuation of a large and potent US nuclear weapons arsenal and have also garnered political support to fund extensive modernization and replacement programs, the future number and composition of nuclear weapons platforms as well as the future direction of US nuclear weapons policy and strategy are yet to be determined and will likely continue to evolve and serve as subjects of debate for many years to come.

References

Broad, W. & D. Sanger (2016, April 16). Race for latest class of nuclear arms threatens to revive Cold War. *New York Times*.

Costlow, M. (2015, December 9). Do more nukes mean more nuclear crises? Not necessarily. *Bulletin of the Atomic Scientists*.

Cox, M. (2016). Pentagon leaders defend nuclear triad overhaul. Military.com. Retrieved from http://www.military.com/daily-news/2016/03/16/pentagon-leaders-defend-nuclear-triad-overhaul.html

Feiveson, H. (ed.) (2010). *The nuclear turning point: A blueprint for deep cuts and de-alerting of nuclear weapons*. Washington, D.C.: Brookings Institution Press.

Foradori, P. (ed.) (2013). *Tactical nuclear weapons and Euro-Atlantic security: The future of NATO*. New York, NY: Routledge.

Friedman, B., C. Preble, & M. Fay (2013). *The end of overkill: Reassessing US nuclear weapons policy*. Washington, D.C.: Cato Institute.

Gard, R. & G. Terryn (2015). American nuclear strategy: The case for a minimal-deterrence policy. *The National Interest*. June 28, 2015.

Garvey, J. (2013). *Nuclear weapons counterproliferation: A new grand bargain*. New York: Oxford University Press USA.

Koren, M. (2013). Top ten cases of nuclear thefts gone wrong. *Smithsonian*. February 4, 2013.

Kulacki, G. (2016). China's military wants to put its nukes on a hair trigger. *Defense One*. Retrieved from http://www.defenseone.com/ideas/2016/03/china-military-nuclear-obama-ICBM/127139/

Lieber, K. and D. Press (2009). The nukes we need: Preserving the American deterrent. *Foreign Affairs*. November/December 2009.

Mehta, A. (2016a). Is the Pentagon's budget about to be nuked? *Defense News*. Retrieved from http://www.defensenews.com/story/defense/policy-budget/budget/2016/02/05/nuclear-option-nuclear-modernization-costs-bomber-icbm-submarine-lrso/79788670/

Mehta, A. (2016b). On nuclear weapons, hill shows bipartisan support. *Defense News*. Retrieved from http://www.defensenews.com/story/defense/air-space/strike/2016/02/26/nuclear-weapons-hill-shows-bipartisan-support-nnsa-triad/80927862/

Murdock, C., T. Karako, S. Brannen, and A. Weaver. (2015). *Project atom: A competitive strategies approach to defining U.S. nuclear strategy and posture for 2025–2050*. New York, NY: CSIS/Rowman and Littlefield.

Norris, R. & H. Kristensen (2010). Global nuclear weapons inventories, 1945–2010. *Bulletin of the Atomic Scientists*. 66(4), 77–83.

Nuclear Threat Initiative (2016). *Country data: United States, 2016*. ttp://2016.ntiindex.org/countries/united-states/?index=theft

Roberts, B. (2015). *The case for U.S. nuclear weapons in the 21st century*. Stanford, CA: Stanford University Press.

Sanger, D. & W. Broad. (2016). As nuclear security summit begins, materials remain vulnerable to theft. *New York Times*. March 29.

Schwartz, S. (2011). *Atomic audit: The costs and consequences of U.S. nuclear weapons since 1940*. Washington, D.C.: Brookings Institution Press.

Spring, B. (2013). Disarm now, ask questions later: Obama's nuclear weapons policy. *Heritage Foundation: Backgrounder*. No. 2826 (July 12, 2013). Retrieved from http://thf_media.s3.amazonaws.com/2013/pdf/bg2826.pdf

Union of Concerned Scientists (2016). "Accidental" and "nuclear war" don't belong in the same sentence. Environmentalists Against War. Retrieved from http://www.envirosagainstwar.org/know/read.php?itemid=17578

United Nations Office of Disarmament Affairs (2016). *Treaty on the Non-Proliferation of Nuclear Weapons (NPT)*. Retrieved from http://www.un.org/disarmament/WMD/Nuclear/NPT.shtml

Walker, D. (2014). *Trends in U.S. military spending*. Council on Foreign Relations website. Retrieved from http://www.cfr.org/defense-budget/trends-us-military-spending/p28855

Weisgerber, M. (2015). Report: US Needs New Small Nuclear Bombs. *Defense One*. Retrieved from http://www.defenseone.com/management/2015/06/report-us-needs-new-small-nuclear-bombs/115765/

Wittner, L. (2009). *Confronting the bomb: A short history of the world nuclear disarmament movement*. Stanford, CA: Stanford University Press.

Wolfsthal, J., J. Lewis, & M. Quint (2014). *The trillion dollar nuclear triad*. James Martin Center for Nonproliferation. Retrieved from http://cns.miis.edu/opapers/pdfs/140107_trillion_dollar_nuclear_triad.pdf

Woolf, A. (2008). U.S. strategic nuclear forces: Background, developments, and issues: *Congressional Research Service Report*. Retrieved from: http://fpc.state.gov/documents/organization/109505.pdf

Woolf, A. (2015). U.S. strategic nuclear forces: Background, developments, and issues. *Congressional Research Service Report*. Retrieved from http://missilethreat.wpengine.netdna-cdn.com/wp-content/uploads/2015/03/CRS-US_Strategic_Nuclear_Forces.pdf

Woolf, A. (2016). Nonstrategic nuclear weapons. Congressional Research Service Report. Retrieved from: https://www.fas.org/sgp/crs/nuke/RL32572.pdf

Zenko, M. (2010). *Toward deeper reductions in US and Russian nuclear weapons*. Washington, D.C.: Council on Foreign Relations.

14 The future of US hegemony in the Middle East

US–Israel–China military technology development and acquisition

Abdelrahman Rashdan

Introduction

According to the Organisation for Economic Cooperation and Development (OECD) report "Education Indicators in Focus," by 2030 the United States (US) will be producing only 8 percent of the OECD and G20's STEM—science, technology, engineering, and mathematics—graduates, while China and India together will account for more than 60 percent ("Education Indicators," 2015). Back in 2010, the US Department of Defense's Defense Advanced Research Projects Agency (DARPA) issued a report signifying the alarming decline ("DARPA-RA-10-03," 2010); by 2012, around 15 percent of tertiary education graduates in the US had STEM degrees while in China, the figure was 40 percent.

The attention given by the Chinese government to the development of science and technology has done much to serve its recent military advancement. China has been able to place its hand over advanced US military technologies, only to improve its own capabilities. It remains in the US strategic interest to limit or deny Chinese access to such defense technologies while encouraging further US technological advancement to support future American interests worldwide. Israel represents one of China's main avenues to access up-to-date US military technology. In addition to harming global US security interests, the rise of cooperation between China and Israel in different fields would allow China an increasingly influential role in one of the most powerful and central Middle Eastern states, posing a potential threat to US regional interests. This study assesses Sino–Israeli relations in different fields with a special focus on the military cooperation, addressing the question as to whether or not Sino–Israeli bilateral relations did develop to a level that could potentially be threatening to the US Middle East hegemony.

In light of the historical background of bilateral relations between China and Israel, this study considers the evidence available on cooperation in the fields of arms, as well as trade, education, technology, and industry. Finally, the study considers the implications of such relation on US hegemony in the Middle East. Based on unclassified data, the assessment of the Sino–Israeli level of military-related cooperation may not include the most recent updates, since there is an intentional information blackout for the sensitivity of the topic.

By ssessing the history and nature of Sino–Israeli relations, it will be clear that the strategic value both sides attach to the different kinds of cooperation represents a long-term threat to the traditional US hegemony in the Middle East. Rapid growth of China's trade relations in the Middle East is coupled with gradual advancement in its own military technology, which, in turn, is transferred to different Middle East client states.

Historical background

Although Israel was the first country in the Middle East and one of the first in the world to recognize the newly established Chinese state in January 1950 (Han, 1993), official bilateral diplomatic relations were not launched until 1992 ("About Sino-Israel," 2009). The Chinese shift towards establishing deeper relations with Israel, before being official in the 1990s, was based on the Chinese aim to achieve rapid economic and industrial development through adopting Western technology (Friedman, 1985).

Sino–Israeli relations witnessed four main stages since 1950 (Institute of Military Research, 1998, pp. 93–103). Only three months after the declaration of the People's Republic of China as a state, Israel recognized it; such a quick Israeli response came out of Israel's own need for allies and international recognition in addition to the need to open up relations with the Third World. However, Israel's decision to establish full diplomatic relations with China was blocked from implementation because of the Chinese involvement in the Korean War. As war dragged on, Israel leant more towards the Western block (Melman & Sinai, 1987). The role of US–Israeli relations was influential at the time in keeping Israel away from China for the fear of the infiltration of Communism into the newly formed Middle Eastern state. Thus, a gap was growing between the two countries, which was widened by the First Afro-Asian Solidarity Conference at Bandung in 1954, in which China became more appreciative of forging bonds with the Arabs and changed its outlook towards the Middle East to conclude that the step of establishing closer relations with Israel could be postponed (Institute of Military Research, 1998, p. 96).

The second stage in Sino–Israeli relations took place between the 1950s and the late 1970s. China developed an inimical diplomatic attitude towards Israel following the 1956 Tripartite Aggression against Egypt and Israel's decision to sell arms to India during the Indian–Chinese war of 1962, which led to all Sino–Israeli contacts being cut by the mid-1960s (Abadi, 2004, p. 74). In addition, China declared the 1967 Arab–Israeli war as "another towering crime against the Arab people committed by U.S. imperialism and its tool Israel, as well as a grave provocation against the people of Asia, Africa and the rest of the world" (Abadi, 2004, p. 74). The end of the 1960s witnessed China's aims to defend its interests against Soviet–American global domination, which included the Middle East. Thus, China gave attention to the support of resistance movements such as the Palestine Liberation Organization (PLO) and the re-establishment of full diplomatic ties with the Arab regimes (Calabrese, 1991,

p. 66). In yet another shift, China had started by the beginning of 1970s to view the Soviet Union as its principle enemy and thus started easing up its relations with the West (Calabrese, 1991, p. 73). The ideopolitical rupture in Sino–Soviet relations led China to seek new alliances and work on limiting the Soviet influence generally and in the Middle East specifically (Melman, 1987, p. 397). Israel, being pro-American, represented a good target for China and was willing to swallow its pride and keep the relationship on a clandestine basis for the vast economic potential carried with China's rush towards modernization (Melman, 1987, p. 397).

Sino–Israeli relations were thus refreshed despite China's firm support of the rights of the Palestinian people. Israel joined in to support China's U.N. membership, opened a Consulate General in Hong Kong in 1972—which closed in 1975 to reopen in 1985—and sought Western mediators to resume diplomatic ties with China (Institute of Military Research, 1998, p. 97). Furthermore, China was showing increasing interest in Israeli military technology, resulting in concrete steps to resume ties.

The Camp David Accords encouraged China to take further steps in its relations with Israel. The third stage of the Sino–Israeli relationship was characterized by increasingly opening secret channels and communication in several fields. The 1980s witnessed a change in the Chinese diplomatic statements towards Israel. In 1982, China's prime minister, Zhao Ziyan, said in Cairo that "all the Middle East countries including Israel" should enjoy "the rights of independence and existence...on the basis of the Israeli withdrawal from the territories occupied in 1967 and the restoration of the rights of the Palestinian people" (Han, 1993, p. 67). With increasing Sino–Israeli contacts and China's policy of opening up with the West and the Soviet Union—to create a "peaceful international environment" as orchestrated by Deng Xiaopin—the Chinese official stance was eased further to become more lenient towards Israel. Such conditions stated in 1982 by Ziyan were dropped 5 years later during a meeting between Israel's foreign minister, Shimon Peres, and his Chinese counterpart in New York (Institute of Military Research, 1998, p. 100). Israel's official diplomatic rhetoric showed parallel willingness for warmer bilateral relations. Addressing the Knesset in 1984, Prime Minister Peres said, "We shall once again knock on great China's door—[for] there is no need for the geographical distance between us to be translated into a pointless diplomatic distance" ("Address in the Knesset," 1984).

During the 1980s, Israeli skills attracted the Chinese government and businessmen especially in the fields of agriculture, solar energy, manufacturing, advanced technology, robotics, construction, and road building, as well as arms manufacturing (Friedman, 1985). Following several business deals where Israeli expertise was invited into various Chinese national projects (Friedman, 1985), initially incited by the military cooperation, the foreign ministers of both countries held a secret meeting in the United Nations in 1987 ("About Sino-Israel," 2009). By 1989, it was clear that a new stage in Sino–Israeli relations had started ("China: Stalemate Continues," 1989).

China's need for stronger relations with Israel was most vivid following the 1989 Tiananmen Square events. China has been trying to modernize its military and was denied US and USSR assistance; such denial was transformed into a European Union (EU) and US arms embargo imposed on China to further restrict its choices on the international market. Israel represented a viable alternative where the technology was developed and modeled after the Western advanced systems. Israel was able to offer China the best technology at the best price (Friedman, 1985).

For Israel, such military cooperation with China was taken counter to US interests (Han, 1993, p. 72); Israeli policymakers were keen not to harm their country's relations with the superpower (Han, 1993, p. 70). As it is explained later in the study, close Sino–Israeli cooperation caused several incidents of friction between Israel and the US on one side and between Israel and China on the other side. Soon after, some steps were taken that resulted in the signature of an agreement in January 1990 to establish diplomatic relations between the two countries. And in January 1992, the Joint Communiqué on Establishment of Diplomatic Relations was signed to announce the establishment of Sino–Israeli relations at an ambassadorial level ("Joint Communiqué," 2000). Since then, several visits by highly ranked officials were exchanged, three Israeli on the prime-ministerial level and four Israeli and one Chinese on the presidential level.

In the same way that the US was a source of pressure in shaping Israel's decision to establish Sino–Israeli relations, Arab states were influencing the Chinese decision. Although China's relations with the Arabs would have been greatly harmed by China's choice to get closer to Israel, the Arabs' official steps to accept the 1967 Israeli borders and the Madrid Peace Conference of 1991 gave China a space for maneuver. Progress in the Middle East peace process determined when the Chinese government would prepare to take steps towards normalizing its relations with Israel ("China: Stalemate Continues," 1989). Moshe Zak wrote in the *Jerusalem Post* in 1993, "Since the core of the discord between Israel and the Arabs has ostensibly been blunted, it is more convenient for the Chinese to act friendly toward Israel as the surrogate of the Jewish people. They thus hope to pave the way for large Jewish capital investments in developing China" (Zak, 1993).

Based on a foreign policy that stems from a pragmatic need to guard its own national interests, China uses a strategy of separating the political from the economic when dealing with the Middle East to be able to escape the complexities of the Arab–Israeli conflict. From the Chinese perspective, its relations with the Arab countries were of a strategic importance that Beijing was not willing to easily risk with official and open relations with Tel-Aviv. To that end, China uses a soft diplomatic rhetoric and asks all the parties of the conflict to settle the conflict through peaceful means. At the same time, it maintains sensitivity to the Arab and Muslim cultures and tries to get closer to them on the cultural level to be able to secure overall cooperation, including economic collaboration. China needs not to lose the support of Arab countries because, most importantly, Arab countries are the main suppliers of oil for the Chinese growing economy, representing the source of 42 percent of China's crude oil imports in 2014 ("China," 2015). China also needs their support in international bodies on Taiwan ("Al-Seen wal-Arab," 2010). In 2009, China launched an Arabic-speaking satellite channel directed at the Arab world along with several cultural centers in some Arab countries ("Al-Seen wal-Arab," 2010).

In light of the mentioned pressures and the four stages the history of the bilateral relation passed through, China and Israel were able to advance their cooperation in several fields. Trade, education, technology, and arms are the main fields of cooperation among the two countries that are critical to evaluate for the purposes of this study. As will be illustrated, Sino–Israeli relations are in an advanced stage of close cooperation and interdependence in several fields, a fact that should be the basis for understanding the Israeli policy of arms technology exportation to China.

Sino–Israeli cooperation

Trade

Israel's overseas arms sales represent a fundamental portion of its international trade. Based on an agreement with the US to share arms technology, the Israeli overseas sales by 1981 had mounted up to $1.3 billion and placed Israel as the seventh largest exporter of military weapons in the world (Bitzinger, 1991). It was arms technology deals that incited the trade relations between China and Israel; since then, the balance of bilateral trade has been growing, and several expensive arms or arms technology deals have been uncovered. For China, Israel was a window for fresh air and a backdoor to the EU and US markets after the trade frictions following the June 1989 Tiananmen Square events, since Israel had free trade agreements with both entities (Han, 1993, p. 73). The Israeli reaction to the Tiananmen events was thus relatively and understandably weak. According to Jerusalem Post, then foreign ministry director-general Reuven Merhav, who also served as consul-general in Hong Kong, expressed "very deep sorrow at the hundreds of people who were murdered" by Chinese authorities at Tiananmen Square (Shalev, 1989).

Among the earliest formal bilateral trade agreements stand the 1992 Trade, Taxation and Investment Protection Agreement and the 1997 Agreement on Customs Mutual Administrative Assistance and Cooperation ("Bilateral Relations," 2011). In the same year of the establishment of the bilateral diplomatic ties, for example, a protocol of cooperation was signed in Tel-Aviv between Israel's largest bank, Bank Hapoalim, and the state-owned Chinese People's Construction Bank ("Banks Sign Deal," 1992). In 1994, the two countries signed a financial protocol, which was updated in 2004 and in 2010, with an aim of increasing Israeli exports to China and allowing the former to finance medium and long-term credit transactions to the latter. The total transactions based on this protocol mounted to more than $550 million ("Financial Protocol").

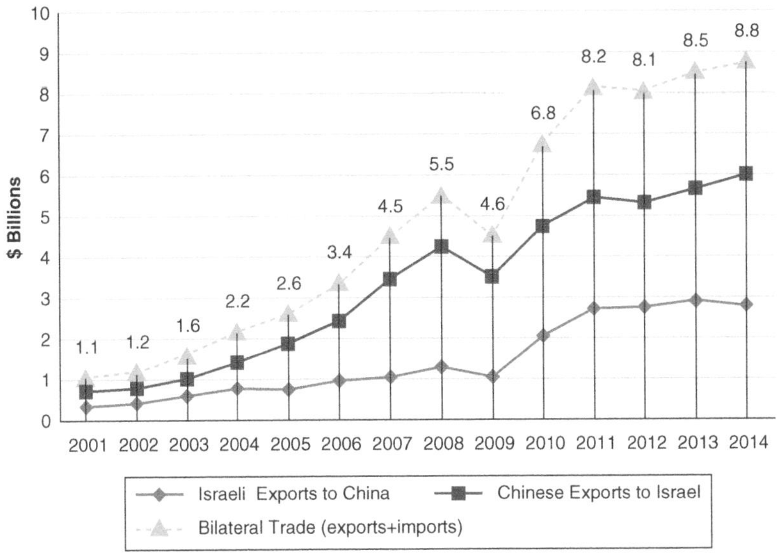

Figure 14.1 China-Israel Bilateral Trade (2001–2014).

Source: UN Comtrade through the International Trade Centre's Trade Map, which I extracted from this website: http://www.trademap.org/Bilateral_TS.aspx

The two-way trade volume was no more than $50 million when formal relations were established in 1992 ("Sino–Israeli Commercial"). However, since then, the overall volume of the bilateral trade increased 175-fold between 1992 and 2014 to become $8.8 billion, as illustrated in the following graph based on the UN Comtrade statistics. Figures also show that during the height of political tension between the two countries, as it was during the 2000 Phalcon crisis described below, trade ties were barely affected, signifying the importance that both sides gave to the economic angle of the relations ("Sino–Israeli Commercial").

Education

The field of education and cultural exchange witnessed the signature of several agreements, among which are the 1993 Agreement on Cultural Exchanges, 1998's Implementation Plan of Sino Israeli Cultural Exchanges 1999–2001, and the 2000 Agreement on Educational Cooperation ("Bilateral Relations," 2011). Chinese students started to be granted scholarships in Israel around 1994, and during the late 1990s, academic exchange programs were active and involved thousands of students ("Sino–Israeli Commercial"). Bar-Ilan University, for example, signed nine cooperation agreements with Chinese universities between 1993 and 2010 ("Cooperation Agreements"), which included more than 100 scholarships to Chinese students (Institute of Military Research, 1998, p. 109).

Based on an agreement signed in 2000, Israel is asked to support "China's program for the development of the West China Region" in "teaching and research in such hi-tech areas as electronics, biotechnology, material science, energy resources and agriculture" ("Educational Exchanges," 2004). One of the direct results of that was the effective transfer of Israeli agriculture expertise to China, which led the crop output in certain regions to be doubled ("Educational Exchanges," 2004). There are more than 20 universities in China with ties with Israeli ones in addition to the signature of nearly 20 Chinese universities twin-city agreements with Israeli cities. In 2011, Sichuan International Studies University (SISU) announced the launch of China's first-ever Israel studies program (Rosenberg, 2011), and in 2015, Tel Aviv University and Beijing's Tsinghua University agreed to establish a research center to research early-stage and mature technologies in biotech, solar energy, water, and environmental technologies (Pevzner, 2015).

Technology and industry

Technology is a field to which Israel has been giving due attention through significant investments. Israel ranks third in innovation among 140 countries studied in the Global Competitiveness Index (Schwab, 2015) and has been number one internationally in 2011, 2012, and 2013 in terms of the percentage of GDP spent on research and development ("Research and Development").

In 2013, machines and technology accounted for 42 percent of the total Israeli exports to China ("What does Israel"). Israeli expertise in agriculture was transferred to China through close technological and industrial cooperation. Among the earliest agreements in this field is the 1993 Memorandum of Understanding on Agricultural Cooperation ("Bilateral Relations," 2011). In the same year, Israel Chemicals Ltd. (ICL) declared the approval of plans to establish a potash plant in western China through its subsidiary company Dead Sea Works ("DSW OKs China," 1998), world's fourth largest producer and supplier of potash ("About Us"). The Israeli know-how is offered (Zak, 1993) in return for a 33 percent share in the $486 million fertilization complex carried out as a joint venture with the Chinese government ("DSW OKs China," 1998). Overall, 29 percent of Israeli exports to China in 2013 were chemical products, which primarily include potash ("What does Israel").

ICL, along with its five subsidiary companies, is placed under the partial control of the Israeli government defined in terms of "special state share." ICL cannot sell more than 14 percent of its capital shares or grant the right of appointment for half or more of its directors to a shareholder without the permission of the state of Israel. In addition, the management and control of ICL cannot be moved outside of Israel, majority of directors and board members must be Israeli citizens and the government of Israel holds the right to access information about ICL in order to maintain "vital state interests" ("Israeli Chemicals Ltd," 2014).

China, the world's largest potash buyer, is dependent on imports to satisfy its needs of potash, which the fertilizer sector depends on. In 2014, for example, China imported about 60 percent of its potash requirements, estimated to be around 12 million tons per year (Tan, 2015). ICL plays a key role in satisfying those Chinese needs. For instance, as part of the cooperation that started in the 1990s, in 2006, ICL signed a contract with a Chinese consumer for at least two million tons of potash over 3 years, which is 30 percent larger than the previous 3-year contract (Rosenberg, 2006), and another contract for 3.4 million tons in 2015 ("ICL Reports Q4," 2016). Another significant deal in the agricultural field was the $2.4 billion Chinese acquisition of an Israeli pesticide company in 2010. China is already the world's largest user, producer, and exporter of pesticides with a significant domestic surplus, which pushes some analysts to argue that the deal is intended to allow China to make use of the Israeli company's distribution channels ("China Spends $2.4," 2011).

Besides being occupied with transferring Israeli expertise in the field of agriculture, a big portion of the technological and industrial Sino–Israeli cooperation is based on the transfer of military technology. There are several incidents that indicate the continuous collaboration between the Israeli government and firms with the Chinese government and firms to transfer cutting-edge technologies including ones with military applications.

Arms

China is the second country in the world in its military expenditures, right after the US, with an estimate of $190 billion in 2014 ("SIPRI Military Expenditure," 2015). China sits second on the list of the world's arms recipients and consumes 5 percent of all the world's arms imports ("International arms transfers," 2014). The People's Liberation Army (PLA) of China underwent major changes since the 1980s towards focusing on quality rather than quantity, which resulted into major cuts in the number of soldiers. However, for more than two decades, China has been able to maintain its position as the world's largest army in terms of the number of active duty military personnel, with an average ratio of 2:1 to its US counterpart since 1989 until present day ("Armed Forces Personnel").

China first showed its interest in Israel's arms technology in 1975. Several Chinese officials visited the Israeli pavilion at the Paris Air Show in June 1975 to see the Kfir fighter jet that was being shown for the first time (Melman, 1987, p. 403). Four years later, the Israeli businessman Shaul Eizenberg arranged a visit to Beijing with representatives of major industrial and military Israeli companies that included Israeli Aircraft Industries (IAI), the manufacturer of Kfir and Lavi fighter jets (Melman, 1987, p. 403).

The history of arms sales and military cooperation between China and Israel has been witnessing ups and downs since the 1980s. Both countries were conducting their military cooperation efforts within a lot of limitations dictated on them by their mutual close relations with other countries, especially the US in the case of Israel and the Arab countries in the case of China. From the early 1980s to the mid-1990s, Israel was the second most important seller of military technology to Beijing ("Procurement, China: Jane's," 2010). The growing Chinese demand in the 1980s was met by the Israeli supplier ("External Affairs,

Israel," 2010), with Israel's arms technology exports to China worth between $1 and $3 billion (Bitzinger, 1991).

Israeli military technology, along with technology from other suppliers, was sold by China to other customers in the Middle East during the 1990s and later. From the early to the mid-1990s, the Middle East represented the most active market for Chinese arms, representing over 50 percent of Chinese arms exports ("Chinese Arms Exports," 2000). Since then, Egypt, Iran, Sudan, and Turkey have represented the main regional clients for China among its 14 other Middle Eastern and Arab clients ("SIPRI Arms Transfers"). Although such a step by Beijing might harm Israel's regional stability, Tel-Aviv was able to indirectly sell its adversaries technology that it can defeat in the case of a confrontation (Bryen & Ledeen, 1997).

One path for arms technology from the US to some of the Arab countries is through Israel and China. Selected US arms technology is transferred to Israel under the annual military aid pack and as part of the "Technical Data Packages" that were laid down in the 1970s Master Defense Development Data Exchange Agreement, allowing Israel the greatest ever technology transfer rights from the US (Bitzinger, 1991). Based on several occasions, detailed later in this paper, Israel would then sell out the technology to China, which, in turn, would pass it on to selected Arab countries. Among the examples of such arms technology transfer is the US-made heat-seeking air-to-air missile AIM-9L Sidewinder.

The US started producing and delivering the AIM-9L in 1976 ("AIM-9 Sidewinder," 2000). By the mid-1980s, Israel was able to deploy its Python 3, which is developed from the AIM-9L ("PL-9/DK-9," 1999), through the Rafa'el Armament Development Authority ("Python 3," 1999). By the late 1980s, Israel had sold its Python 3 to China ("PL-9/DK-9," 1999) and in 1991, China was able to unveil its copy of the Python 3 ("EP-3E Incident," 2001) through Beifang Industrial Company, calling it the PL-9 ("PL-9/DK-9," 1999). Among the early candidates to receive the Chinese helmet-sighted PL-9 was Iran (Lilley & David, 1999) and today the air-to-air missile shows in the Iranian missile arsenal through a Chinese manufacturer in the name of China Precision Machinery Import–Export Corporation (CPMIEC) ("Iran, Air Force," 2010). Thus, almost a decade passed between the Israeli deployment of the Python 3 and the Iranian acquisition of the same missile, which allowed Israel the time needed to develop its missile to overcome the capabilities of the earlier technology transferred to Iran.

Israel was thus able to win the financial leverage from selling the expensive technology to China while passing to its adversaries certain arms technology it knew very well, including knowing its weaknesses and how to defeat it (Bryen, 1997). In addition, Israel was benefiting—indirectly but controllably—from strengthening its Arab and regional adversaries, because it allowed Tel-Aviv to ask for more US military aid to counter the growing regional threat (Neff, 1997).

However, Israel is confronted with the US in its arms relations with China. Since the US remains the main supplier of arms and military technology to Israel—accounting for 80 percent, followed by Germany (17 percent) and

Italy (2 percent) of all Israeli arms imports over the past 20 years ("SIPRI Arms Transfers")—it has the power to cancel Israel's deals with China when it deems them to be threatening. The US–Israeli agreement on technology transfer is guarded by US laws and regulations that firmly prohibit the transfer of the technology to a third party without the approval of the US ("Arms Export Control," p. 372). Yet, several incidents surfaced about such deals leading to the embarrassment of the Israeli government and in some cases the eventual cancellation of the deal and occasionally causing damage to Israel's relations with China.

Following are some specific cases that revealed the mentioned Sino–Israeli illegal arms cooperation. These cases are mainly focused on missiles technology, fighter jets, early warning systems, and unmanned aerial vehicles.

Patriot and SAM systems

A declassified CIA 1992 report revealed the existence of "very specific details" about the transfer of US Patriot missile technology by the Israeli government to China ("Israeli Patriot Missile"). A joint US State-Defense Department team failed to find any evidence of the unauthorized transfer (Ehrlich, 2008, p. 260), while at the same time the State Department inspector-general released a report suggesting that Israel had transferred other US arms technology without US permission (Mark, 2005). In 1997, the US Office of Naval Intelligence suggested that "technology from advanced Western systems may be incorporated into the HQ-9" ("HQ-9/FT-2000," 2000).

With the transfer of this technology beginning in 1987, Israel was aiming to gain trade rights with China in exchange for the technology. China needed the Patriot technology to improve its surface-to-air missile (SAM) systems and to develop its countermeasures against the US Patriot ("China's Missile Imports," 2003). With successful trials as early as 1999 ("China, Self-propelled," 2010), China was able to display its advanced HQ-9 SAM at the Africa Aerospace and Defense Exhibition in 2009. Despite the fact that the HQ-9 is partially based on the best US Patriot system guidance technology ("China, Self-propelled," 2010) passed to China through Israel's assistance ("China's Missile Imports," 2003), its range against conventional air targets exceeds that of the Patriot to the extent that it can be compared to the US THAAD theater-level defense system ("China Gearing up," 2009). With the proliferation of this advanced air defense missile/system, depending on its exportation price, US will be facing an extending challenge to its air dominance if the missile reaches militaries in the third world ("China Gearing up," 2009).

A recent challenge that the HQ-9 is posing to the US regionally is the Turkish interest in acquiring the Chinese long-range air and missile defense system without integrating it into the infrastructure of NATO. The $3.4 billion deal that was to be signed in 2013 between Turkey and China Precision Machinery Import and Export Corp (CPMIEC) prompted Western concerns since CPMIEC has been under US sanctions for selling items related to weapons of mass destruction (WMD) to Iran, Syria, or North Korea (Coskun & Karadeniz, 2015).

Cruise missiles

In addition to the Patriot system guidance technology, Israel helped China develop its cruise missiles. Reports refer to several kinds of cruise missiles involved in such technology transfer, including the YJ-12A (anti-ship/medium range/supersonic); YJ-62 (anti-ship/long range); YJ-91 (air-launched/ anti-radiation); and sharing technology of Israeli land-attack cruise missiles based on the Star-1 attack drone ("China's Missile Imports"; Twing, 1996; "Procurement, China: Jane's," 2010; Gurtov & Hwang, 1998, p. 228).

Other missile-related technology transfer includes a cooling system manufactured by Ricor, which can be used for missiles. In December 2013, a joint US–Israeli investigation concluded that instead of selling it only to a French company as agreed with the US, Israel retransferred the miniature cooling system to China (Sharp, 2015). Meir Shalit, then head of Israel's Defense Export Control Agency (DECA), resigned right after.

Lavi and fighter jets

Israel was able to reach an advanced stage in developing its multipurpose jet-fighter "Lavi" in the 1980s (Ehrlich, 2008, p. 260) using more than $1.3 billion of US funds along with US technical assistance (Mark, 2005). Lavi shared a lot of characteristics with the US-built F-16 in Israel's arsenal (Mark, 2005). Yet, because the designer— IAI—failed to accomplish a design that is 100 percent Israeli-built (it was all Israeli up to 95 percent), which would expose the jet to the vulnerabilities of dependence on foreign suppliers, the project was frozen; the design was made use of by selling the Lavi technology to China (Ehrlich, 2008, p. 261).

Consequently, by 1993, China possessed an all-metal mockup of its single-seat light multirole fighter the J-10 ("Chengdu J-10"), which turned out to be heavily based on the Israeli prototype Lavi (Heginbotham, 2015, p. 31; "Worldwide Challenges;" "External affairs, Israel," 2010). It was revealed in early 2010, according to Jane's, that the engine intake and the vertical stabilizer shape of the early design concepts for the J-10 were based on one of Lavi's prototypes ("Procurement, China: Jane's," 2010). In addition, the J-10 was dependent on the fly-by-wire technology that was first developed by Lockheed-Martin for its F-16 (Fisher, 2007). When the J-10 was revealed for the first time in 2006, it was clear that it had the potential to become one of the most significant fighters in the next few decades ("Chengdu J-10"). By the end of 2014, 294 J-10s were built on a production rate of about 30 aircraft per year (Heginbotham, 2015).

Phalcon

The Phalcon incident, which involved four Israeli prime ministers up to Ariel Sharon, is considered one of the most serious rifts in the US-Israeli relations

(Ehrlich, 2008, p. 263). Following the successful design and delivery of the Israeli-designed Airborne Early Warning and Control (AEW&C) radar Phalcon system to Chile in 1994, Israel signed a $1.3 billion agreement with China to install the Phalcon reconnaissance platforms on a number of large jet aircrafts that were later imported from Russia (Ehrlich, 2008, p. 262). The Israeli-developed system is claimed to be superior to its US counterpart E-3A Sentry Airborne Warning and Control System (AWACS) (Mark, 2000). In 2000, the Clinton administration expressed its concerns about the Phalcon deal and forced Israel to cancel the deal with China at its late phases before delivery, and, later in 2002, Israel was forced to pay China $350 million in compensation (Barzilai, 2002). The US had concerns about the Chinese possible usage of the Phalcon over Taiwan and the Taiwan Straits in addition to the negative impact on US regional allies such as Japan and South Korea. Moreover, there was a US fear that China might be able to reverse-engineer the system and solely develop it in the future (Barzilai, 2002). However, the Israeli spokesman at the time did not rule out future transfers when he announced the cancellation of the deal on July 12, 2000 (Barzilai, 2002). The incident caused considerable turbulence in the US–Israeli and Sino–Israeli relations.

The Phalcon incident was one that made Israel "on the brink of endangering its strategic alliance with the United States," according to NYTimes columnist William Safire writing in 2000 (Zacharia & Harman, 2000). The then chairman of the US House Appropriations Subcommittee on Foreign Operations, Republican Congressman Sonny Callahan, who controls foreign aid, threatened "to find a mechanism to withhold about $250 million, which is the equivalent of the sale of proceeds going to Israel" (Zacharia, 2000).

As for Sino–Israeli relations, although the commercial side was not as highly impacted ("Sino–Israeli Commercial"), the Phalcon incident caused a diplomatic rift between the two sides. The cancellation of the deal came right after the first visit of a Chinese president to Israel, accompanied by several guarantees and promises by the Israeli Prime Minister Ehud Barak that the deal would go through regardless of all the pressures (Shai, 2009, p. 27). Through the Phalcon deal, Israel was targeting hard currency from China as well as the Chinese diplomatic support internationally. Israel also aimed at gaining leverage in influencing Chinese policy when it came to Chinese arms transfer to Iran. Some Israelis believed that this deal could also make China play a more productive role in the Middle East peace process (Shai, 2009).

Harpy and UAVs

Resembling the Phalcon incident to a degree, the Harpy was also based on Israeli technology and resulted in public criticism directed against Israel from the US administration and eventually led to the cancellation of the Israeli deal with China. The nature of the Chinese need for the antiradiation unmanned aerial vehicle (UAV) was also similar to that for the Phalcon, which was to be able to monitor and control Taiwan and the Taiwan Straits. Yet, UAVs—commonly known as drones—represent a field that China is rapidly catching up in for domestic and exportation needs. Chinese cyber attacks against US defense contractors have been spotted since 2011 with the aim of acquiring the advanced US technology behind the military UAVs (Wong, 2013). Chinese efforts to develop UAVs began back in 1958 with the establishment of China's first UAV research and production center at the Northwestern Polytechnical University. Since then, university-based programs have been representing the backbone of UAV development and among the most established in China. In addition, private and state-owned firms play a key role in research and development of this flourishing sector in China (Hsu, 2013, p. 12; Wong, 2013). The total number of Chinese entities working in UAV-related technology is believed to be between 75 and 100 (Rawnsley, 2016). At the same time, between 1985 and 2014, Israel accounted for 60.7 percent of the drone exports worldwide, surpassing the US's 23.9 percent (Arnett, 2015).

IAI's Harpy is a part-UAV and part-missile—falling under the category of unmanned combat aerial vehicles (UCAV)—in which the entire aircraft acts as a missile when locking onto a target. Israel was said to have taken the Clinton administration's permission to sell an espionage version of the drone to China in 1994 (Ehrlich, 2008, p. 263). Yet, Israel then took the initiative to "repair" the drones for China with no due permission from the US. The repair process was a cover for the development and upgrade of the drone (Ehrlich, 2008, p. 263). In 2004, Bush's administration forced Israel to return a Harpy drone to China with no upgrading ("U.S. Pressure Threatens," 2005) in a tension that reached the degree of making the US undersecretary of defense for policy, Douglas Feith, demand the resignation of the top bureaucrat in the Israeli Defense Ministry, Amos Yaron (Ehrlich, 2008, p. 264). Under American pressure, Israel was forced to sign an agreement that would allow the US to review Israel's future weapons transactions (Wilson, 2005).

The Chinese PLA first revealed its interest in UAVs when it conducted its first pilotless successful flight in 1972. Chinese forces were able to shoot down dozens of the US AQM-34 drones—also called Ryan Firebee—during the Vietnam War in the 1960s and to acquire Soviet UAVs in the 1950s. The American drones were reverse-engineered to produce the Chinese WuZhen-5—also referred to as WZ-5 and ChangHong-1 CH-1—that entered into service in 1981 and is still part of China's UAV fleet ("Wu Zhen 5 (WZ-5)," 2011). In 1994, the first Chinese-developed advanced tactical UAV, the twin boom tail pusher ASN-206, came to light ("China, World Armies," 2010). Although unconfirmed sources refer to the involvement of the Israeli company Tardiran Spectralink Ltd. in the development of the ASN-206 (Kospoth, 2009), the large magnitude of Sino–Israeli cooperation in the field of UAV technology was later to become clear.

Resemblances between the Chinese UAVs and the Israeli ones are obvious; Chinese companies like the Nanjing Research Institute on Simulation Technique (NRIST) produce tactical UAVs that are similar to the Israeli Hunter developed by the IAI ("China, World Armies," 2010). In 2006, the Israeli aviation technology company EMIT was caught shipping UAV technology to China ("China Closes the UAV," 2010).

There are several reasons to believe that Israeli companies are still sharing UAV technology—among other kinds of military technologies, such as the aforementioned 2013 Ricor incident—with their Chinese counterparts under governmental approval on both sides. The Chinese reaction to the cancellation of the upgrade of the Harpy in 2004 was alarming although not made on an official level. As mentioned before, the UAVs are of strategic importance for Beijing for the vast territories it has to control and monitor, especially Taiwan and the Taiwan Straits; hence, an article written by the fellow of the Chinese government-linked Institute for International Studies during the height of the Harpy crisis can only be understood in this context. Jin Liangxiang wrote in 2005 that "if Israel does not meet its commitments to upgrade the drones, contractual terms would lead Beijing to launch sanctions on Israeli enterprises not only on the Chinese mainland but also in Hong Kong" (Ehrlich, 2008, 264).

IAI—the largest government-owned aerospace and defense company in Israel ("Company Profile")—was the one that sold the Harpy to China in 1994 and still has major investments in the Chinese market with five airline companies on its unclassified customers list ("Bedek Aviation Group"). In addition, 2016 witnessed the establishment of a joint venture between IAI and the state-owned Chinese Lingyun Science and Technology Group Co., Ltd. to establish its first local Chinese maintenance, repair, and overhaul (MRO) enterprise in the Hubei Province ("IAI to Cooperate," 2016).

One of the most important possible pieces of evidence that needs investigation is the nature of the UAVs/UCAVs that appear in Chinese air shows. In November 2010, Chinese companies revealed 25 new different UAV models during the 8th China International Aviation and Aerospace Exhibition in Zhuhai. The increase in the number of the UAVs in that year's show, along with the sophistication of the unmanned civil and military aircrafts, caused surprise among Western defense officials and experts (Page, 2010) because it means that China is catching up with the US and Israel, who are the world's leading countries in UAV production and technology.

The models revealed the increasing attention the PLA is giving to UAV technology (Arthur, 2010). One model that was especially interesting if compared to the US models in the same category is the jet-powered WJ-600. It can—in theory, since most of the UAVs exhibited are still in production and not on the market yet—fly faster than the US Predator and Reaper drones (Page, 2010). The appearance of such models coincides with the release of a report on China's military to the US Congress by the US–China Economic and Security Review Commission that referred to the Chinese work on developing a variety of medium and high-altitude long-endurance UAVs with options for long-range reconnaissance and strike (Minnick, 2010). The WJ-600 was developed further and reappeared in the 2012 Chinese Zhuhai air show (Hsu, 2013).

The 2014 Zhuhai air show revealed an even more open Chinese appetite for UAVs. The US Department of Defense, based on watching the air show and evaluating the development of the Chinese UAVs, warned that China will acquire nearly 42,000 UAVs between 2014 and 2023, worth $10.5 billion, with some of these systems expected to be strike capable. China already began incorporating its UAVs in military exercises in 2013 and conducted intelligence, surveillance, and reconnaissance (IRS) over the East China Sea with the BZK-005 UAV ("US DoD Annual," 2015; "Annual Report to Congress," 2015).

As expected, Chinese UAVs are making their way onto foreign markets. During the 2010 Zhuhai air show alone, Chinese aviation companies signed 28 contracts totaling $9.3 billion (Grevatt, 2010), and Iran is a potential customer (Page, 2010). Until the present, the countries purchasing Chinese UAVs are either the country's traditional allies or the Middle Eastern oil exporting countries ("China Starting," 2015).

In 2014, China demonstrated its ASN-209—a medium-altitude, medium-endurance aircraft with an upgraded version (ASN-209G) able to carry two laser-guided missiles—at the Africa Aerospace and Defense Expo held in South Africa in an attempt to attract foreign customers. "We've been contacting many countries, especially from Africa and Asia," Guo Qian, a director at a division of the state-owned China Aerospace Science and Technology Corporation (CASC), said in 2012 when discussing the company's interest in marketing the military-grade UAVs. Countries "are quite interested in the intermediate and short-range UAVs because they are portable and low-cost" (Standaert, 2012).

Among the Middle Eastern countries that have cooperation with China in UAVs is Egypt. The year 2012 witnessed Egypt's signature of an agreement with China to build the ASN-209 domestically under Chinese license ("Warplanes: Egypt Builds," 2012) and negotiations to buy a "whole package of air defense system" that also includes a medium-range CH-91 UAV (Standaert, 2012).

The Chinese Wing Loong UAV—also called the Pterodactyl I—was sold to Egypt, the United Arab Emirates (UAE) (Lin, 2015), Saudi Arabia ("Saudi Arabia Signs," 2015), and possibly Jordan (Gady, 2015). The medium-altitude long-endurance UAV can be fitted with missiles and bears a striking resemblance to the US Reaper/Predator while standing at about one-fourth of the price (Bodeen, 2013). The more advanced version of the UAV—Wing Loong II—appeared in the 2015 Beijing Air Show and is shown to be capable of carrying 12 air-to-surface missiles (Fisher & Wong, 2015). Overall, the Chinese UAV industry is witnessing diversification and expansion in production and exportation, especially given that Chinese defense firms are not subject to the same export restrictions imposed on their American and Israeli counterparts and enjoy a lower production cost (Hsu, 2013).

Another related field that has been witnessing significant Chinese development is cyber security. China's PLA seems to have developed its new concept of "information confrontation" as an update to its older "Integrated Network

Electronic Warfare" doctrine that used to guide its computer-based operations. Since 2008, according to IISS's 2016 Military Balance report, PLA has been integrating cyber and information-operations components, both defensive and offensive, to its military exercises.

Over the past few years, Chinese cyber attacks against the US have been growing with systemic exfiltration of data from American companies. Since this sector of technology falls in a grey area between the civilian and military applications and Israel enjoys a global leadership in it, a new debate arises as to whether there is cooperation between China and Israel that would harm US interests (Chester, 2013). University-based collaboration between the two countries can safely put this area outside of the American radar and save Israel from directly violating its 2005 agreement with the US.

Implications

With the rapid growth that the Chinese economy is showing, along with the increasing attention given to the military, China could soon catch up on certain strategic weapons and overtake US technology, as can be inferred from the examples above, including UAVs and missiles, through advanced Israeli technologies and Tel Aviv's access to US advanced weaponry. In such a case, through utilizing the growing technically educated manpower and with the Sino–Israeli ongoing arms trade and technical cooperation, China would be able to increase its arms exports to the Middle East because of the advantage that China has in terms of cheap labor and overall production costs.

Chinese foreign policy is relatively limited to securing domestic needs for prosperity and economic growth. In the Middle East, the main Chinese interests involve energy security and market access to the region and through its strategic location in connecting the markets in Asia, Africa, and Europe. Towards those ends, and with its change in foreign policy towards a proactive rather than responsive one following the 2003 US Iraq invasion, China is utilizing two strategies; first, establishing what is called the New Silk Road, and secondly projecting its military strategy from being "coastal defense" to "far seas defense." Tools used to achieve the mentioned strategies are military, political, and economic in nature (Lin, 2013).

China, understandably, is seeking to modernize and develop its military power. The close relations between China and Israel can significantly harm the US regional hegemony by allowing China an increasing political leverage to shape actors' decisions in the Middle East, including Israel, which would threaten the strategic and special US–Israeli relations. Furthermore, China's exportation of military technology, which is partly based on Israeli technical assistance, would help proliferate advanced military technology that would destabilize US-orchestrated regional power balances in the Middle East.

Israel's illegal transfer of US arms technology to China gives the latter a long-sought technological leverage, which, combined with cheap labor, allows it a big push ahead in the global arms market. The obvious repercussions include shifting the Arab countries' arms markets and political preferences more in China's favor. Currently, China is the world's third-largest arms exporter after the US and Russia; over the next decade, China is expected to become a global supplier of advanced weapons platforms (Lin, 2013).

Until 2014, based on the available data, there is no increase in the Chinese conventional arms exports to the Arab countries (Cordesman, 2016). Yet, frequent news about Chinese–Arab UAV exportation deals has been surfacing recently with clients including Saudi Arabia, the United Arab Emirates, Egypt, and Iraq (Rawnsley, 2016). Part of the reason why those kinds of deals do not all appear in official reports is that the Chinese UAV marketing strategy revolves around "price, privacy and product," since the clients and the supplier might face Western opposition when disrupting the Western-imposed regulation on the proliferation of such advanced technology (Rawnsley, 2016). Such advantages that the Chinese UAVs enjoy in terms of privacy of deals, price, relative quality, and easier sales approvals make them the preferred option over the US ones, even though the US State Department loosened its export restrictions in February 2016 (Rawnsley, 2016).

The dominant belief since Operation Desert Storm that information superiority and precision weapons can easily defeat mass forces is now being questioned. China's ability to produce and export smart, small, and cheap technically advanced weapons, especially UCAVs, dwarfs precision weapons and allows developing nations to acquire capabilities that were previously reserved for great powers (Hammes, 2016). Such development would open up some policy options for some Middle Eastern countries—as well as nonstate actors—that were previously under the classical US regional dominance. The acquisition of "anti-access/area denial" capabilities through China would also make an American military intervention in the Middle East a more costly decision to take. Moreover, with the existence of a high level of violence in many countries in the Middle East, the proliferation of Chinese weapons would certainly set the whole region on fire and push it out of control. The scene becomes even more complicated when considering the growing power of nonstate actors and their various regional and international alliances.

The tension in US–Israeli relations with regards to espionage and the illegal transfer of sensitive technology has been present in the background for some time, yet it gets portrayed in public only occasionally. In a Snowden-leaked NSA 2007 Strategic Mission List, Israel was named as the "third most aggressive intelligence service against the U.S.," coming right after China and Russia (Stein, 2014). Israel was also listed as a leading perpetrator of "espionage/intelligence collection operations and manipulation/influence operations… against U.S. government, military, science & technology and Intelligence Community" organs (Stein, 2014). Another 2008 NSA top-secret document revealed that the CIA regards Israel as a "priority threat country," along with China, Russia, Iran, Pakistan, and Cuba (Greenwald & Fishman, 2015).

At the same time, there is no evidence that the Chinese or Israeli sides have incentives to give up their special relationship. Israel is gaining well financially from trading with one of the world's fastest growing economies, one with apparent and rising military needs. The bigger the Chinese economy grows, the bigger the Chinese appetite will be in order to protect its stretching interests worldwide, and the more it will be willing to pay to acquire the latest military technologies from the best developer, which happens to be Israel in key fields. Listed incidents show that Israel is willing to take the risk as long as it thinks it will be able to keep secretly selling US technology or its own technology that is based on US assistance. At the same time, China is gaining international political influence, besides protecting its interests, through developing and re-selling the technology acquired from Israel. In fact, as P. R. Kumaraswamy rightfully argues, Israel will head towards a more open cooperation with China as American economic and political power weakens (Kumaraswamy, 2013).

It remains important to say that, based on the listed incidents, a solution in the form of an agreement to limit the Israeli ability to transfer US military technology to China might not be the best choice for the US to make. As discussed, different agreements and incidents have failed to realize that end. More solid measures such as the revision of the US–Israeli 1970 Master Defense Development Data Exchange Agreement should be discussed. In addition, the US should continue investing in edgy military technologies and not share them with any other country, whether through research and development or exports of the final product. It is in the American vital interest to maintain its technical superiority in order to protect its interests from the growing Chinese dragon that is feeding off America's own technology.

References

Abadi, J. (2004). *Israel's Quest for Recognition and Acceptance in Asia: Garrison State Diplomacy*. London: Frank Cass.

About Sino–Israel Relations. (2009, September 7). *China.org.cn*. Retrieved from http://www.china.org.cn/culture/expriencechina/2009-09/07/content_18475777.htm

About Us. *Dead Sea Works Ltd*. Retrieved from http://www.iclfertilizers.com/FERTILIZERS/DSW/Pages/AboutUs.aspx

Address in the Knesset by Prime Minister Designate Peres, Upon the Presentation of His Government, 13 September 1984. (1984, September 14). *Israel Ministry of Foreign Affairs*. Retrieved from http://www.mfa.gov.il/MFA/Foreign+Relations/Israels+Foreign+Relations+since+1947/1984-1988/3+Address+in+the+Knesset+by+Prime+Minister+Designa.htm

AIM-9 Sidewinder. (2000, April 23). *Federation of American Scientists (FAS)*. Retrieved from http://www.fas.org/man/dod-101/sys/missile/aim-9.htm

Al-Seen wal-Arab: Men Tareeq al-Hareer ila Hareer al-Deplomaceyyah (China and the Arabs: From the Silk Road to the Silk of Diplomacy). (2010, September 26). *AlJazeera.net*.

Annual Report to Congress: Military and Security Developments Involving the People's Republic of China 2015. (2015, Apr. 7). *Office of the Secretary of Defense*. Retrieved from http://www.defense.gov/Portals/1/Documents/pubs/2015_China_Military_Power_Report.pdf

Armed Forces Personnel, Total. *The World Bank*. Retrieved from http://data.worldbank.org/indicator/MS.MIL.TOTL.P1/countries/1W

Arms Export Control Act, P.L. 90-629. *Federation of American Scientists (FAS)*. Retrieved from http://fas.org/asmp/resources/govern/aeca00.pdf

Arnett, G. (2015, March 16). The Numbers behind the Worldwide Trade in Drones. *The Guardian*. Retrieved from http://www.theguardian.com/news/datablog/2015/mar/16/numbers-behind-worldwide-trade-in-drones-uk-israel

Arthur, J. (2010, November 19). UAVs Showcased at Airshow China: Jane's Defense Weekly. *IHS Jane's*.

Banks Sign Deal In Israel, China. (1992, November 23). *San Francisco Chronicle*.

Barzilai, Amnon. China to Get $350M for Lost Phalcon Deal. (2002, March 14). *Haaretz*. Retrieved from http://www.haaretz.com/china-to-get-350m-for-lost-phalcon-deal-1.50444

Bedek Aviation Group. *Israeli Aerospace Industries (IAI)*. Retrieved from http://www.iai.co.il/sip_storage/FILES/7/38717.pdf

Bilateral Relations: Bilateral Treaties and Agreements. (2011, July 24). *Consulate General of Israel in Guangzhou*. Retrieved from http://embassies.gov.il/guangzhou-en/Relations/Mashav/Pages/event.aspx

Bitzinger, R. A. (1991). Chinese Arms Production and Sales to the Third World, *RAND Corp*. Retrieved from http://www.rand.org/pubs/notes/2009/N3334.pdf

Bodeen, C. (2013, May 3). China Emerging as New Force in Drone Warfare. *The Associated Press*. Retrieved from http://phys.org/news/2013-05-china-emerging-drone-warfare.html

Bryen, S. & Ledeen, M. (1997, June 27). China-related Challenges. *Federation of American Scientists (FAS)*. Retrieved from http://www.fas.org/news/china/1997/970627.htm

Calabrese, J. (1991). *China's Changing Relations with the Middle East*. London and New York: Printer Publishers.

Chengdu J-10 (Jian-10 Fighter Aircraft 10) / F-10. *Global Security*. Retrieved from http://www.globalsecurity.org/military/world/china/j-10.htm

Chester, S. (2013, June 28). As Chinese-Israeli Relations Enjoy a Second Honeymoon, America Frets. *Tablet Magazine*. Retrieved from http://www.tabletmag.com/jewish-news-and-politics/136348/china-israel-united-states

China Closes the UAV Gap. (2010, November 19). *Strategy Page*. Retrieved from http://www.strategypage.com/htmw/htairfo/articles/20101119.aspx

China Gearing up to Export HQ-9 Anti-Air Missiles. (2009, March 8). *Defense Industry Daily*. Retrieved from http://www.defenseindustrydaily.com/China-Gearing-up-to-Export-HQ-9-Anti-Air-Missiles-05319/

China, Self-propelled Surface-to-air Missiles: Jane's Land-Based Air Defence. (2010, March 15). *IHS Jane's*.

China's Missile Imports and Assistance from Israel. (2003, March 28). *The Nuclear Threat Initiative (NTI)*. Retrieved from http://web.archive.org/web/20090328203127/http://www.nti.org/db/china/imisr.htm

China Spends $2.4 Billion…On Pesticides? (2011, January 11). *The Wall Street Journal*.

Retrieved from http://blogs.wsj.com/chinarealtime/2011/01/11/china-spends-24-billionon-pesticides/

China: Stalemate Continues; China/Israel: Confidential, Secretary's Morning Summary, June 2, 1989. *Digital National Security Archive*.

China Starting to Export its Military Drones. (2015, January 11). *iHLS*. Retrieved from http://i-hls.com/2015/01/china-starting-export-military-drones/

China, World Armies. (2010, October 15). *Jane's World Armies, IHS Jane's*.

China. (2015, May 14). *US Energy Information Administration*. Retrieved from https://www.eia.gov/beta/international/analysis.cfm?iso=CHN

Chinese Arms Exports: Policy, Player and Process. (2000, August). *The Strategic Studies Institute*. Retrieved from http://www.fas.org/nuke/guide/china/doctrine/chinarms.pdf

Company Profile. *Israeli Aerospace Industries (IAI)*. Retrieved from http://www.iai.co.il/2013/10285-en/CompanyInfo-CompanyProfile.aspx

Cooperation Agreements. *Bar-Ilan Univeristy Research Authority*. Retrieved from http://research.biu.ac.il/content/cooperation-agreements

Cordesman, A. H. (2016, February 16). The Changing Patterns of Arms Imports in the Middle East and North Africa. *Center for Strategic and International Studies*. Retrieved from http://csis.org/files/publication/160205_cordesman_changing_arms_race_MENA.pdf

Coskun, O., & Karadeniz, T. Turkey Goes Back to Other Missile System Bidders as China Drags Feet: Officials. (2015, February 26). *Reuters*. Retrieved from http://www.reuters.com/article/us-turkey-defence-china-idUSKBN0LU1KD20150226

DARPA-RA-10-03: Computer Science, Science, Technology, Engineering, and Mathematics (CS-STEM) Education Research Announcement (RA). (2010, January 29). *DARPA*. Retrieved from http://www.darpa.mil/ipto/solicit/baa/RA-10-03_Mod01.pdf

DSW OKs China Venture. (1998, September 11). *Jerusalem Post*. Retrieved from https://www.highbeam.com/doc/1P1-17260085.html

Educational Exchanges. (2004, September 21). *Embassy of the People's Republic of China in the State of Israel*. Retrieved from http://il.china-embassy.org/eng/jyjl/t159822.htm

Education Indicators in Focus. (2015, April). *OECD*. Retrieved from http://www.oecd.org/education/EDIF%2031%20(2015)--ENG--Final.pdf

Ehrlich, M. A. (2008). *The Jewish-Chinese nexus: A Meeting of Civilizations*. New York, NY: Routledge.

EP-3E Incident Confirms Chinese PL-9 is a Python 3 Copy: Jane's Missile and Rockets June 1, 2001. (2001, May 29). *IHS Jane's*..

External Affairs, Israel: Jane's Sentinel Security Assessment – Eastern Mediterranean. (2010, October 26). *IHS Jane's*.

Financial Protocol. *Israel Trade & Economic Office, Embassy of Israel in China*. Retrieved from http://itrade.gov.il/china-en/doing-business-with-china/financial-protocol/

Fisher J., Richard D., & Wong, K. (2015, September 18). China unveils next-gen Wing Loong II UAV. *IHS Jane's Defense Weekly*. Retrieved from http://www.janes.com/article/54648/china-unveils-next-gen-wing-loong-ii-uav

Fisher, R. (2007, January 16). China's J-10 Jet Fighter: How Much Do We Know? *International Assessment and Strategy Center*. Retrieved from http://www.strategycenter.net/research/pubID.140/pub_detail.asp

Friedman, T. L. (1985, July 22). Israel and China Quietly form Trade Bonds. *The New York Times*. Retrieved from http://query.nytimes.com/gst/fullpage.html?sec=technology&res=9B02E4DF1339F931A15754C0A963948260&partner=rssnyt&emc=rss

Gady, F. (2015, May 16). Will China Sell Armed Drones to US Ally? *The Diplomat*. Retrieved from http://thediplomat.com/2015/05/will-china-sell-armed-drones-to-us-ally

Greenwald, G. and Fishman, A. (2015, March 25). Netanyahu's Spying Denials Contradicted by Secret NSA Documents. *The Intercept*. Retrieved from https://theintercept.com/2015/03/25/netanyahus-spying-denial-directly-contradicted-secret-nsa-documents/

Grevatt, J. (2010, November 23). AVIC Reports 28 Contracts Signed at Zhuhai Airshow. *IHS Jane's: Jane's Defense Industry*.

Gurtov, M. & Hwang, B. (1998). *China's Security: The New Roles of the Military*. London: Lynne Rienner Publishers.

Hammes, T. X. (2016, March 29). Cheap Technology Will Challenge U.S. Tactical Dominance. *Joint Force Quarterly*, 81. Retrieved from http://ndupress.ndu.edu/Media/

News/NewsArticleView/tabid/7849/Article/702039/cheap-technology-will-challenge-us-tactical-dominance.aspx

Han, X. (1993). Sino-Israeli Relations. *Journal for Palestine Studies*, 22(2). Retrieved from http://www.jstor.org/stable/2537269

Heginbotham, E. (2015). The U.S.-China Military Scorecard. *RAND*. Retrieved from http://www.rand.org/content/dam/rand/pubs/research_reports/RR300/RR392/RAND_RR392.pdf

HQ-9/FT-2000. (2000, June 23). *Federation of American Scientists (FAS)*. Retrieved from http://fas.org/man/dod-101/sys/missile/row/hq-9.htm

Hsu, K. (2013, June 13). China's Military Unmanned Aerial Vehicle Industry. *U.S.-China Economic and Security Review Commission*. Retrieved from http://origin.www.uscc.gov/sites/default/files/Research/China's%20Military%20UAV%20Industry_14%20June%20 2013.pdf

IAI to Cooperate with Chinese Lingyun (Yichang) Science and Technology Group Co. LTD on Establishing a Local MRO Enterprise in the Hubei Province. (2016, January 12). *Israeli Aerospace Industries (IAI)*. Retrieved from http://www.iai.co.il/2013/32981-46662-en/MediaRoom_News.aspx

ICL Reports Q4 2015 Results. (2016, February 9). *PR Newswire*. Retrieved from http://www.prnewswire.com/news-releases/icl-reports-q4-2015-results-300217290.html

Institute of Military Research. (1998). *Al-'Alāqāt al-Isrā'īlīyah ma'a kull min Turkiyā, al-Sīn wa-al-Hind*. Cairo: Egyptian Ministry of Defense.

International Arms Transfers, 2009–2013. (2014). *Stockholm International Peace Research Institute (SIPRI)*. Retrieved from http://www.sipri.org/yearbook/2014/05

Iran, Air Force: Jane's Sentinel Security Assessment - The Gulf States. (2010, November 23). *IHS Jane's*.

Israeli Chemicals Ltd. (2014, September 12). *U.S. Securities and Exchange Commission*. Retrieved from https://www.sec.gov/Archives/edgar/data/941221/000104746914007594/a2221184zf-1.htm

Israeli Patriot Missile Transfers to China: Confidential, Intelligence Information Report. Mar. 26, 1992. *Digital National Security Archive*.

Joint Communiqué of the Government of the People's Republic of China. (2000, November 15). *Ministry of Foreign Affairs, the People's Republic of China*. Retrieved from http://www.fmprc.gov.cn/mfa_eng/wjb_663304/zzjg_663340/xybfs_663590/gjlb_663594/2828_663646/2829_663648/t16344.shtml

Kospoth, N. (2009, October 14) China's Leap in Unmanned Aircraft Development. *Defence Professionals*. Retrieved from http://www.tolerance.ca/Article.aspx?ID=60875

Kumaraswamy, P. R. (2013). China, Israel and the US: The Problematic Triangle. *China Report*, 49(1).

Lilley, J. R., & David L. S. (1999). *China's Military Faces the Future*. New York, NY: American Enterprise Institute.

Lin, C. Y. (2013, March 18). China's Strategic Shift Towards the Region of the Four Seas: The Middle Kingdom Arrives in the Middle East. *Rubin Center*. Retrieved from http://www.rubincenter.org/2013/03/chinas-strategic-shift-toward-the-region-of-the-four-seas-the-middle-kingdom-arrives-in-the-middle-east/

Lin, J. E. (2015, March 20).China's Weapons of Mass Consumption. *Foreign Policy*. Retrieved from http://foreignpolicy.com/2015/03/20/chinas-weapons-of-mass-consumption/

Mark, C. (2000, July 13) Israel's Sale of Airborne Early Warning Aircraft to China: CRS Report for Congress. *Congressional Research Service*. Retrieved from http://www.lexis-nexis.com/congcomp/getdoc?CRDC-ID=CRS-2000-FDT-0033

Mark, C. (2005, April 26). Israel: U.S. Foreign Assistance. *Congressional Research Service.* Retrieved from https://www.fas.org/sgp/crs/mideast/IB85066.pdf

Melman, Y., & Sinai, R. (1987, April). Israeli-Chinese Relations and Their Future Prospects: From Shadow to Sunlight. *Asian Survey*, 27(4). Retrieved from http://www.jstor.org/stable/2644679

Minnick, W. (2010, November 24). China Developing Armed/Recon UAVs. *Defense News.* Retrieved from http://www.defensenews.com/story.php?i=5101322

Neff, D. (1997, June/July). U.S. Had to Wage Long Battle Against Israel's Technology Transfers to China. *Washington Report on Middle East Affairs (WRMEA).* Retrieved from http://www.wrmea.com/backissues/0697/9706070.htm

Page, J. (2010, November 18). China's New Drones Raise Eyebrows. *The Wall Street Journal.* Retrieved from http://online.wsj.com/article/SB10001424052748703374304575622350604500556.html

Pevzner, A. B. (2015, January 21). Israel-China Bilateral Relations Already 'Mainstream'. *Jerusalem Post.* Retrieved from http://www.jpost.com/Opinion/Israel-China-bilateral-relations-already-mainstream-388492

PL-9/DK-9. (1999, August 10). *Federation of American Scientists (FAS).* Retrieved from http://www.fas.org/man/dod-101/sys/missile/row/pl-9.htm

Procurement, China: Jane's Sentinel Security Assessment – China and North East Asia. (2010, November 8). *IHS Jane's.*

Python 3. (1999, August 10) *Federation of American Scientists (FAS).* Retrieved from http://www.fas.org/man/dod-101/sys/missile/row/python3.htm

Rawnsley, A. (2016, January 14). Meet China's Killer Drone. *Foreign Policy.* Retrieved from http://foreignpolicy.com/2016/01/14/meet-chinas-killer-drones/

Research and Development Expenditure (% of GDP). *The World Bank.* Retrieved from http://data.worldbank.org/indicator/GB.XPD.RSDV.GD.ZS?order=wbapi_data_value_2012+wbapi_data_value&sort=desc

Rosenberg, D. (2006, September 27) Israel Chemicals inks potash deals with China, India. *Jerusalem Post.* Retrieved from http://www.jpost.com/Business/BusinessNews/Article.aspx?id=36092

Rosenberg, D. (2011, June 7). Israel-China Relations Growing Deeper. *Jerusalem Post.* Retrieved from http://www.jpost.com/Diplomacy-and-Politics/Israel-China-relations-growing-deeper

Saudi Arabia Signs Deal for China's Pterodactyl Drone. (2015, May 6). *Want China Times.* Retrieved from https://web.archive.org/web/20150320050301/http://www.wantchinatimes.com/news-subclass-cnt.aspx?cid=1101&MainCatID=11&id=20140506000088

Schwab, K. (2015). Global Competitiveness Report 2015-2016. *World Economic Forum.* Retrieved from http://www3.weforum.org/docs/gcr/2015-2016/Global_Competitiveness_Report_2015-2016.pdf

Shai, A. (2009, September). Sino-Israeli Relations: Current Reality and Future Prospects. *The Institute for National Security Studies.* Retrieved from http://www.inss.org.il/upload/(FILE)1255339864.pdf

Shalev, M. (1989, June 11). Merhav: China Will Close Off from World. *Jerusalem Post.* Retrieved from http://pqasb.pqarchiver.com/jpost/doc/320952390.html

Sharp, J. M. (2015, June 10). U.S. Foreign Aid to Israel. *Congressional Research Service.* Retrieved from https://www.fas.org/sgp/crs/mideast/RL33222.pdf

Sino-Israeli Commercial Relations: An Historical Perspective. *Israel Chamber of Commerce in China.* Retrieved from http://www.ischam.org/index.php/business-in-china/commercial-history.html

SIPRI Arms Transfers Database. *Stockholm International Peace Research Institute (SIPRI)*. Retrieved from http://www.sipri.org/databases/armstransfers

SIPRI Military Expenditure Database. (2015, November 3). *Stockholm International Peace Research Institute (SIPRI)*. Retrieved from http://www.sipri.org/research/armaments/milex/milex_database/milex-data-1988-2014

Standaert, M. (2012, November 15). China Unveils New Drones Aimed at Buyers in Developing Countries. *Global Post*. Retrieved from http://www.globalpost.com/dispatch/news/regions/asia-pacific/china/121114/china-unveils-new-drones-developing-economies

Stein, J. (2014, August 4). Israel Flagged as Top Spy Threat to U.S. in New Snowden/NSA Document. *Newsweek*. Retrieved from http://europe.newsweek.com/israel-flagged-top-spy-threat-us-new-snowdennsa-document-262991

Tan, L. (2015, December 14). China Will Keep Exporting Nitrogen and Phosphate, Importing Potash. *The Progressive Farmer*. Retrieved from http://kticradio.com/agricultural/china-will-keep-exporting-nitrogen-and-phosphate-importing-potash/

Twing, S. L. (1996, November/December). Congress Calls for Sanctions If Israeli Technology Transfer to China Is Proven. *Washington Report on Middle East Affairs*. Retrieved from http://www.wrmea.org/1996-november-december/congress-calls-for-sanctions-if-israeli-technology-transfer-to-china-is-proven.html

U.S. DoD Annual Report Highlights China's Growing UAV Strike Capabilities. (2015, May 14). *IHS Jane's 360*. Retrieved from http://www.janes.com/article/51430/us-dod-annual-report-highlights-china-s-growing-uav-strike-capabilities

U.S. Pressure Threatens Israel-China Trade. (2005, January 7). *IHS Jane's: Jane's Defense Weekly*.

Warplanes: Egypt Builds Chinese UAVs. (2012, May 24). *Strategy Page*. Retrieved from http://www.strategypage.com/htmw/htairfo/articles/20120524.aspx

What Does Israel Export to China? *The Observatory of Economic Complexity, MIT*. Retrieved from http://atlas.media.mit.edu/en/visualize/tree_map/hs92/export/isr/chn/show/2013/

Wilson, S. (2005, June 27). Israel Set to End China Arms Deal Under U.S. Pressure. *The Washington Post*. Retrieved from http://www.washingtonpost.com/wp-dyn/content/article/2005/06/26/AR2005062600544.html

Wong, E. (2013, September 20). Hacking U.S. Secrets, China Pushes for Drones. *The New York Times*. Retrieved from http://www.nytimes.com/2013/09/21/world/asia/hacking-us-secrets-china-pushes-for-drones.html

Wu Zhen 5 (WZ-5) / Chang Hong (CH-1). (2011, July 11). *Global Security*. Retrieved from http://www.globalsecurity.org/military/world/china/ch-1.htm

Zacharia, J. & Harman, D. (2000, April 7). U.S. Congressman: We'll block Israeli aid unless China deal cancelled. *Jerusalem Post*.

Zak, M. (1993, October 15). China's Path to 'Jewish Power.' *Jerusalem Post*. Retrieved from http://pqasb.pqarchiver.com/jpost/doc/321135367.html

15 Veblen and the impacts of stealth technology in war

Ronald Lorenzo and Tony Tai-Ting Liu

Introduction

Stealth technology demonstrates the civilized barbarism of modern warfare. Even though stealth technology and its uses are highly innovative, they are geared toward destructive physical efficiency. Unlike other forms of weaponry and military technology, stealth technology's characterizing feature is the use of deception to gain an advantage over an adversary. In the example of stealth technology found in contemporary military technologies such as stealth fighter aircraft and bombers, the use of deception accompanies the use of destructive force. Although this is the case for stealth technology in general, this chapter emphasizes these aspects especially concerning the F-35 Lightning II.

In sociology, one theorist who discusses the use of force and deception is Thorstein Veblen. His social theory can be used to explore many of the social features of stealth technology. Thorstein Veblen was a Norwegian–American scholar who began his work as an economist but is remembered as a social critic and as a sociologist. Although other influential sociologists such as David Riesman and C. Wright Mills recognize Veblen's contribution to the discipline as important, sociology textbooks rarely mention him (Mestrovic, 2003: 2). The term *conspicuous consumption* is a widely used concept both within and outside academia, and even though it comes from Veblen's best known work, *The Theory of the Leisure Class*, most people use the term unaware of its origin. In *The Theory of the Leisure Class*, Veblen describes conspicuous consumption as a feature of a modern society whose veneer of civilization conceals much older and darker human urges. Veblen indicated that modern society betrayed signs of barbarism, despite the trappings and appearance of civilized, modern society.

The barbarian bent of modern society has implications for the use of stealth technology. Nations acquire stealth technology in part as displays of status and wealth. The barbaric motivation towards the acquisition of status symbols corresponds with what Veblen termed *invidious distinction*. Likewise, the acquisition of technology as a display of a country's economic devotion to military pursuits corresponds with Veblen's idea of *pecuniary damage*, or the ability to spend lavishly without harming one's economic position. The drive towards acquisition of stealth technology and the tremendous financial commitment to stealth technology, makes nations vulnerable to counter-stealth technology. Finally, the use of deception entailed in stealth technology has moral implications in regard to international law. What Veblen termed as *chicanery*, or a very specific form of barbaric fraud, has similarities to the principle of perfidy in the law of war. As such, stealth technology may ensnare barbaric societies with strategic and legal dilemmas.

Force and deception as barbaric traits

Veblen stated that among other signs of barbarism, modern society still had impulses toward the use of force and deception. Veblen writes that, "as it finds expression in the life of the barbarian, prowess manifests itself in two main directions—force and fraud. In varying degrees, these two forms of expression are similarly present in modern warfare, in the pecuniary occupations, and in sports and games" (Veblen, 1994 [1899]: 273). Contemporary warfare's main directions of force and fraud are made concrete in examples of its material culture, such as military stealth aircraft. In many ways, stealth technology as a material object encapsulates many of the values and beliefs of the modern society that created it. Veblen's theoretical perspective also makes other facets evident when examining stealth technology as a social phenomenon.

In *Theory of the Leisure Class*, Veblen discusses the characteristics of barbaric or predatory culture. Veblen's perspective is in stark contrast with common conceptions of modern society as ever enlightened and constantly improving in its morality and behaviors. The idea that society is still barbaric, only in a way that is transformed or disguised from its distant past, stands in sharp contrast to common ideas of progress and modernity. Within modernity, Veblen saw society as promoting a predatory culture that promotes barbaric values and habits. For Veblen, "predatory culture" stood in contrast with what he termed *peaceable culture*. In addition to the use of force and deception in modern warfare, Veblen also discusses other elements in modern society that are related to predatory culture. Such associated predatory elements include the fascination with novelty, the habit to spend wastefully in order to draw attention, a belief in luck, and an exemption from morality other the barbarian's own. These barbaric elements are inseparable, and where one is present or detected in a society, one will soon find all the others as well.

By contrast, the opposite of a predatory culture is a peaceable culture. According to Veblen, peaceable cultures are marked by elements that he refers to as the *instinct of workmanship*, the *parenting instinct*, and *idle curiosity*. Similarly, when one element is present in a peaceable culture, the other elements will be present also. For example, if one is acting on the impulse of workmanship by creating a piece of art for its own sake, one becomes like the parent of that artwork. Through the creation of artwork and through the exploration of ideas for no ulterior end, one is idly curious, or pursuing ideas merely for the joy of doing so and for no other ulterior motive.

Veblen admits that in societies, one can find elements of both predatory culture and peaceable culture. What is important to note is that the peaceable activities and elements of a culture become subsumed into its predatory elements and predatory culture. In such a manner, the peaceable segments of a society, such as the arts, to draw from one of many examples, become incorporated into the predatory culture. Art no longer becomes the beloved project of an artist who creates the work because he or she loves doing so and merely creates out of joy. The artwork enters the predatory segment of the society when it is assigned monetary value and when the artist no longer creates out of the joy of creating but because of other impulses, such as commissions or the pursuit of recognition. Artwork may remain an object of beauty, but it also becomes a predatory object when it becomes an object of distinction in the collection of a predatory individual or museum.

Stealth technology encompasses a wide array of military technologies. A definition of stealth technology is technology that is "intended to render aircraft, missiles, and ships harder to detect by sight, sound, and heat as well as radar," although a primary focus is on countering detection by radar (Hacker and Vining, 2007: 161). Stealth technology by its character seems like the present day catching up to science fiction. Emerging technologies that currently exist as prototypes seem to have been ripped from the movie screen or the comic-book page. A company even claims to have created a material that renders the person using it nearly invisible by bending rays of light (Gayle, 2012). The real material does what fictional magic rings, enchanted cloaks, or alien technology have done in books such as *The Hobbit* or the *Harry Potter* series, or in the *Predator* movie series. The broader definition of stealth technology would encompass "ordinary" military technologies not considered as stealth technologies, such as submarines (themselves ripped from the pages of a Jules Verne novel), decoys, and camouflage uniforms. Other technologies such as encrypted communications, including steganography (or a message secretly embedded within another message, such as one sound recording secretly embedded, and normally inaudible, within another sound recording) could be considered stealth technology. The quality of deception in order to gain an advantage over a potential adversary is the defining feature of stealth technology.

Stealth airplanes, ironically, are a very visible example of stealth technology in production and use by modern militaries. If Veblen could apply the phrase *conspicuous consumption* to the wasteful, public spending of money in order to draw attention, one could say that the stealth fighter as a cultural object is a nation's attempt at conspicuous secrecy. Despite its status as cutting-edge technology, the stealth fighter is also a popular culture object: one can buy stealth fighter toys and models or view videos of stealth fighters in flight on the Internet. The stealth fighter provides a good focus for discussions of the sociology of stealth technology.

Stealth technology and the US pursuit for military supremacy

Since the denotation of Little Boy and Fat Man in Japan in August 1945, the United States (US) instantly achieved global military superiority. While Washington and, later, Moscow would soon realize that the atomic bomb is a weapon that demonstrates more threatening capability than practical use, "the bomb" opened up a new round of the arms race in the world. Nuclear weapons, in a sense, became the focal point of military competition among great powers in the twentieth century, to the extent that the international community recognized the insanity in mutually assured destruction (MAD) and agreed to limit the proliferation of nuclear technology.

While the US lost its nuclear monopoly in less than a decade after 1945, competition soon moved on to other forms of competition. In the new century, besides the incessant fear of global thermonuclear war, the latest competition among states has taken on the form of stealth technology. When combined with conventional military vehicles such as tanks, ships, and planes, military pundits and strategists have argued that in theory, stealth technology may lead the way toward dominance on the battlefield for states. In other words, the key to military superiority in the twenty-first century rests largely on the successful application of stealth technology to warfare. The ability to penetrate enemy frontlines, to deceive without being detected, is widely believed to be the game changer in future war.

Comparable to Hiroshima and Nagasaki, the staging of the US F-117 striker planes in the opening phase of Operation Desert Storm in 1991 marked a turning point in technology and modern warfare. This section will briefly discuss US military modernization efforts in recent years related to stealth technology before describing the implications and limitations of such developments. A particular emphasis is given to the development of stealth aircraft—fighters and bombers—as the field has generated much more discussions and debates than other fields in recent years.

The stealth priority in US military modernization

Besides the US achievement of quick, overwhelming victories in Operation Desert Storm in 1991 through the massive display of force, the mission also witnessed the use of stealth technology. The F-117A Nighthawk, a stealth fighter commenced for development by the US more than a decade ago, breached the defenses of a heavily guarded Baghdad and destroyed or crippled numerous strategic targets, including electrical power stations, military headquarters, communications sites, air defense operation centers, airfields, ammo bunkers, and chemical, biological, and nuclear weapons plants (Myers, 1992). F-117's successful operation in Iraq was not unnoticed. Since then, the integration of stealth technology in combat became an important priority for the US military.

Paradoxically, while the application of stealth technology to combat literally implies the shrouding or disappearance of military vehicles in the event of war—a condition that may sound baffling and perhaps lunatic to the common ear—in essence, the aim of stealth is not to vanish into thin air but to become intractable through radar. The chances for potential detection can be greatly reduced by significantly lowering the radar signature of beamed objects and generating confusions so as to present moving objects as something otherwise. In addition to mechanical fine tuning that reduces the sound level of moving vehicles, as noted by Pietrucha (2016), stealth mainly relies on the material, color, and shape adopted by the object. For example, metal generally reflects signature better than fabric-covered plywood, which becomes problematic if the objective is to conceal. However, a vehicle made of plywood—if tenable—would simply be a vulnerable target in modern warfare. The tradeoffs among material, color, and design of the military vehicle define the basic challenges in stealth technology.

Since the Gulf War, the US has made many efforts to apply stealth technology to a variety of military vehicles with some success. In the maritime arena, notwithstanding the fact that submerging underwater has long been an effective tactic used to avoid enemy detection, stealth technology is applied to submarines of different classes to further enhance their intractability in water. In naval applications, current discussions focus on the USS *Zumwalt*, a guided missile destroyer with stealth capability that first took to the water in December 2015 (Prigg, 2015). On land, researchers and engineers in national defense are contemplating the possibility of producing stealth armored vehicles, a concept that is naturally handicapped by spatial limitations (Peck, 2014). On the other hand, discussions also center on the possibility of developing stealth cruise missiles, a formidable advancement should the concept be realized (Majumdar, 2016). In the air, aside from fighters, efforts have been made to employ stealth technology in helicopters. The Sikorsky UH-60 Black Hawk helicopters used in the raid against Osama bin Laden's compound in 2011 may have incorporated stealth technology in order to evade radar detection in Pakistan (Geoghegan and Shenker, 2011).

Nonetheless, following the success of Operation Desert Storm, advancements in stealth aircraft remain the central focus of the US military, abetted by the idea that gaining control over airspace is a vital step toward victory. At the Air Force Association's Air Warfare Symposium held in Orlando on February 26, 2016, Air Force Secretary Deborah Lee James unveiled the first rendering of the Long Range Striker Bomber (LRS-B) designated as the B-21, a new stealth bomber that will replace the B-2 Spirit that has been operationalized in 2003 (Kimmons, 2016). The revelation came merely weeks after James and the chief of staff of the US Air Force, General Mark A. Welsh, called for an increased budget before the Senate Appropriation Committee to expand and modernize the US Air Force (Martin, 2016). Militaries have invested the improvement of stealth technology in fighters and bombers.

1 Stealth fighters: As an early user of stealth technology in military vehicles, the US was the first country in the world to be in possession of stealth fighters. In November 1978, Washington awarded a contract to aviation giant Lockheed Corporation to commence full-scale development of stealth aircraft. After few years of tests, Lockheed delivered the first F-117A Nighthawk to the US Air Force (USAF) in 1982 and made its final delivery of

the model in 1990. In total, the USAF took delivery of 59 F-117A stealth fighters. The F-117A first gained public recognition in Operation Desert Storm in Iraq and went forth to see action in Operation Allied Force in 1999, Operation Enduring Freedom in 2001, and Operation Iraqi Freedom in 2003. In 2008, the F-117A was retired and replaced by the fifth-generation F-22 Raptor.

Currently, the F-22 serves as the main stealth fighter for the USAF, with the F-35 Lightning II Joint Strike Fighter waiting in line to become America's newest mainstay of fighters. First introduced in 2005, the F-22 is capable of carrying nuclear missiles and bombs and plays an important role in Washington's deterrence strategy ("US F-22 stealth fighters fly over South Korea in show of force toward DPRK," 2016). As of September 2015, the USAF has an inventory of 183 F-22 fighters in service (United States Air Force, 2015). On the other hand, development of the F-35 seeks to complement the F-22. Introduced in late 2015, the F-35 is set to establish three main models: the F-35A conventional takeoff and landing variant, the F-35B short-takeoff and vertical-landing variant and the F-35C carrier based variant ("Lockheed Martin F-35 Lightning II Joint Strike Fighter Specification & Technical Data," 2013). While discussions have emerged for the possibility to invest in the establishment of sixth-generation fighters, dissenters have argued that the F-22 and F-35, if regularly maintained and upgraded, may well serve Washington's military needs into the near future (Forbes and Wynne, 2016; Weisgerber, 2016).

2 Stealth bombers: In 1981, the B-2 stealth bomber went under development when the USAF awarded the contract for the "Advanced Technology Bomber" to Northrop Grumman Corporation. Several years lapsed before the world witnessed the first flight of the B-2 in 1989 and the first operational use of the bomber in Operation Allied Force in Kosovo and Serbia in 1999. Despite initial plans by the USAF to contract for production 127 stealth bombers from Northrop, following the collapse of the Soviet Union in 1991 and a cut in the US defense budget, the order for B-2s eventually comprised 21 aircraft, a significantly lower number compared to stealth fighters. Nonetheless, with its bat-like and all-wing design (no fuselage and vertical tail) that evades radar detection and ability to carry air-launched stealth cruise (nuclear) missiles, the B-2 proved to be a game changer in modern warfare (McArthur, 1990). As Northrop describes in a phrase, "[the B-2] changed the fundamental calculus of air combat from 'number of air sorties required to destroy a target' to 'number of targets that could be destroyed by a single sortie" ("The B-2 Spirit Stealth Bomber turns 25," 2014).

Currently, with the US facing a host of global security challenges from international terrorism to North Korea to Syria, the USAF has argued for an expansion to the existent air force program and the development of a second generation of stealth bombers in response to national security demands. As Deborah James concisely put it, "the aircraft represents the future of [US] Airmen" (Martin, 2016). However, with an ironically similar design as its predecessor, commencement of the B-21 project has generated controversies no less than the high production price and the new model's capabilities. It remains to be seen whether the B-21 can successfully replace the B-2 in the future.

The implications and limitations of stealth technology

While success in Iraq spurred the US onto a path of more than two decades of interest and research on stealth technology, one cannot equate Washington's engineering achievements with military supremacy. Rather, as many observers note, stealth technology has opened up a new round of arms competition among states. In terms of stealth aircraft, the demonstration of US military capability through the F-117, F-22, and F-35 have encouraged countries such as Russia and China to come up with matching models such as the Sukhoi T-50 (PAK-FA), Chengdu J-20, and Shenyang J-31 Falcon Hawk. Besides great power competition, the race for stealth aircraft has spilled over to countries such as Japan, India, Turkey, Iran, and Finland, each with aspirations to introduce new vehicles in the near future. In short, the arms race over stealth aircraft is growing steadily, though incrementally.

Aside from the potential for new military innovations to destabilize international security, in terms of stealth technology, a major implication for states comes from transformations to the ways future wars may be conducted on the tactical level. While air combat and targeted bombing can be expected from stealth aircraft, the true power of stealth lies in its combined use with other technologies. For example, besides increasing the difficulty of detecting its aircraft, Russia is currently investigating the possibility of equipping its next generation of stealth bombers, the Tupolev PAK-DA, with hypersonic weapons (Majumdar, 2015). On the other hand, Russia has also equipped its Lada-class submarine with specialized launchers for cruise missiles (Kazianis, 2016); missiles, as noted above, have the potential to become stealthy as well. In such sense, competition in stealth technology is no longer confined to a single field arms race when other factors are taken into account.

Meanwhile, on the defensive end of the battlefield, corresponding developments in anti-stealth technologies may further complicate competitions and conflict scenarios. As Carlo Kopp (2010, 90) warns, in any near future conflict, US forces will have to confront a complex spectrum of air defense systems ranging from legacy Soviet systems to newly built Russian and Chinese systems, with various hybrid mixes of Cold War and new systems being possible and likely. In addition to improved radar detection systems that seek to expose stealth vehicles, Zachary Keck (2015) also notes that both Russia and China have undergone research into the development of unmanned aerial vehicles (UAV) or drones that have the goal of finding, detecting, and possibly even eliminating enemy vehicles. The Divine Eagle, China's anti-stealth vehicle, is touted to have the ability of detecting stealthy enemy aircraft, missiles and ships well before they come in range of the mainland (Lin and Singer, 2015).

In short, in the quest for military supremacy through stealth technology, the parallel development of anti-stealth technologies persistently seeks to offset or cancel the advantages gained through advancements in stealth. Beyond tactical implications, the application of stealth technology to military affairs also raises both practical and moral concerns:

1 Myth of military supremacy: Supported by successful campaigns such as Operation Desert Storm, the application of stealth technology to military vehicles may be the next-best thing to the direct use of nuclear weapons in ensuring victory. Yet many observers have questioned the truth of such belief (de Briganti, 2012a). For example, in 2015, after a mock skirmish between the F-35 Joint Strike Fighter and the F-16—a soon-to-be-replaced older model of F-class fighters—the test pilot for the F-35 stealth aircraft released a controversial statement claiming that errors in design have led the aircraft to be outmaneuvered by its F-16 counterpart (Axe, 2015). While supporters of the F-35, including the F-35 Joint Program Office, later argued in defense of the aircraft, responses have not been able to quell ever-growing doubts and criticisms against stealth technology (Axe, 2015). Furthermore, even if stealth aircraft do come out on top in simulations and tests, there is no guarantee that similar results can be reproduced in actual combat.

2 Decreasing lifespan of stealth technology: Notwithstanding disillusions surrounding the potential dominance afforded by stealth technology, the initiation of state competition over research and development into stealth vehicles essentially entails that even if military superiority could be achieved, such advantage would be short lived as other states invest efforts to develop counterbalancing vehicles and devices. Moreover, as Mike Pietrucha (2016) notes, the fact that adversaries may be able to acquire high resolution photos of new stealth models and study the new designs several years in advance effectively shortens the lifespan of breakthrough technologies by allowing competing states substantial time to ponder over countermeasures.

3 High price of technology: Connected with fleeting advantages, in order to stay ahead of the game, large investments need to be placed in research and development (R&D). In terms of stealth technology, the demand for funds has grown alongside the contracted period of technological lead, to the extent that skeptics question the utility of stealth vehicles. For example, as noted above, while the USAF is keen to upgrade to the B-21 stealth bomber, recalling extremely high costs induced by past upgrade to the F-35, Senator John McCain, among others, voiced concerns about cost overruns (Swarts, 2016; Korb, 2015). Official estimates put the cost of B-21 development at 23 billion USD (Drew, 2016). Regarding the F-35 Joint Strike Fighter program commenced earlier, skeptics issued a conservative estimate of 400 billion USD on development and maintenance (Cohen, 2015). Meanwhile, the high price tag associated with the development of stealth aircraft forced the Netherlands, Italy and Canada—countries that were once interested in developing stealth vehicles—to re-evaluate their

decisions. In Italy, acquisition of the F-35 entailed an increase to more than 50 percent of taxpayer income (de Briganti, 2012b), an immense price to pay in any event.

4 Legal gray zone: Beside practical concerns, a less discussed aspect of stealth technology relates to its legal use in military combat. While war may be a chaotic event, the Geneva Convention introduced in 1949 makes an effort to regulate the conduct of war. The application of stealth technology to war, nonetheless, raises questions concerning the legality of its use in combat. Former air commodore and military lawyer Bill Boothby points out that the use of stealth technology in warfare may breach international laws that demand weapons and military equipments to be openly discernable (Bowcott and Ross, 2016). Boothby's suggestion that camouflage for disguising as a non-combatant in order to deceive the enemy and cause death may be unlawful represents a type of fraud in warfare discussed in the Geneva Convention clause on the "prohibition of perfidy," (Bowcott and Ross, 2016). Therefore, in the present circumstance that stealth technology has yet to be fully mature, the legal and moral questions surrounding the topic may continue to be left in the dark. While policy makers and military pundits may rejoice when stealth technology can live up to its potential and become fully applied in battle in the near future, similar to the debate on drones, legal and moral scruples may demand reconsiderations on going stealth.

The predatory culture of stealth technology

Stealth technology is the product of a predatory culture that has subsumed the peaceable segments of its society. Two of the main elements of barbarism are force and fraud (Veblen, 1994 [1899]: 236). The barbarian is not a person who creates. The creative process is part of peaceable society, and Veblen states that industry and industriousness is a peaceable activity. The barbarian, because he cannot create, takes or steals. However, the manner in which a barbarian takes or steals from another is not only through the use of brute force but also by fraud or chicanery. Chicanery is simply the use of trickery or deceit to achieve some ulterior motive.

In modern societies, barbarians also develop a sense of status that transforms the way in which they force or chicane from others. Veblen, for instance, states that in modern barbaric society, the barbaric professions that reflect predatory values and methods are those related to finance and law. For example, Veblen states:

> The lawyer is exclusively occupied with the details of predatory fraud, either in achieving or in checkmating chicane, and success in the profession is therefore accepted as marking a large endowment of that barbarian astuteness which has always commanded men's respect and fear.
>
> (1994 [1899]: 231)

What Veblen states is that because of status, which is another element of barbarism, cultures develop intricate esthetics of barbarism. The barbarian in modern society is not someone who stands in the middle of the road with an axe robbing passersby of their wealth. The modern-day barbarian is most likely the graduate of a top-ranked business or law school and is more likely to be found in the board room of investment banks or vulture funds, manipulating the law in order to chicane the wealth of others. The modern-day barbarian metaphorically kills his victim not with the single swing of an axe but through hundreds of precise, yet immoral, punctures of a scalpel.

The esthetics of stealth technology are highly refined barbarian esthetics. Stealth fighters, for example, are incredibly forceful. The variants of the F35 can carry around 15,000 pounds of payload (i.e., missiles and bombs) to drop on an enemy ("F-35 Joint Strike Fighter (JSF) Lightning II: Specifications"). The esthetic of the stealth fighter is that it should conform to principles of high technology stealth, or the ability to use trickery and deceit in order to carry out the deadly blow. The phrase uttered by Macbeth to Lady Macbeth to carry out an assassination with a "bloody but invisible [or unseen] hand" (Act 3 Scene 2 from Macbeth) comes to mind when describing the esthetic of stealth technology. Chicanery, in warfare and in other predatory pursuits, is used to gain strategic advantage by tricking an opponent.

The esthetic of force and fraud through stealth (in the case of stealth meaning high technology) is only possible when a predatory culture incorporates the segment of peaceable workmen and workmen from its society into predatory pursuits. The instinct of workmanship of otherwise peaceable engineers and designers are subsumed into the predatory exploits of fraud and deception. Both the predatory and peaceable segments coexist within the culture that produces stealth technology, but the barbarian instinct ultimately takes preeminence.

The idea can be encapsulated by the fact that stealth fighters and other military technologies are designed by mathematicians and engineers who studied at reputable schools, belong to professional associations with codes of professionalism, live in orderly neighborhoods, and work in respectable office buildings. These are not early people from long ago chipping away at flint to make arrowheads, but people who consider themselves civilized applying modern science and ideas in the manufacture of cutting edge, near-science-fiction, cutting-edge technologies. However, the same peaceable individuals spend their days on designs for machines that can almost instantaneously kill large numbers of their fellow human beings. The engineers and designers who work on the technologies focus on the workmanship of their efforts and leave the predatory aspect of their endeavors for others to worry about. Engineers and scientists who design the technology of war often anthropomorphize their work, and projects become "my baby" and "our baby" in discussions among themselves.

National symbols reflect both the predatory and peaceable aspects of a nation. The US has as one of its national symbols the eagle, a predatory bird, even though one should remember that Benjamin Franklin originally proposed the turkey (which is not a predatory bird) as the national symbol. Likewise, the heraldry and symbolism of stealth technology reflects both the predatory and peaceable aspects of its culture. However, the same heraldry and symbolism demonstrates the superiority of predatory instincts over the peaceable instincts of the culture that produces stealth technology. For example, two of the most current stealth fighter planes are named after predatory totems. The F-22 is given the name of Raptor. A raptor is a bird of prey, or a bird that hunts and feeds on other living animals. The colloquial use of the word *raptor* is also culturally linked to the movie *Jurassic Park* and its sequels, which introduced the word and image of the velociraptor (depicted as a predatory dinosaur) into popular culture. Another stealth fighter aircraft named after a predatory totem is the F-35, which is given the name of Lightning II (in order to not confuse it with an earlier World War II era aircraft, the P-38 Lightning). Lightning still is a predatory totem because it is a destructive force of nature.

The heraldry of another aircraft, the F-117, also demonstrates the presence of both predation and peaceableness within the culture that produces stealth aircraft, as well as highlighting the ultimate superiority of its predatory, or barbaric, instincts. A helpful book on this subject is *Nighthawks: Insider's Guide to the Heraldry and Insignia of the Lockheed F-117A Stealth Fighter* by Blazek (2004). The book catalogs the embroidered patches produced by various groups of people involved in the testing, development, and operation of that stealth aircraft.

The patches for personnel involved in the creation or care of the aircraft (the workmanship and parenting of the aircraft, respectively) are of a whimsical nature. One patch was created to commemorate *flutter tests* of the aircraft; in lay terms, flutter tests are a series of performance tests early in the development stages. The "Dynamics Duo" patch is thus described:

> This cartoon patch was created by Rick Wyllie and Dan Fuller to commemorate the flutter tests performed on the F-117A used during these dynamic response tests and their respective pilots. However, some of the Flight Test members of the time like to say that the name applied to two engineers, Stephanie McNay and Bill Anderson, who were specifically responsible for monitoring these tests and analyzing the data.
>
> (Blazek, 2004: 9)

The patch shows two cartoonish World War II propeller airplanes with oversized pilots' heads sticking out of the open canopies. The airplanes fly over a similarly cartoonish flat, desert landscape populated with a single cactus and a house-like structure. Similar patches exist for the A.G.E. (aircraft ground equipment) personnel responsible for maintaining equipment. In Veblenian terms, maintenance is seen as a parenting, peaceable function; when one is looking after an object and taking care of it, one is essentially acting as a parent figure. The A.G.E. patch shows a cartoon dog from the comic strip "Mother Goose & Grimm" grinning humorously. Another patch, one made for the Product Excellence Committee of the company that manufactured the F-117, shows a cartoonish skunk (a nonpredatory animal that drives away potential prey with chemicals) inside a similarly cartoonish wooden fort with the words "Workmanship," "Quality," and "Performance" underneath the image. In the same patch, the outside borders are made to resemble the analog gauge dial of flight instruments.

The patches created much later in the history of the F-117A program reflect the predatory nature of the plane's operation and use. For example, the patch for the 415 Tactical Fighter Squadron, which flew the plane in combat missions, features an eagle with a halo over its head. Patches for other flight squadrons show similar totemic, predatory symbols: the 416 Tactical Fighter Squadron features a rider cloaked in black with red eyes riding a black horse with similar red eyes. In that patch, the rider is carrying a lightning bolt, another totemic, predatory symbol. Similarly, patches for test-flight teams show dragons, arrows on a bull's-eye, eagles with lightning bolts emitting from their talons, and swords with eagles' wings attached to them, all of which are predatory, totemic symbols of destructive forces.

Another aspect of barbarism that Veblen noted as a common feature was its close association with sports, and this is evident in the culture of stealth technology. Veblen states that sports becomes a preoccupation for the predatory segments of society because at both of the nonindustrial nature of sporting (Veblen, 1994 [1899]: 2). Veblen writes:

> The men of the upper classes are not only exempt, but by prescriptive custom they are debarred, from all industrial occupations. The range of employments open to them is rigidly defined. As on the higher plane already spoken of, these employments are government, warfare, religious observances, and sports.
>
> (1994 [1899]: 2)

At the other end of the range of employments, manual labor, industry, and the everyday work of production is relegated to the peaceable segments of a society (Veblen, 1994 [1899]: 2). The culture of sports and the culture of stealth technology intersect in several ways.

In a very visible and direct way, the conspicuous secrecy of stealth technology is sometimes on display at sports events. Professional and collegiate sporting events in the US are almost always preceded by rituals that connect elements of culture that Veblen would characterize as predatory. There is often a color-guard that occupies the field and then, after its ritual is concluded, cedes it to the sports teams. Almost always, the national anthem is played before an official match (Veblen thought of patriotism as "partisan solidarity in respect of prestige," associating it with barbarism). There might also be an invocation. For high-profile events, there might be a military fly over stadiums prior to the start of the game. For example, every Super Bowl, except for Super Bowl I in 1967, has had a pregame military flyover ("Navy spent $450,000 of taxpayer money on Super Bowl flyover… when the stadium roof was shut," 2011).

In Veblenian terms, military flyovers have many of the elements of barbarism, including a high-profile waste of money. The per-hour cost of flying a military aircraft can exceed $10,000, and it is estimated that in some years the military performs over 1000 flyovers for funerals, sporting events, and airshows (Durkin, 2012). Recently, the 2015 Rose Parade and Rose Bowl had military flyovers from B2 stealth bombers (Brantley, 2015). The estimated cost of flyovers is disputed, with a journalist reporting that the military flyover of the closed-roof Dallas Cowboy stadium cost taxpayers over $450,000, while a spokesperson for the US Navy countered that the maneuver amounted to a sum in the range of $100,000 (Rovell, 2011). An argument in defense of the cost of military flyovers of events is that they serve as recruiting and public relations efforts for the military, and that the hours the pilots spend flying are logged in their records as training (Milhoces, 2013). The social effect of the military flyover in conjunction with the other pregame rituals is to associate all of the elements Veblen would have characterized as predatory or barbaric to each other: militarism, patriotism, and sports. Thereafter, sports become patriotic and martial in character, while airplanes used in flyovers become sporty and patriotic. The aura from each individual element extends into the others.

The character of sports extends into the process of awarding manufacturing contracts for military technology. In various documentaries and news stories about the JSF (Joint Strike Fighter), filmmakers often include a clip of the workers from the company that ultimately won the contract to manufacture the F-35. The workers sit in rows of chairs in a large room, watching a television broadcast. The television broadcasts the decision that their company will be selected over a competitor to build the JSF based on the results of a competition between the companies' prototypes for the stealth plane. The initial contract for the first order of planes would eventually amount to $200 billion ("Lockheed awarded contract: Defense Dept. taps Lockheed-Martin for $200B Joint Strike Fighter," 2001). When the employees hear that it is their company that has been chosen, the employees rise from their seats and cheer, as if reacting to watching their team score the winning point after the last, agonizing, and tense moment of a game. The moment reveals that the competition for the contract takes on a dimension similar to that of a contest of athletic prowess between the two companies. The F-35 manufacturer's otherwise peaceable workmen and workwomen are swept up into vicarious barbaric prowess. The awarding of the contract takes on the characteristic of sport, and the image of the company's employees celebrating is similar to images of people celebrating the announcement that their city has been selected to host the Olympics.

The sporting character of winning contracts extends beyond the shores of the US into the competition for contracts in other countries. The F-35's maker has been aggressive in pursuing contracts around the world to replace aging fleets of aircraft such as the F-18, which is flown by Canada and Australia. In choosing the American manufacturer over British and French competitors, a list of nations have agreed to pay almost $5 billion to participate in the development and manufacture of the stealth fighter ("F-35 Joint Strike Fighter (JSF) Lightning II International Partners").

Conspicuous national spending and stealth technology

Veblen associates the wasteful consumption of goods with predatory behavior, a hallmark feature of the barbaric segment of society that he terms the *leisure class*. Often, the wasteful consumption is of a very public nature, which Veblen terms as *conspicuous consumption*. Veblen states of wasteful consumption:

> Unproductive consumption of goods is honourable, primarily as a mark of prowess and a perquisite of human dignity; secondarily it becomes substantially honourable to itself, especially the consumption of the more desirable things. The consumption of choice articles of food, and frequently also of rare articles of adornment, becomes tabu to the women and children; and if there is a base (servile) class of men, the tabu holds also for them. With a further advance in culture this tabu may change into simple custom of a more or less rigorous character; but whatever be the theoretical basis of the distinction which is maintained, whether it be a tabu or a larger conventionality, the features of the conventional scheme of consumption do not change easily. When the quasi-peaceable stage of industry is reached, with its fundamental institution of chattel slavery, the general principle, more or less rigorously applied, is that the base, industrious class should consume only what may be necessary to their subsistence. In the nature of things, luxuries and the comforts of life belong to the leisure class. Under the tabu, certain victuals, and more particularly certain beverages, are strictly reserved for the use of the superior class.
>
> (1994 [1899]: 68)

Stealth technology comes to be associated with status among countries. Similar to "choice" and "rare" articles of things, stealth technology acquires cachet and distinction.

Veblen states that the conspicuous consumption of valuable goods is a means of reputability to the gentleman of leisure (1994 [1899]: 75). The consumption of goods is seen as evidence of wealth and comes to be seen as an activity that brings honor to the person consuming. As such, the gentleman of leisure consumes "freely and of the best, in food, drink, narcotics, shelter, services, ornaments, apparel, weapons and accoutrements, amulets, and idols or divinities." The list brings to mind the distinction between what is consumed in modern society by the leisure class of our own day: specialty food from high-class supermarkets, high-end wines and spirits, brand-name clothing from exclusive stores, private colleges and universities, property ownership in historic city districts (poor districts, apparently, have no history), and membership of exclusive churches.

The countries of the world have societies that are a mixture of predatory and peaceable individuals, classes, habits, and social structures. In some countries, the predatory aspects are more exaggerated and pronounced than in others. The predatory habits that Veblen describes in individuals and classes of society are apparent also at a national level. The large expenditures in stealth technology can be seen from a Veblenian perspective as conspicuous consumption on a national level.

Stealth aircraft are expensive. The F-117 Nighthawk had a cost of $122 million per unit in 1998 or around $170 million in 2016. The B-2 Spirit bomber has a unit cost of $2.1 billion ("B-2 Spirit" 2016). The F-22 Raptor had a unit cost exceeding $150 million (United States Air Force, 2010: 1–3). The F-35 Lightning II has a unit cost that ranges between $98 million and $116 million for the three variants of the plane (the Navy, Air Force, and Marines each have their own variant of the F-35).

Critics of the F-35 have targeted the cost of the program, one aspect of the stealth fighter that is highly conspicuous. The US anticipates purchasing over 2400 F-35s for a sum of $382 billion (Tierney, 2011). Furthermore, the General Accounting Office estimates that an additional $650 billion will be needed to operate and maintain the fleet of F-35 airplanes (Tierney, 2011). With a notable exception of a few countries (France and Germany prominently among them), the list of countries that have committed to purchasing F-35s reads like a list of the leisure class and aspiring-to-be leisure class countries aligned with US foreign policy and interests: the United Kingdom, Italy, the Netherlands, Canada, Australia, Denmark, Norway, and Turkey (Tierney, 2011). With the exception of the US, the other countries participating as customers in the F-35 program are divided into three levels of participation based on financial stake in the program, subcontracts open for bidding by their national companies, and, commitment of technology transfer ("F-35 Joint Strike Fighter (JSF) Lightning II: International Partners"). The classification of levels almost acts as a proxy for classifying the countries on the list by degree of predation with the United Kingdom as the most predatory as the only Level 1 "partner" on the list, followed by the predatory countries of Italy and the Netherlands as Level 2 "partners," and Canada, Turkey, Australia, Norway, and Denmark as the least predatory "partners" at Level 3. The levels of "partners" on the list calls to mind the boarding classes for flights at the airports, with gradations of classes from business class and first class (the leisure class of passengers) down to coach class (the more workman and workwoman class of passengers).

On the international stage, a country that buys a fleet of F-35 airplanes is on a par with the individual consumer who buys an assortment of Gucci, Armani, or Chanel products. The F-35 has become a status symbol, proof that a country is capable of some degree of pecuniary damage, or the ability to waste without sustaining privations. The least predatory of the countries on the list show greater reticence at their commitment to the F-35 program. The Liberal Party in Canada came into office in 2015 in part because of their promise to consider other options to upgrade their air force:

> We will not buy the F-35 stealth fighter-bomber. We will immediately launch an open and transparent competition to replace the CF-18 fighter aircraft. The primary mission of our fighter aircraft should remain

> the defence of North America, not stealth first-strike capability. We will reduce the procurement budget for replacing the CF-18s, and will instead purchase one of the many, lower-priced options that better match Canada's defence needs.
>
> (Liberal Party of Canada, 2015: 70)

Political dissent in Australia has also centered on the purchase of the F-35 to replace aircraft reaching the end of their service lives ("Warring over aircraft," 2007). In a denunciation of Australia's participation in the JSF program, an Australian government backbencher, Dennis Jensen, characterized his country's support for the F-35 as a "great national scandal" and "worse than a disgrace" (Doherty, 2014).

Novelty and stealth technology

Stealth technology, by its very nature, can be described as cutting edge with an almost science-fiction-like quality. The novelty of this technology is one of its appeals to the predatory segments of society, and this appeal is related to a second feature, which is the aspect of the technology's mystique. Veblen notes that both a fondness for novelty and a belief in luck are present in the barbarian character. The mystique of stealth technology, as critics have pointed out, can be associated with a predatory belief in the efficacy of luck.

In modern society, Veblen notes that the barbarian temperament has not been eliminated but transformed. The modern day barbarians of the leisure class use status and esthetics as a way of cloaking their uses of force and chicanery. Novelty becomes a distinction in the status of objects:

> The canons of beauty must be circumvented by some contrivance which will give evidence of a reputably wasteful expenditure, at the same time that it meets the demands of our critical sense of the useful and the beautiful, or at least meets the demand of some habit which has come to do duty in place of that sense. Such an auxiliary sense of taste is the sense of novelty; and this latter is helped out in its surrogateship by the curiosity with which men view ingenious and puzzling contrivances.
>
> (Veblen, 1994 [1899]: 152)

The technologies that are part of the F-35 have been touted as revolutionary and groundbreaking.

As an example, the F-35 is equipped with sensors that can project a three-dimensional, virtual-reality representation of the world around the plane for its pilot. The project is based on cameras on the exterior of the airplane, which transmit their images to the pilot's helmet. Pilots effectively are in virtual reality when flying the airplane, with the ability to "see through" the floor of their cockpit on a screen that is projected on the helmet's visor. The helmet is also a computer onto itself that networks into the airplane's main computer. Each helmet carries a price tag of $770,000, and it must be custom made for each pilot. The process of sitting and being fitted for the helmet takes two days so that the screen on the helmet's visor can be accurately aligned with the pilot's eyes. Critics note that if the helmet becomes damaged or if the software malfunctions, the pilot is incapable of controlling the airplane (Hutcheon, 2015).

Pierre Sprey, an airplane designer who was one of the designers on the F-16 and A-10 airplanes, which are both planes that the F-35 is supposed to replace, has been critical of its technologies (Ciralsky, 2013). In an interview with *Vanity Fair* magazine, Sprey was critical of the helmet. The display screen on the helmet's visor was jittery and latent (or slow in transmitting live images on the visor, with noticeable lag) (Ciralsky, 2013). Sprey was also critical of the resolution of the images transmitted to the helmet's visor, contending that it was inferior:

> "Right from the start, they should have known there would be a huge computation problem and a huge resolution problem," says Sprey. "Why do drones shoot up wedding parties in Afghanistan? Because the resolution is so poor. That was knowable before the helmet was built." The helmet-mounted display, says Sprey, is "a total fuckup from start to finish."
>
> (Ciralsky, 2013)

The helmet is at the front of a list of problems that critics find with the F-35's novel features.

In addition to the helmet, other critics have noted a long list of defects with current versions of the airplanes. Among the critics was the general who had been the head of the military's Joint Strike Fighter program, who compiled a list of "50 top parts of the airplane that break more often than we expect them to" (Ciralsky, 2013). Another noted problem is that the ejector seat could snap a pilot's neck through an unfortunate combination of the ejector seat's speed and the weight of the F-35's expensive, heavy helmet (Wroe, 2015). The F-35's onboard intelligence, ALIS, is made up of 24 million lines of programming code that sometimes refuses to accept the plug-and-play installation of spare parts or weapons ("The $1 trillion F-35 tries to be all things, but succeeds at few, say critics. But is Australia's new weapon now too big to fail?" 2014). ALIS's rigidity in obeying commands from its human users has drawn comparisons with the fictional, ultimately murderous computer HAL from the movie 2001 (McGarry, 2014). Even the central selling feature of the F-35 is prone to failure, as the coating of stealth material at the back of the plane starts to disbond after continuous heat from plane's engine (Ciralsky, 2013).

Against the litany of problems related with any one particular stealth fighter, the apologists for stealth have countered that the technologies are revolutionary and game changing. In general, the promoters of stealth technology have created an aura of mystique around it. This belief in a technology that has its many critics is associated with a barbarian belief in luck. Veblen states that the barbarian temperament is one in which the impulse towards gambling, and of luck, is strong:

> While the belief in luck is the basis of the gambling habit, it is not the only element that enters into the habit of betting. Betting on the issue of contests of strength and skill proceeds on a further motive, without which the belief in luck would scarcely come in as a prominent feature of sporting life. This further motive is the desire of the anticipated winner… to heighten his side's ascendancy at the cost of the loser
>
> (Veblen, 1994 [1899]: 277)

A nearly magical belief in the ultimate superiority of stealth technology, including fighter aircraft like the F-35, is akin to placing a one trillion dollar bet.

Conclusion

Stealth technology embodies the cultural values of civilized barbarism. Veblen noted that when one element of predatory culture, or barbarism, was present, the other elements were present also. Stealth technology conforms to the esthetics of status: the futuristic, front-line aspects of the science of stealth technology cloak the two hallmark features of barbarian aggression, which are force and chicanery. Stealth technology directly speaks to the barbaric or predatory trait to deceive and trick an adversary. However, stealth technology marries the two barbaric traits: one is able to deceive in order to destroy an adversary. In addition to fusing these two traits in a single weapon, stealth technology feeds into other predatory impulses as well. By being expensive and wasteful, stealth technology allows the predatory segment of society that champions it to engage in unbounded conspicuous consumption amounting to hundreds of billions of dollars. The cutting-edge aspect of stealth technology gives status in its conspicuous consumption. The leisure-class nations who indulge in the conspicuous consumption of stealth technology draw invidious distinction from their expenditures.

In its implications for the future of warfare, the application of "stealth" or literally "the ability to become unnoticed" in combat will increasingly become a defining concept of war in the new century. A review of the development of stealth technology in combat suggests that the way war is fought on the battlefield has evolved considerably. Besides the enlargement of the spatial confine of warfare, or the expansion of warfare from land, sea, and air to the atmosphere and beyond, stealth technology is set to become a ubiquitous feature of military arsenals.

Such development hints at two ideas that may expose the nature and orientation of future wars. First, while firepower and payload may continue to be a dominant aspect of arms competitions, stealth technology implies that war may be taken to a different level as a potential counter alternative against traditional weapons. The application of "stealth" in combat opens up the possibilities of achieving first-strike ability or closing in and striking while the enemy is off guard—tactical level measures that sometimes prove to be more critical in war than objective military capability.

Second, with the downsizing of war in recent years through the combined use of drones and technologies such as the Internet and global positioning systems (GPS), stealth technology adds another dimension to combat that may or may not bode well for future warfare. For states in possession of stealth technology, infinite possibilities may be available for its application to various weapons and armed and unmanned vehicles, which further increases the military prowess of a state. Nonetheless, for state or nonstate actors without the technology, asymmetric measures such as terrorist attacks or concentrated attacks on stealth weapons and vehicles (for example, targeting bases and sites where such weapons are housed, or hacking into their systems) may be adopted as potential countermeasures. In essence, stealth technology may fuel a new round of tensions between the "haves" and "have nots" in the future that may give rise to implications that cut back on potential advantages.

Stealth technology is a high-technology concept resulting from barbaric or predatory impulses. Wherever the predatory instinct is present in a segment of a society, there is also an accompanying belief in the efficacy of luck. Proponents of stealth technology have imbued it with an aura of luck and mystique, and this has been noted by their critics. Stealth technology is marketed with the almost magical belief that regardless of the circumstances, the use of the technology can only be an advantage and will result in success. According to that belief, stealth technology leaves no possibility for unintended consequences, be they diplomatic, political, or legal.

In its political implications, nations who wield stealth technology may well be tempted to misplace their trust in it, overvaluing its effectiveness and dependability. Thus, nations that use stealth technology may be more likely to engage in combat because of a misplaced trust in its dependability. The fact that such technology is bestowed with such omnipotent power signals a more disturbing idea among the nations that possess it: that diplomacy is redundant. If so, stealth technology is an element that makes war—in whatever shape it takes—more likely.

References

Axe, David. 2015. "Test pilot admits the F-35 can't dogfight." *Warisboring.com* June 29. Retrieved April 30, 2016. (https://warisboring.com/test-pilot-admits-the-f-35-can-t-dogfight-cdb9d11a875#.qs1t2t1fw).

"B-2 Spirit." *Federation of American Scientists*. Retrieved March 31, 2016. (http://fas.org/nuke/guide/usa/bomber/b-2.htm).

"The B-2 Spirit Stealth Bomber turns 25." 2014. *Northrup Grumman* July 9. Retrieved April 30, 2016. (http://www.northropgrumman.com/Capabilities/B2SpiritBomber/Documents/pageDocuments/B-2_25th_Anniversary_Fact_Sheet.pdf).

Blazek, Patrick Allen. 2004. *Nighthawks: Insider's Guide to the Heraldry and Insignia of the Lockheed F-117A Stealth Fighter*. Atglen, Pennsylvania: Schiffer Publishing, Ltd.

Bowcott, Owen and Alice Ross. 2016. "Military 'invisibility cloak' could breach Geneva conventions." *The Guardian* March 14. Retrieved April 30, 2016. (https://www.theguardian.com/science/2016/mar/14/military-invisibility-cloaks-stealth-could-breach-geneva-conventions).

Brantley, Briggitte N. 2015. "B-2s fly over Tournament of Roses Parade, Rose Bowl." *Inside Whiteman AFB.*

Ciralsky, Adam. 2013. "Will it fly?" *Vanity Fair* September 16. Retrieved March 31, 2016. (http://www.vanityfair.com/news/2013/09/joint-strike-fighter-lockheed-martin).

Cohen, Zachary. 2015. "The F-35: Is the world's most expensive weapon's program worth it?" *CNN* July 16. Retrieved April 30, 2016. (http://edition.cnn.com/2015/07/16/politics/f-35-jsf-operational-costs/).

De Briganti, Giovanni. 2012a. "F-35 reality check ten years On—Part 1: 'Fifth-generation' and other myths." *Defense-Aerospace.com* May 19. Retrieved April 30, 2016. (http://www.defense-aerospace.com/articles-view/feature/5/135080/f_35-reality-check-10-years-on-%28part-1%29.html).

De Briganti, Giovanni. 2012b. "Ballooning F-35 costs questioned in Netherlands, Italy and Canada." *Defense-Aerospace.com* October 29. Retrieved April 30, 2016. (https://web.archive.org/web/20151004160959/http://www.defense-aerospace.com/articles-view/feature/5/139745/f_35-faces-new-hurdles-in-netherlands%2C-italy-and-canada.html).

Doherty, Bob. 2014. "F-35 Joint Strike Fighter purchase 'a great national scandal' says Coalition MP." *Sydney Morning Herald* June 16. Retrieved March 31, 2016. (http://www.smh.com.au/national/f35-joint-strike-fighter-purchase-a-great-national-scandal-says-coalition-mp-20140616-zs9po.html).

Drew, James. 2016. "Penalty for breaking Northrop's B-21 contract is '$300 million'." *Flight Global* March 9. Retrieved April 30, 2016. (https://www.flightglobal.com/news/articles/penalty-for-breaking-northrops-b-21-contract-is-3-422960/).

Durkin, Ali and Medill News Service. 2012. "Military does more than 1,000 flyovers per year but doesn't track cost" *Washington Guardian* November 26. Retrieved March 31, 2016. (http://www.washingtontimes.com/news/2012/nov/26/military-does-more-than-1000-flyovers-per-year-but/?page=all).

"F-117A Nighthawk." *Federation of American Scientists.* Retrieved March 31, 2016. (http://fas.org/man/dod-101/sys/ac/f-117.htm).

"F-35 Joint Strike Fighter (JSF) Lightning II: International Partners." *Globalsecurity.org.* Retrieved March 31, 2016 (http://www.globalsecurity.org/military/systems/aircraft/f-35-int.htm).

"F-35 Joint Strike Fighter (JSF) Lightning II: Specifications." *Global Security.org.* Retrieved March 31, 2016 (http://www.globalsecurity.org/military/systems/aircraft/f-35-specs.htm).

Forbes, J. Randy and Michael W. Wynne. 2016. "US air power needs an F-22 upgrade." The Wall Street Journal April 28. Retrieved April 30, 2016. (http://www.wsj.com/articles/u-s-air-power-needs-an-f-22-upgrade-1461884661).

Gayle, Damien. 2012. "The camouflage fabric 'that can make soldiers invisible': Company claims it has Pentagon backing for miracle material" *The Daily Mail* December 10. Retrieved March 31, 2016 (http://www.dailymail.co.uk/sciencetech/article-2245935/The-camouflage-fabric-make-soldiers-INVISIBLE-Company-claims-Pentagon-backing-miracle-material.html).

Geoghegan, Tom and Sarah Shenker. 2011. "Stealth helicopters' used in Bin Laden raid." *BBC News* May 6. Retrieved April 25, 2016 (http://www.bbc.com/news/world-us-canada-13297846).

Hacker, Brian C. and Margaret Vining. 2007. *American Military Technology: The Life Story of a Technology*. Baltimore, Maryland: John Hopkins University Press.

Hutcheon, Stephen. 2015. "Revolutionary F-35 Joint Strike Fighter pilot's smart helmet will cost a bomb." *The Sydney Morning Herald* February 25. Retrieved March 31, 2016. (http://www.smh.com.au/technology/technology-news/revolutionary-f35-joint-strike-fighter-pilots-smart-helmet-will-cost-a-bomb-20150220-13ko9d.html#ixzz44UOMyTm4).

Kazianis, Harry. 2016. "This could 'sink' the US Navy: Lethal stealth submarines." *The National Interest* January 27. Retrieved April 30, 2016. (http://nationalinterest.org/blog/the-buzz/could-sink-the-us-navy-lethal-stealth-submarines-15034).

Keck, Zachary. 2015. "This is how China and Russia plan to crush America's stealth aircraft." *The National Interest* August 26. Retrieved April 30, 2016. (http://nationalinterest.org/blog/the-buzz/how-china-russia-plan-crush-americas-stealth-aircraft-13708).

Kimmons, Sean. 2016. "State of AF: Modernizing for next offset strategy." *Air Force News Service* Feb 26. Retrieved May 1, 2016. (http://www.af.mil/News/ArticleDisplay/tabid/223/Article/674159/james-modernizing-for-next-offset-strategy.aspx).

Korb, Lawrence. 2015. "New US stealth bomber: Costly and unneeded." *The Japan Times* November 9. Retrieved April 30, 2016. (http://www.japantimes.co.jp/opinion/2015/11/09/commentary/world-commentary/new-u-s-stealth-bomber-costly-unneeded/#.VyLmUeafewk).

Liberal Party of Canada. 2015. *Real Change: A New Plan for a Strong Middle Class*. Retrieved March 31, 2016. (https://www.liberal.ca/files/2015/10/New-plan-for-a-strong-middle-class.pdf).

Lin, Jeffrey and P. W. Singer. 2015. "Divine Eagle, China's enormous stealth hunting drone, takes shape." *Australian Popular Science* May 29. Retrieved April 30, 2016. (http://www.popsci.com.au/robots/drones/divine-eagle-chinas-enormous-stealth-hunting-drone-takes-shape,404616).

"Lockheed awarded contract: Defense Dept. taps Lockheed-Martin for $200B Joint Strike Fighter." 2001. *CNN* October 26. Retrieved March 31, 2016. (http://money.cnn.com/2001/10/26/companies/strikefighter/).

"Lockheed Martin F-35 Lightning II Joint Strike Fighter specification & technical data." 2013. Defenceaviation.com. Retrieved April 30, 2016. (http://www.defenceaviation.com/2013/07/lockheed-martin-f-35-lightning-ii-joint-strike-fighter.html).

Majumdar, David. 2016. "Why America needs lethal nuclear stealth cruise missiles." *Nationalinterest.org* March 3. Retrieved April 30, 2016. (http://nationalinterest.org/blog/the-buzz/why-america-needs-lethal-nuclear-stealth-cruise-missiles-1539).

Majumdar, David. 2015. "Russia to build lethal PAK-DA stealth bomber—with hypersonic weapons?" *The National Interest* April 18. Retrieved April 30, 2016. (http://www.nationalinterest.org/blog/the-buzz/russia-build-lethal-pak-da-stealth-bomber%E2%80%94-hypersonic-15821).

Martin, Mike. 2016. "Air Force Reveals B-21 long range strike bomber." *Air Force News Service* February 26. Retrieved May 1, 2016. (http://www.af.mil/News/ArticleDisplay/tabid/223/Article/673784/air-force-reveals-b-21-long-range-strike-bomber.aspx).

McArthur, David. 1990. "Hidden capabilities of B-2 stealth bomber threaten START talks." *Peace Magazine* June–July, Vol. 6, No. 3.

McGarry, Brendan. 2014. "Is the F-35's computer R2-D2 or HAL?" *Defense Tech* February 19. Retrieved March 31, 2016. (http://www.defensetech.org/2014/02/19/is-the-f-35s-computer-r2-d2-or-hal/).

Mestrovic, Stjepan G. 2003. *Thorstein Veblen on Culture and Society*. Thousand Oaks, CA: Sage Publications.

Milhoces, Gary. 2013. "Federal budget cuts would end flyovers at sports events." *USA Today*, February 28. Retrieved March 31, 2016 (http://www.usatoday.com/story/sports/2013/02/27/federal-budget-cuts-sequestration-flyovers-air-force-navy/1952415/).

Myers, Harold P. 1992. *Nighthawks Over Iraq: A Chronology of the F-117A Stealth Fighter in Operations Desert Shield and Desert Storm*. Special Study 37FW/HO-91-1, Office of History Headquarters 37th Fighter Wing Twelfth Air Force Tactical Air Command.

"Navy spent $450,000 of taxpayer money on Super Bowl flyover… when the stadium roof was shut." 2011. *Daily Mail* February 10. Retrieved March 31, 2016. (http://www.dailymail.co.uk/news/article-1355727/Navy-spent-450-000-Super-Bowl-flyover-shut-roof.html).

Peck, Michael. 2014. "How in the world do you make a tank stealthy?" *Warisboring.com* August 19. Retrieved April 30, 2016. (https://warisboring.com/how-in-the-world-do-you-make-a-tank-stealthy-5dada62af8b9#.x6svcfy4k).

Pietrucha, Mike. 2016. "The US Air Force and stealth: Stuck on denial part I." *War On The Rocks* March 24. Retrieved April 28, 2016. (https://warontherocks.com/2016/03/stuck-on-denial-part-i-the-u-s-air-force-and-stealth/).

Prigg, Mark. 2015. "Watch the US Navy's largest ever 'stealth' destroyer put through its paces in first sea trials." *Daily Mail* December 9. Retrieved April 30, 2016. (http://www.dailymail.co.uk/sciencetech/article-3351736/Watch-Navy-s-largest-stealth-destroyer-paces-sea-trials-radical-design-fitted-ultrapowerful-RAILGUN.html).

Rovell, Darren. 2011. "How much did you pay for the Super Bowl flyover?" *CNBC.com* February 9. Retrieved March 31, 2016. (http://www.cnbc.com/id/41495109).

Swarts, Phillip. 2016. "Lawmakers worry new B-21 bomber will be another F-35 money pit." *The Air Force Times* March 7. Retrieved April 30, 2016. (http://www.airforcetimes.com/story/military/2016/03/07/lawmakers-worry-new-b-21-bomber-another-f-35-money-pit/81319112/).

"The $1 trillion F-35 tries to be all things, but succeeds at few, say critics. But is Australia's new weapon now too big to fail?" 2014. *News Limited Company* June 11. Retrieved March 31, 2016. (http://www.news.com.au/technology/the-1-trillion-f35-tries-to-be-all-things-but-succeeds-at-few-say-critics-but-is-australias-new-weapon-now-too-big-to-fail/story-e6frfrnr-1226950254330).

Tierney, Dominic. 2011. "The F-35: A weapon that costs more than Australia." *The Atlantic* March 15. Retrieved March 31, 2016. (http://www.theatlantic.com/national/archive/2011/03/the-f-35-a-weapon-that-costs-more-than-australia/72454/).

United States Air Force. 2010. *FY 2011 Budget Estimates: Aircraft Procurement, Air Force. Volume I*. Retrieved March 31, 2016. (http://www.saffm.hq.af.mil/shared/media/document/AFD-100128-072.pdf).

United States Air Force. 2015. "F-22 Raptor." September 23. Retrieved April 30, 2016. (http://www.af.mil/AboutUs/FactSheets/Display/tabid/224/Article/104506/f-22-raptor.aspx).

"US F-22 stealth fighters fly over South Korea in show of force toward DPRK." 2016. *Xinhua* February 17. Retrieved April 30, 2016. (http://www.defencetalk.com/us-f-22-stealth-fighters-fly-over-south-korea-in-show-of-force-toward-dprk-66760/).

Veblen, Thorstein. 1994 [1899]. *The Theory of the Leisure Class*. New York, NY: Penguin Books.

"Warring over aircraft." 2007. *Australian Broadcast Corporation* March 4. Retrieved March 31, 2016 (http://www.abc.net.au/radionational/programs/nationalinterest/warring-over-aircraft/3395374).

Weisgerber, Marcus. 2016. "Skunk Works chief: How to keep America's airborne advantage." *Defenseone.com* March 15. Retrieved April 30, 2016. (http://www.defenseone.com/technology/2016/03/skunk-works-chief-how-keep-americas-airborne-advantage/126699/?oref=d-dontmiss).

Wroe, David. 2015. "Joint Strike Fighter ejector seat could break lightweight pilot's neck." *Sydney Morning Herald* October 23. Retrieved March 31, 2016. (http://www.smh.com.au/federal-politics/political-news/joint-strike-fighter-ejector-seat-could-break-lightweight-pilots-neck-20151023-gkh480.html).

16 Stealth combat aircraft and US warfare

Shih-Yueh Yang

Introduction

Stealth combat aircraft will play an increasing and crucial role in United States (US) warfare. Although stealth technology has many applications in almost every major category of combat platforms, such as warships and tanks, it is mainly associated with combat aircraft, which are exactly the foremost striking power in the US way of war. Currently, only a small fraction of US combat aircraft is stealthy, namely, 187 F-22 Raptor fighters and 20 B-2 Spirit bombers. Within a decade or so, around 2400 F-35 Lightning II Joint Strike Fighters (JSF) and 100 newly conceived Long Range Strike Bombers (LRS-B), presently designated B-21 (Tirpak, 2016), will make up the bulk of US combat strength. These two are the only US new combat aircraft projects underway and will cost more than 1.5 trillion (US Department of Defense, 2016) and 100 billion (Harrison and Hunter, 2015) USD respectively. They are essentially the core of future US warfare.

Stealth combat aircraft are by no means invisible. They are only "less observable" in certain directions against electromagnetic waves of a certain wavelength spectrum, and they can still be detected as long as the sensors are close enough (Grant, 1998). Even so, stealth combat aircraft do have clear advantages over their nonstealthy counterparts. Since stealth aircraft are less observable and can only be detected at shorter distances, an air defense network designed to detect nonstealthy aircraft will be full of loopholes against stealth aircraft. Resuming the network with more sensors is often too expensive to be feasible, so the stealth aircraft can go through with fewer supports from fighter escorts and electronic warfare aircraft. One stealth fighter/bomber is said to be worth of 75 nonstealthy aircraft (Ruehrmund and Bowie, 2011: 77). An all-stealth combat aircraft fleet will be a great leap forward in US combat effectiveness.

However, there are challenges ahead. Many potential adversaries are now trying to defeat the US's increasing reliance on stealth technology. After all, stealth technology had its combat debut in the late 1980s and is thus no longer on the cutting edge. With the diffusion of related knowledge and technologies, various tactics and actual weapon systems to defeat US stealth technology may be ready within a decade. Although stealth technology cannot be rejected altogether, alternatives are required. This chapter, then, explores all these emerging challenges and possible alternatives in more detail.

Fighters

In the short term, the major challenge for US stealth fighters is that potential adversaries can simply overrun them by sheer number (Stillion and Perdue, 2008). The F-22 Raptor is arguably the best fighter in the world, but even the best fighter cannot appear in two places simultaneously. With only 187 copies on hand for three major theaters (Europe, Asia-Pacific, and the Middle East), one theater could have around 60 at most. Optimists may suggest that the exchange rate between F-22 and other latest nonstealthy fighters (so-called 4th or 4.5th-generation fighters, which have state-of-the-art avionics and engines similar to F-22s installed on fighter designs of the 1970s) is 1 to 100, which means 60 F-22s equal 6000 of the latest nonstealthy fighters, a number that no country has. However, things may not be as easy as these optimists suggest (Gilbert, 2011: 49–51).

F-22 carries only six beyond visual range (BVR) missiles. Even if they are 100 percent accurate, the seventh enemy fighter will still survive and get through. Stealthy F-22s have no advantage in the subsequent within visual range (WVR) combat against other nonstealthy fighters, and the modern high-off-boresight missiles simply make everyone equal in WVR combat (Sweetman, 2001: 38–45). Thus, the eighth enemy fighter will prevail. Even worse, it is highly possible that BVR missiles will be far less than 100 percent effective. Until now, all BVR kills were achieved against easy targets. If the targets have advanced electronic countermeasures (ECM), two missiles per target is a reasonable and necessary policy. Thus, the best an F-22 can achieve in BVR combat is three kills. Then, the fourth enemy fighter and F-22 perish together in WVR combat, and the fifth enemy fighter prevails.

This hypothetic picture duplicates the air-to-air combat in the Vietnam War, when the heavy, well-equipped Phantoms were often defeated by the swarm of light and rather primitive Migs. Ironically, the Vietnam War era Mig-21s, which are still in service in many countries in large quantities, will be even more effective in this scenario than those latest 4th or 4.5th generation nonstealthy fighters. With adequate updates of ECMs and high-off-boresight WVR missiles, Mig-21s will be a much cheaper option if someone wants to overrun F-22s with sheer number. One counterargument is that F-22s can conduct hit-and-run tactics, that is, evade safely after achieving three BVR kills. Nevertheless, this is not always feasible. In an escort mission, which should be one of the most important missions of F-22s, evade means mission failure (Cliff, Fei, Hagen, Hague, Heginbotham, and Stillion, 2011: 219–223). Furthermore, high-value assets like tankers and early warning aircraft simply cannot evade like F-22s or other striking fighters that can give up their weapon loads. Without the support from these assets, how can F-22s fight? In the mid-term, 2400 F-35s will gradually ease up the F-22s' inferior in quantity from 2019, but the challenge remains. After all, the delivery of all 2400 F-35s is to be spread over a decade. It takes many years to accumulate enough quantity (US Department of Defense, 2015: 29).

In the long term, the challenge points directly to stealth technology. As mentioned, stealth technology only makes aircraft "less observable" against electromagnetic waves of a certain wavelength spectrum. For fighter-sized aircraft like the F-22 and F-35, they are stealthy only against centimeter waves and still subject to the detection of radars using decimeter and meter waves (Groves, 2004: 53–58). These radars are huge, heavy, and cumbersome to operate (the longer the wavelength, the larger the antenna) and are largely phased out of service in advanced militaries, but stealth technology, ironically, gives these radars a new life. In fact, during the First Gulf War in 1991, the US army dispatched nonstealthy helicopters to take out old Iraqi early warning radars of this kind, and this kind of radar helped the Serbians to shoot down an F-117 in the 1999 Kosovo War (Lambeth, 2001: 116–119). Upgrade kits are now available to these older radars to extend their service life. In addition, new designs also appeared. They have foldable antennas and can be deployed or relocated within 30 minutes, making them much more mobile than the older ones, which are largely static (Kopp, 2008: 32–36).

Networks also play an important role in defeating stealth technology (Friedman, 2009). Ordinary radars have their emitter and receiver placed together, so the basic principle of stealth technology is to deflect the incoming radar waves to other directions and thus prevent the waves from being received by the radar. These waves do not disappear and can still be received if the receivers of radar are located in different places. Networks coordinate these receivers and can thus achieve better detection. In addition, networks enable another way to defeat stealth aircraft, passive detection, as well. Stealth aircraft have their own electromagnetic emissions from their onboard radars and radios. Although these intermittent signals are still hard to identify and track, this difficulty will be overcome sooner or later, and then many receivers can be coordinated by networking (Sweetman, 2006a: 59–63). Supporters of stealth technology often argue that the detection of these networks is still not as accurate as conventional centimeter wave radars and cannot achieve a firing solution. However, as mentioned earlier, even against centimeter waves, stealth fighters can still be detected "as long as the sensors are close enough." Networks are valuable in cueing fighters and missiles with centimeter wave radars to score a hit (Stillion and Perdue, 2008). This combined method will eventually make stealth aircraft not stealthy anymore (Majumdar, 2014; Sweetman, 2015).

F-35s, the US premier fighter through the 2040s, would suffer most in such a situation. The F-35 was originally a small, light fighter in great numbers like the F-16 but was subsequently evolved into a big, heavy fighter bomber (maximum take-off weight at 31 tons, as heavy as the famous B-17 Flying Fortress heavy bomber in the 1930s and 1940s)—the largest single-engine combat aircraft ever built in history. To fulfill the requirement of striking missions, F-35 now has two large weapon bays and a huge fuel reserve inside its bloated fuselage, let alone the bulky lift fan and other associated mechanisms in the short take-off and vertical landing (STOVL) version. In contrast to F-22, which is stealthy, fast, and highly maneuverable, F-35 is stealthy but slow and poorly maneuverable, a "strike" fighter indeed (Hunter, 2015: 34–35). This is not a good news for a fighter. Maneuverability may count little in the age of high-off-boresight WVR missiles, but speed is still the life of fighters. F-35 depends too heavily on stealth and is too slow to evade the enemy fighters. Once its stealth breaks, it is doomed (Gertler, 2012).

Airbase survivability is another challenge (Tol, Gunzinger, Krepinevich, and Thomas, 2010: 11–13). There are always imperatives to strike airbases to destroy or downgrade enemy air forces, but stealth technology provides even more imperatives to launch such a strike (Krepinevich, 2010). If stealth aircraft remain difficult to find, the best timing to get them is when they are still on the ground. Stealth aircraft are also more vulnerable than nonstealth aircraft in this regard. Stealth aircraft spend more time on the ground because they need more maintenance and other logistic supports. The facilities and personals required are also ideal targets for an enemy strike. Ground base air/missile defense systems and bunkers are hardly the solution, because they could be overrun by massive, accurate attacks (Stillion and Orletsky, 1999; Shlapak, Orletsky, Reid, Tanner, and Wilson, 2009: 32–34, 50–51). This is particularly troublesome for the US because the US's forward deployments in its allies are largely concentrated in a very limited number of airbases (Easton, 2014: 6–18).

Bombers

Stealth bombers seem to be the cure to all the challenges listed above. Given their range, bombers can locate in secure bases thousands of miles away from enemy threats. Because of their size, stealth bombers can accommodate more sophisticated designs and structures that make them still "less observable" against decimeter and meter wave radars. Stealth bombers actually look much smaller than stealth fighters on radar screens, although bombers are visually much larger. That's why 20 B-2 bombers are considered the most lethal aircraft in US arsenal and the pride of US military power. These aircraft are so valuable that each of them has its own name. The only weakness of B-2s is that they are really too scarce, so 100 B-21 LRS-B are conceived to enter service in the coming decade. Regrettably, stealth bombers also have their own challenges to be addressed.

Airbase survivability remains to be a serious challenge. The logic is simple: since stealth bombers are even more difficult to detect than stealth fighters, there are greater incentives to strike the bombers at their bases. Furthermore, bombers are much more limited in numbers, and to take them all out at one base is an attractive and reasonable option. Stealth bombers are also more complex then stealth fighters and require even more maintenance and other logistic supports. They spend even more time on the ground, making them ideal targets to strike. The long range of stealth bombers seems to be the strongest shield of their base, but to strike a distant base is not as difficult as one might think at the first glance. In fact, this illusory sense of security is exactly the reason why Pearl Harbor was successfully attacked in 1941 (Prange, Goldstein, Dillon, 1991: 443). Surely, few nations have the ability to conduct a similar all-out attack against the US today like Japan did 75 years ago. Nevertheless, as late as 1945, when Japanese forces were largely destroyed, they still launched successful attacks on US bomber bases in the Marianas (Cate, 1953).

Modern technology provides more options for such attempts. While a small number of nuclear tipped long-range ballistic missiles can be defeated by missile defenses, cruise missiles pose a real threat (Easton, 2009: 6). One might wonder how cruise missiles can succeed in this scenario. There are indeed cruise missiles with a 3000-km range, but they take several hours to cover such a great distance. Many of them could be intercepted during this lengthy flight. Cruise missiles are small and can have stealth features, but their size makes them similar to stealth fighters, which are still subject to decimeter and meter wave radars, as mentioned. Nevertheless, cruise missiles can be carried by other platforms in large quantities and can thus strike a distanced bomber base as a fighter base nearby at short notice. Submarines are ideal carriers for cruise missiles (Patton, 2008: 18–23). In addition, asymmetric tactics also provide good cover to other missile carriers. For example, cruise missiles can be packed into commercial containers (Watts, 2011). Every container ship can be an arsenal ship. Even a fishing ship can carry tens of cruise missiles. How, then, can anyone identify these covered missile vessels from so many ordinary ships on the high sea? Special operations forces and agents are also capable of attacking bombers' bases, especially for those bases close to civilians like those in Guam or Hawaii.

Stealth technology as regards bombers also has its challenges. Networking and passive detection can also defeat stealth bombers. In addition, although stealth bombers are still stealthy against decimeter and meter wave radars, there are radars using even greater wavelengths at 10–100 meters (decameter). Indeed, this kind of radar is still in its infancy, and the resolution is very poor even at its best. The errors are measured in kilometers if not tens of kilometers. However, this sort of information is still enough to cue fighters to search inside this killing box (Perrett, 2014). In close range, stealth bombers are no longer stealthy and can be spotted by centimeter wave radars, infrared sensors, and the naked eye in daytime. Stealth bombers are hopeless in air-to-air combat against fighters. If F-35s depend too heavily on stealth, stealth bombers depend almost entirely on stealth. Fighter escorts are not an option here. Fighters do not have the range and endurance to accompany bombers all the way in and out (Grant, 2010: 44; Tirpak, 2011: 24–30). Additionally, stealth fighters are less stealthy than stealth bombers as mentioned. How can stealthier bombers be protected by less stealthy fighters, which have to struggle for their own survival in hostile airspace?

Defensive armament for stealth bombers would not work. Stealth bombers can have BVR missiles, but these missiles will have difficulty in engaging stealth targets. One should keep in mind that a potential enemy can also acquire stealth fighters for air defense (Tiezzi, 2015). Invading US stealth bombers cannot have the initial cuing from decimeter and meter wave radar networks like the enemy, because these radar systems can only be ground-based due to their size. Taking out enemy decameter wave radars won't work either. These radars are huge, larger than the Eiffel Tower, due to the ultralong wavelength they use, and are thus immobile, which seems to make them easy prey. However, their detection range is also very long, at thousands of miles (Air Combat Command, 2012). Stealth bombers will be detected and intercepted well before they could destroy those radars.

Alternatives

Given these challenges, although stealth technology is still a necessity, the overreliance on this technology should be adjusted. Unfortunately, available options are limited and questionable from a technological and financial perspective. Quantity is the first to be considered. Originally, there should be around 700 F-22s, but the number was subsequently cut and cut after the end of the Cold War. The US authorities asserted that the F-22 is not relevant to the requirement of the new era, where the air-to-ground capabilities are much more necessary (Schanz, 2012: 36–37). Then it became clear that this judgment may be too risky, the re-opening of the F-22 production line was considered. However, the F-22 is too luxurious. The cut in number does contribute to the rise of the F-22's unit cost, but the F-22's projected high unit cost is an important reason for the cut as well. In any case, the F-22 is a 38-ton, twin-engine fighter. Like its predecessor, the F-15, it cannot be cheap, and stealth technology only makes it even more expensive. Because of the same reason, any futuristic fighters beyond F-22s (known as 6th-generation fighters, while F-22s are 5th-generation fighters) will be very limited in number if they are still big fighters, let alone the fact that these fighter designs are just at the beginning of conceptual development (Harrington, 2010: 5); they are still too fantastic to be true (Majumdar, 2015; Tadjdeh, 2016).

This is exactly why there are F-16s in addition to F-15s, and F-22s should also have their own partners. Since F-35s have grown so big and heavy that can no longer be a fighter in the same class as F-16s, how about a slimmed-down F-35? It should still be BVR-capable but smaller, lighter, faster, more maneuverable, and less stealthy to keep its price down. This approach would be an ideal solution from a technological and operational perspective, but the budget reality and tight schedule prevent this from happening (Schanz, 2010: 32–34). This conceptual fighter can share many important subsystems like engine and avionics with F-35s but essentially represent a new design. There are no funds for such a new project. From the beginning, the F-35 is the "Joint" Strike Fighter, a compromised fighter that combines the requirements for all US service branches to minimize the overall cost in a shrinking-budget environment. If there is a new fighter design parallel to F-35 and F-35's production significantly cut, the entire "joint" concept breaks down. Additionally, F-35 is approaching its full rate production (Butler, 2013), and axing the entire program in exchange for a new one will be a waste as well as time-consuming.

The only hope that remains is the B-21 LRS-B bomber, which is not yet finalized and is thus alterable, but this is not a realistic perspective either. Given the budgetary constraints, LRS-B is now an off-the-shelf solution only, which utilizes mainly those matured technologies (Tirpak, 2012: 30–35). Thus, it will have a very similar layout and will be largely identical in terms of performance to the B-2, being essentially a smaller B-2 with less payload (Gunzinger, 2010). In other words, LRS-B will face the same challenges as B-2. Stand-off weapons like air-launched cruise missiles may be a way out, but their relatively high price limits their use. The US has global reach and various missions and cannot rely on missiles like those potential regional opponents who can concentrate their efforts in just one limited conflict with the US. Furthermore, missiles are ideal in striking fixed targets like airbases as mentioned, but these regional opponents do not depend on these facilities to the same extent as the US since they do not operate as many stealth aircraft as the US. Their rugged air defense networks, cruise missile launchers and nonstealth aircraft are much easier to scatter into highways, towns, or forests, which complicates the targeting.

Given all this, can LRS-B's flight characteristics be improved? Altitude is the most achievable aspect. B-2 has a service ceiling of about 50,000 ft. If a larger and longer wing is adopted, LRS-B could have a service ceiling of about 70,000 ft (Cook, 2006), away from enemy air defenses. Nevertheless, altitude as a means to aircraft's survivability was already superseded in the 1960s when high-flying spy planes like U-2s were shot down by surface-to-air missiles (SAM). Fighter-interceptors in the 1960s could also have a service ceiling of more than 60,000 ft. Armed with rocket-boosted missiles, they could easily shoot down targets flying at 70,000 ft. If a stealth bomber is to be spotted, altitude will not significantly improve its chance of survival. After all, modern SAMs can even intercept ballistic missiles outside the atmosphere. How can a stealth bomber be secure simply because it flies higher?

High speed is another approach for an aircraft's survivability but is much more expensive to achieve than high altitude. Besides, this approach was also superseded in the 1960s. The B-58 bomber could fly two times the speed of sound but had only a very brief service life. The XB-70 bomber could fly three times the speed of sound but was too expensive to be mass produced. Combining high speed, high altitude, and stealth seems to be the answer, but the technologies required are far from maturity. High-speed flying needs more stabilizers, but these tail fins will undermine an aircraft's stealth. The B-2 is so stealthy because it is a tailless flying wing, but this kind of layout cannot go supersonic. In addition, high speed may ironically be very hazardous to a bomber's survivability. Even if a supersonic bomber can be as stealthy as a B-2 on radar screens, sustained high speed will produce a lot of infrared signature from the aircraft's heated skins due to the strong friction with the air, which can be easily detected hundreds of miles away at cloudless high altitude. Sonic booms are just another signature produced by supersonic flight. Acoustic waves may not be a reliable source for tracking and weapon guidance but are still valuable for initial cueing (Sweetman, 2006b: 19).

The last possible option is to push the limits to the edge of the atmosphere, a hypersonic near-space aircraft. It flies at 7–11 times the speed of sound and 200,000 ft high, vastly faster and higher than any aircraft (Hewish, 2004: 55). At such an extreme speed and altitude, no SAMs or fighters have the ability to reach these aircraft. Ballistic missile defenses may have the speed and altitude but are still short of the maneuverability required to engage this kind of targets (Scott, 2009: 40–46). Only peers can intercept hypersonic near-space aircraft, but no country except the US (the People's Republic of China [PRC] and Russia, perhaps) has both the financial and technological potential to develop such aircraft (Gady, 2015; Fisher, 2015; Perrett and Norris, 2015). Nevertheless, even the US will have a hard time accomplishing such an ambition. The earliest time that such aircraft can enter service may be well beyond the 2040s, too late to be a solution (Sweetman, 2004: 54). Moreover, such aircraft will be extremely expensive, limited in number, and low in sortie generation. It is still not quite, but almost, a space shuttle. Whether it is really cost-effective is open to question. Even the US is now refocusing hypersonic research efforts to build missiles rather than reusable aircraft (Malenic, 2015).

Conclusion

US warfare is increasingly relying on airstrikes using stealth combat aircraft, and this dependence on stealth technology will create insurmountable difficulties. The dilemma of today is the result of those shortsighted decisions in the 1990s and 2000s. Based on the mindset that the US remains the preeminent military power, the US armed forces have been lost in the so-called asymmetric scenarios and have wasted efforts on this kind of operation. The US assumed that it would not encounter significant resistance in the air and thus decided to acquire mainly strike fighters and bombers at the expense of real fighters with a balanced design between quality and quantity. This is just the epitome of the entire US military after the Cold War: based on the assumption that the US would continue controlling the sea without challenge, US naval forces shifted their focus from naval battle and convoy escort to inland strikes and littoral counterinsurgence operations. Ships designed and built for these purposes, DDG-1000 Zumwalt class destroyers and Littoral Combat ships (LCS), which are also stealthy, are reprehensible to maintain the US supremacy on the global commons (Patch, 2011; O'Rourke, 2012: 55–58). In the same way, due to its confidence in a conventional mechanized warfare, the US ground forces invested first in rapid deployment and later in counterterrorist operations in the past two decades (Bonsignore, 2003: 62–66; Zenko, 2004: 11), the consequence being that its main armor force remained underfunded (Tucker, 2004: 4). Thousands of newly built Strykers and armored trucks (mine-resistant ambush protected [MRAP]) are useless in countering Russian aggressions in eastern Europe. In any case, only after the advantage in symmetric conflict is secured and sustained can asymmetric missions be successfully addressed. Without the former as the pillar, the whole structure of the latter will collapse. Unfortunately, only a major defeat can wake the US up.

References

Air Combat Command. 2012. "Over the Horizon Backscatter Radar: East and West." *U.S. Air Force*. <http://www.acc.af.mil/AboutUs/FactSheets/Display/tabid/5768/Article/199120/over-the-horizon-backscatter-radar-east-and-west.aspx>.

Bonsignore, Ezio. 2003. "Armour Deployment Common Sense, Anyone?" *Military Technology*, 27, 11: 62–66.

Butler, Amy. 2013. "Lockheed Looks to Push F-35 Production Rate." *Aviation Week & Space Technology*. <http://aviationweek.com/defense/lockheed-looks-push-f-35-production-rate>.

Cate, James L. 1953. "Volume Five, The Pacific: Matterhorn To Nagasaki June 1944 to August 1945, The Army Air Forces in World War II." *Hyper War*. <http://www. ibiblio.org/hyperwar/AAF/V/index.html>.

Cliff, Roger, John Fei, Jeff Hagen, Elizabeth Hague, Eric Heginbotham, John Stillion. 2011. *Shaking the Heavens and Splitting the Earth: Chinese Air Force Employment Concepts in the 21st Century*. Santa Monica: RAND.

Cook, Nick. 2006. "Skunk Works Unveils Secret Polecat UAV," *Jane's Farnborough 2006 News Briefs*. <http://farnborough.janes.com/docs/farnborough2006/sections/ daily/day3/skunk-works-unveils-secre.shtml>.

Easton, Ian. 2009. "The Assassin under the Radar: China's DH-10 Cruise Missile Program." *Project 2049 Institute*. <http://project2049.net/documents/assassin_under_radar_china_cruise_missile.pdf>.

Easton, Ian. 2014. "China's Evolving Reconnaissance-Strike Capabilities Implications for the U.S.–Japan Alliance." *Project 2049 Institute*. <http://www.project2049.net/documents/Chinas_Evolving_Reconnaissance_Strike_Capabilities_Easton.pdf>.

Fisher, Richard D. 2015. "US Officials Confirm Sixth Chinese Hypersonic Maneuvering Strike Vehicle Test." *IHS Jane's 360*. <http://newsmilitary.com/pages/63489417-us-officials-confirm-sixth-chinese-hypersonic-manoeuvring>.

Friedman, Norman. 2009. *Network-Centric Warfare: How Navies Learned to Fight Smarter through Three World Wars*. Annapolis: US Naval Institute Press.

Gertler, Jeremiah. 2012. "F-35 Joint Strike Fighter (JSF) Program." *Congressional Research Service*. <http://www.fas.org/sgp/crs/weapons/RL30563.pdf>.

Gilbert, Ronald E. 2011. "Strategic Implications of US Fighter Force Reductions: Air-To-Air Combat Modeling Using Lanchester Equations," *Defense Technical Information Center*. <http://www.dtic.mil/cgi-bin/GetTRDoc?AD=ADA546639>.

Gady, Franz-Stefan. 2015. "China Tests New Hypersonic Weapon." *The Diplomat*. <http://thediplomat.com/2015/11/china-tests-new-hypersonic-weapon/>.

Grant, Rebecca. 1998. *The Radar Game: Understanding Stealth and Aircraft Survivability*. Arlington: IRIS Independent Research.

Grant, Rebecca. 2010. "Countering the Missile Threat." *Air Force Magazine*, 93, 12: 42–44.

Groves, Jim. 2004. "Signature Management and Reduction: A Technological Review." *Military Technology*, 28, 3: 53–58.

Gunzinger, Mark A. 2010. "Sustaining America's Strategic Advantage in Long-Range Strike." *Center for Strategic and Budgetary Assessments.* <http://www.csbaonline.org/4Publications/PubLibrary/R.20100914.Sustaining_America/R.20100914.Sustaining_America.pdf>.

Harrington, Caitlin. 2010. "Boeing Unveils Concepts for Sixth-generation Fighter." *Jane's Defence Weekly*, May 19, 5.

Harrison, Todd, and Andrew Hunter. 2015. "Seven Ways to Make the LRS-B Program a Success." *Center for Strategic and International Studies* (CSIS). <http://csis.org/files/publication/151027_Harrison_LRS-B_Program_Commentary.pdf>.

Hewish, Mark. 2004. "Making Space Fast and Cheap." *Jane's International Defense Review*, 37, 2: 53–58.

Hunter, Jamie. 2015. "Thunderbolt to Lightning: USAF Aims to Borden Close Air Support Capability." *Jane's International Defense Review*, 48, 12: 34–35.

Kopp, Carlo. 2008. "Russian VHF Counter Stealth Radars Proliferate." *Australian Airpower.* <http://www.ausairpower.net/SP/DT-Rus-VHF-Radar-2008.pdf>.

Krepinevich, Andrew F. 2010. "Why AirSea Battle?" *Center for Strategic and Budgetary Assessments.* <http://www.csbaonline.org/4Publications/PubLibrary/ R.20100219.Why_AirSea_Battle/R.20100219.Why_AirSea_Battle.pdf>.

Lambeth, Benjamin S. 2001. *NATO's Air War for Kosovo: A Strategic and Operational Assessment.* Santa Monica: RAND.

Majumdar, Dave. 2014. "Ways to Track Low-Observable Aircraft." *Aviation Week & Space Technology.* http://aviationweek.com/defense/ways-track-low-observable-aircraft.

Majumdar, Dave. 2015. "The U.S. Military's 6th-Generation Fighter: 5 Ways To Dominate the Sky." *The National Interests.* <http://nationalinterest.org/blog/the- buzz/the-us-militarys-6th-generation-fighter-5-ways-dominate-the-14091>.

Malenic, Marina. 2015. "USAF Using X-51 Lessons Learned to Weaponise Hypersonic Vehicles." *IHS Jane's Defence Weekly.* <http://www.janes.com/article/51472/usaf-using-x-51-lessons-learned-to-weaponise-hypersonic-vehicles>

O'Rourke, Ronald. 2012. "Navy DDG-51 and DDG-1000 Destroyer Programs: Background and Issues for Congress." *Congressional Research Service.* <http:// www.fas.org/sgp/crs/ weapons/RL32109.pdf>.

Patch, John 2011. "The Wrong Ship at the Wrong Time." *US Naval Institute Proceeding.* <http://www.usni.org/print/7140>.

Patton, James. 2008. "The SSGN—Not Your Father's Oldsmobile Submarine." *Naval Forces*. 29, 1: 18–23.

Perrett, Bradley. 2014. "Australia's Jindalee Radar System Gets Performance Boost." *Aviation Week & Space Technology.* <http://aviationweek.com/technology/australia-s-jindalee-radar-system-gets-performance-boost>

Perrett, Bradley, and Guy Norris. 2015. "China's Scramjet Claim Puzzles U.S. Hypersonic Researchers." *Aviation Week & Space Technology.* <http://aviationweek.com/technology/china-s-scramjet-claim-puzzles-us-hypersonic-researchers>

Prange, Gordon W., Donald M. Goldstein, Katherine V. Dillon. 1991. *Pearl Harbor: The Verdict of History.* Collingdale: Diane Publishing Co.

Ruehrmund, James C., Jr. and Christopher J. Bowie. 2011. "Arsenal of Airpower," *Air Force Magazine*, 94, 2: 76–78.

Scott, Richard. 2009. "Setting Higher Standards: Missile Line Takes Aim at New Targets." *Jane's International Defense Review*, 42, 5: 40–46.

Schanz, Marc V. 2010. "F-35 at End Game." *Air Force Magazine*, 93, 8: 30–34.

Schanz, Marc V. 2012. "The Last Raptor." *Air Force Magazine*, 95, 2: 36–37.

Shlapak, David A., David T. Orletsky, Toy I. Reid, Murray S. Tanner, and Barry Wilson. 2009. *A Question of Balance: Political Context and Military Aspects of the China-Taiwan Dispute.* Santa Monica: RAND.

Stillion, John, and Scott Perdue. 2008. "Air Combat Past, Present and Future." *RAND.* <http://www.mossekongen.no/downloads/2008_RAND_Pacific_View_Air_Combat_Brie fing.pdf>.

Stillion, John, and David Orletsky. 1999. *Airbase Vulnerability to Conventional Cruise Missile and Ballistic Missile Attacks: Technology, Scenarios, and US Air Force Responses.* Santa Monica: RAND.

Sweetman, Bill. 2001. "Fighter Tactics." *Jane's International Defense Review*, 34, 6: 38–45.

Sweetman, Bill. 2004. "Back to Bomber." *Jane's International Defense Review*, 37, 6: 52–55.

Sweetman, Bill. 2006a. "Skunk Works Confident over Supersonic design for LRS." *Jane's Defense Weekly*, August 2, 19.

Sweetman, Bill. 2006b. "Worth the Cost?" *Jane's Defence Weekly*, July 19, 59–63.

Sweetman, Bill. 2015. "New Radars, IRST Strengthen Stealth-Detection Claims: Counter-stealth Technologies Near Service Worldwide." *Aviation Week & Space Technology.* <http://aviationweek.com/technology/new-radars-irst-strengthen-stealth-detection-claims>.

Tadjdeh, Yasmin. 2016. "Military Eyes Sixth-Generation Fighter." *National Defense.* < http://www.nationaldefensemagazine.org/archive/2016/March/Pages/MilitaryEyesSixthGenerationFighter.aspx>.

Tiezzi, Shannon. 2015. "China's First Stealth Fighter Is About to Enter Production." *The Diplomat*, <http://thediplomat.com/2015/12/chinas-first-stealth-fighter-is-about-to-enter-production/>.

Tirpak, John A. 2011. "Rising Risk in the Fighter Force." *Air Force Magazine*, 94, 2: 24–30.

Tirpak, John A. 2012. "Time To Get Started." *Air Force Magazine*, 95, 2: 30–35.

Tirpak, John A. 2016. "Back to the Future Bomber." *Air Force Magazine*, <http:// www.airforcemag.com/DRArchive/Pages/2016/February%202016/February%2029%202016/Back-to-the-Future-Bomber.aspx>.

Tol, Jan van with Mark Gunzinger, Andrew Krepinevich, and Jim Thomas. 2010 "AirSea Battle: A Point-of-Departure Operational Concept." *Center for Strategic and Budgetary Assessments.* <http://www.csbaonline.org/4Publications/PubLibrary /R.20100518.Air_Sea_Battle__A_/R.20100518.Air_Sea_Battle__A_.pdf>.

Tucker, Terry L. 2004. "Abrams Tank: Centerpiece of the Combined Arms Team." *Armor*, 113, 6: 4.

US Department of Defense. 2015. "F-35 Joint Strike Fighter Aircraft (F-35)As of FY 2016 President's Budget." *US Department of Defense.* <http://www.dod.mil/pubs/foi/Reading_Room/Selected_Acquisition_Reports/15-F-0540_F-35_SAR_Dec_2014.PDF>

US Department of Defense. 2016. "F-35 Lightning II Program Fact Sheet Selected Acquisition Report (SAR) 2015 Cost Data." *Joint Strike Fighter.* <http://www.jsf.mil/news/docs/20160324_Fact-Sheet.pdf>

Watts, Barry D. 2011. "The Maturing Revolution in Military Affairs." *Center for Strategic and Budgetary Assessments.* <http://www.csbaonline.org/wp-content/uploads/2011/06/2011.06.02-Maturing-Revolution-In- Military -Affairs1.pdf>

Zenko, Micah. 2004. "Saving Lives with Speed: Using Rapidly Deployable Forces for Genocide Prevention." *Defense & Security Analysis*, 20, 1: 3–19.

Conclusion

Lessons of the past for future war

Scott N. Romaniuk

The contributors of this book collectively build on the common theme of the future direction of United States (US) warfare, which currently stands as a new and daunting point of inflection, and points to areas of theory and practice that appear to be more distorted than ever before. Each chapter in this book finds a more nuanced position, addressing specific areas that we have identified as some of the main trajectories of the US on this subject matter. As outlined in the table of contents and introduction, neither the chapters nor the categories in which they have been placed represent an exhaustive list of the areas of warfare that are still to be comprehensively explored, the future of warfare, and the future of US warfare. Indeed, these areas demand yet further attention in terms of the depth and breadth of exploration and analysis, as well as a more potent problematizing of the concepts, issues, and challenges contained within the theme of the book if we are to better understand where the US has been heading and where it is heading during the course of the twenty-first century.

In discussing the challenges of the future, we have to turn to the past. As Francis Grice noted in the introduction of this book, the past provides valuable lessons, and although the events of the past have elapsed in time, many of its implications either remain with us or are recurring. The Cold War's lessons and contests are much the same as those facing the US today. It is worth noting that Michael Lind, co-founder of New America, former editor/staff writer for the New Yorker, Harper's, and the National Interest, and author of *The American Way of Strategy*, identified the greatest threats to US power an influence around the world stemming other states willing to spar for positions of power alongside the US (perhaps even replacing the US), rather than from nonstate actors. In doing so, those states can be expected to engage the US indirectly through the use of, for instance, hybrid warfare, economic warfare, propaganda and proxy warfare, among other forms of indirect confrontation. Lind advocates the need for a military with the capacity to engage the enemy indirectly through low-level competition styled after Cold War engagements as opposed to less likely conventional confrontation between two or more economically and militarily powerful states.

With this in mind, we have attempted to show in this book the types of warfare that the US is likely to encounter in the coming years, as well as how changes in war generally both can and will impact the US. The rapid transformations that have occurred throughout the world over the past several decades have shaped, and even extremely altered, many facets of contemporary society, from the social and cultural to the political and economic dimensions of societies around the world. Profound changes in these areas are mirrored to no less extent in the realm of war, which, as illustrated by the 16 chapters in this book, is in a perpetual state of change and even in the midst of revolutionary change. The future of US warfare—or more precisely, the future of the way that the US practices, engages, and responds to conflict, security, and warfare—is fraught with controversy, dispositions toward describing their character in a way that supports the civilian and military efforts of a particular actor or group of actors, and questions of strategy, tactics, legitimacy, and morality, and so on.

Today, the topic of military affairs, war and conflict, and the future of warfare, including (advanced) technology and questions of the robotics way of war, leads to dozens of new articles, chapters, and books every year. Readers will have noticed that technology is a persistent matter in the previous chapters. The world is currently in the center of a monumental change in the way that war is conceptualized and waged, and technological innovation has occupied a central position in warfare overall. The development of new and sophisticated technologies shares an intimate relationship with new and dynamic assessments of the contemporary spatial and temporal battlespace, not to mention what is considered acceptable in war. Scholars of military studies like Martin Van Creveld (1991: 1) have placed a great deal of emphasis on technology. He notes that:

> war is completely permeated by technology and governed by it. The causes that lead to wars, and the goals for which they are fought; the blows with which campaigns open, and the victories with which they (sometimes) end; the relationship between the armed forces and the societies that they serve; planning, preparation, execution, and evaluation; operations, and intelligence and organization and supply; and objectives and methods and capabilities and missions; command and leadership and strategy and tactics; even the very conceptual frameworks employed by our brains in order to think about war and its conduct—not one of these is immune to the impact that technology has had and does have and always will have.

Yet the assertion of Van Creveld can be seen as placing too much emphasis on the role of technology. Rather, technology, while certainly playing a considerable role in the advancement of warfare, remains just a single factor in the future of US warfare. The US, like any other country, has come to embrace the advantages accompanying technological development, particularly in the field of military affairs and warfare, but it will also work hard to keep ahead of adversaries and imminent threats also seeking to acquire the latest and most sophisticated weapons of war. On the other hand, the most advanced weapons of war are not at all prerequisites for inflicting large-scale damage on a powerful state like the US, as was shown on September 11, 2001, and in smaller terrorist attacks on US soil in the past 15 years. Nor can technology act as a substitute for the role of military doctrine and the fundamental principles of agency and interagency cooperation and coordination. In spite of what the US holds in its arsenals and how the US is able to apply its advanced weapons systems, autonomous military technologies and impressive military size is subject to knowing the enemy, understanding and skillfully implementing joint doctrine, and employing other essentials of operational art. Because such aspects of warfare are not constant, they have to be learned and relearned with the passing of time. Reinterpreting US power in an increasingly globalized world, and in a world in which joint action and reaction can mark the difference between security and disaster, is a critical but sometimes overlooked aspect of the future of warfare in scholarship.

Given that fundamental changes in warfare are not based principally on technology, and that the practice of warfare and behavior of states in world politics is not defined by technology alone, there is value in reminding readers of a recent speech delivered by US Deputy Secretary of Defense Bob Work, in which he highlighted the importance of "offsetting" and "offset strategy" where technology is not everything, nor is the concept about a revolution in military affairs (RMA). Work spoke of a "strong technological component, but it is strategy-based, technologically oriented, and you want operational and organizational constructs" that also provide advantages (US Department of Defense [DoD], 2015). Thus, the strategies and practices, including operational and organizational superiority, put into practice by the US are equally important in the context of the changes in technology that both shape the nature of warfare and profoundly alter warfighting.

Many view war as a slowly evolving practice, but it is also one that make great leaps and bounds due to sudden and massive changes that take place due to technological, doctrinal, or strategic innovation; this "fundamentally alters warfare so as to render previous conventions of war obsolete" (Springer, n. d.). Gunpowder (or the sixteenth-century century gunpowder revolution), airpower, and nuclear weapons stand out as three seminal revolutions in military affairs (RMAs) over previous decades and centuries (Murray & Knox, 2001). The term is omnipresent in the field of security and military studies as a concept that can forever change the ways in which war is orchestrated, but the term fails to encapsulate a single idea. Instead, the term *RMA* has come to be a "somewhat nebulous and mutable concept that has had many different interpretations, even among its most ardent proponents" (Chapman, 2003: 2).

Speed is a distinctive part of RMAs and the nature of US warfare. Paul Virilio drew special attention in 2006 to the NASA (National Aeronautics Space Administration) slogan, "Faster, smaller, cheaper," reasoning that it "could shortly become the watchword of globalization itself." He noted one nuance, however, "since the speed and smallness in question would no longer refer to devices designed to conquer extra-terrestrial space, but to our geography at the moment of its sudden temporal compression." Virilio's quotes make clear that speed has become one of the most important factors in modern warfare. This of course has been a common feature in wars of the past, most during the years leading up to the Second World War (inclusive of the element of surprise). Speed was the core component of the concept of victory for German generals, who organized and planned attacks with their Panzer columns in concert with coordinated artillery bombardments from the rear and from the air (i.e., support from the *Luftwaffe*, Nazi Germany's Air Force).

Speed shares a relationship with the interests, efforts, and the need to control time and space of a state the former leaving an individual or state less empowered, as in the case of the latter. In his chapter "Toward Robotic Warfare," Łukasz Kamieński writes disquietingly about how the role of speed has combined with the technological innovations of war and war-making to have a disconcertingly potent effect on the soldier, citing Virilio's (2006) reference to the "violence of speed." "A murderous 'operational tempo'," writes Kamieński, "requires military personnel to perform at the edge of human capability. That is why the US military continually searches for biotechnological means to finely adjust humans to super-efficient machines. As high-tech unmanned systems can perform faster, longer, and more efficiently, soldiers must maintain persistent combat readiness and the ability to regenerate quickly." In Antônio Sampaio's chapter entitled "The Dark Dimension of War," he notes that states like the US have led the charge in speed-based technology and military applications but have simultaneously failed to master the element of speed in modern combat and warfare. "Despite the military search for increased speed and lethality," argues Sampaio, "dense, chaotic, and sprawling developing-world cities have often defeated modern armies." Thus, the future of US warfare looks problematic, as it is described as either unprepared or unfit to fight some of the battles and battlespaces that will likely define warfare in the twenty-first century.

It is no embellishment to say that the while the very idea of war has become deeply contested, conventional understandings of peace have also become the subject of intensive debate and reconceptualization. The distinctions between war and peace have become blurred so much that the US, it has been claimed, is in a perpetual state of war at home and abroad. Taken together, the chapters in this book aid in formulating news ideas and fuel ongoing debate about the nature of warfare, how it can be conceptualized, and what it means for the US military branches at all levels of command. As noted previously, they serve to build on current discussions of how the US engages with war and warfighting, where its future wars will likely take place or have already begun to emerge, and what role the US might occupy in a new and possibly unfamiliar era of warfare. As noted by this book's contributors, the US has already entered into a new era of warfare and is in some ways struggling to adapt to a world of new asymmetric and unfamiliar (and even old but forgotten) threats. Despite turbulence in this domain, the US is playing a leading role in the so-called "robotic turn" and is learning to be on-site in times of war, able to wage war at a distance (i.e., as part of the risk-free/casualty-free dimension of contemporary warfare). The contributors expertly depict, as a final note, that the old or traditional threats from state powers, not just nonstate or terrorist actors, are very much a part of the future of US warfare.

References

Chapman, G. (Sept. 2003). "An Introduction to the Revolution of Military Affairs," XV Amaldi Conference on Problems in Global Security, Helsinki, Finland. [Online]. Available at: <http://www.lincei.it/rapporti/amaldi/papers/XV-Chapman.pdf>.

Murray, W. & Knox, M. (eds.) (2001). *The Dynamics of Military Revolution, 1300–2050.* Cambridge, UK: Cambridge University Press.

Springer, P. J. (n. d.). "Thinking about Military History in an Age of Drones, Hackers, and IEDs." [Online]. Available at: <http://www.fpri.org/docs/springer1.pdf>.

United States (US) Department of Defense (DoD) (Nov. 7, 2015). "Reagan Defense Forum: The Third Offset Strategy," Deputy Secretary of Defense Speech—As Delivered by Deputy Secretary Bob Work, Reagan Presidential Library Simi Valley, CA. [Online]. Available at: <http://www.defense.gov/News/Speeches/Speech-View/Article/628246/reagan-defense-forum-the-third-offset-strategy>.

Van Creveld, M. (1991). *Technology and War: From 2000 BC to the Present.* New York, NY: The Free Press.

Virilio, P. (2006). *The Information Bomb (Radical Thinkers).* London, UK and New York, NY: Verso.

Index

For Product Safety Concerns and Information please contact our EU representative GPSR@taylorandfrancis.com Taylor & Francis Verlag GmbH, Kaufingerstraße 24, 80331 München, Germany